Vibration of linear mechanical systems

Vibration of linear mechanical systems

H. McCallion

Longman

LONGMAN GROUP LIMITED
London

Associated companies, branches and representatives throughout the world

First published 1973

ISBN 0 582 44301 6 cased
0 582 44302 4 paper

Made and printed in Great Britain by William Clowes & Sons, Limited London, Beccles and Colchester

Contents

3 Systems having distributed mass and elasticity

4 Piecewise representation of continuous systems

5 Natural frequencies—on finding numerical values

Preface

Discomfort resulting from close association with vibrating mechanical systems is common experience: damage to components arising from oscillating stresses is common knowledge. Prime examples of delinquent systems cover the range from diesel-engined marine propulsion systems through building frames and transformer cores, panels of aircraft, turbo-alternator shafts, gas turbine rotors to gyroscopic instruments.

It is also well known that vibratory phenomena may be utilised in specially designed systems, for example to convey material, to drive foundation piles, or to filter out unwanted frequencies in electronic circuits.

Minimisation of discomfort and damage, and optimisation of systems which utilise vibration, are desirable goals: to achieve them it is necessary for engineers to understand how modifications to a system affect its oscillatory behaviour.

In the present trend towards more and more specialisation, engineers in this field appear to concentrate either on experimental techniques associated with the development of equipment to a trouble-free state, or on theoretical techniques associated with technical design problems. Both occupations require a basic understanding of the theory of mechanical vibration; the former in order that the experimental results may be interpreted and changes in system parameters made in the correct direction; the latter as a basis for the development and use of methods for predicting the response of real systems, which generally are more complicated than necessary for a basic understanding.

As a field of study, vibration theory has wider application than indicated above. Oscillatory phenomena arise in many other areas of human activity, one of the most recent to become prominent being the control of management systems; fortunately much of the underlying eigenvalue theory is common. Therefore, although developed in this book in relation to the oscillations of mechanical systems, in studying the subject one has the satisfaction of learning a basic subject of relevance to a wide range of disciplines.

Essentially engineering is the use of a combination of art and science to solve problems which arise in harnessing the forces of nature on behalf of mankind. As an example of part of the art, successful engineers are assumed to have an instinct for correct proportions.

The increasingly used scientific approach to the discovery of knowledge may be expressed in terms of four steps:

(i) define the problem, or refine its definition;
(ii) model the relevant aspects of the real world situation, usually mathematically;
(iii) use the model to discover its relevant behaviour;
(iv) check the validity of the model by comparing this behaviour with that of the real system, and return to (i) continuing the sequence as often as necessary.

That is, in scientific discovery the model is continually refined until eventually it predicts, without exception, the behaviour of the real system.

Engineering design involves steps similar to the first three, but instead of step four the results of step three must be interpreted in a design. Model refinement, if it is necessary, arises at the development stage and may result in costly hardware modifications. Hence problem-solving in an engineering design environment involves a considerable knowledge of the state of the art in that particular field of application at the problem definition and modelling stages, if previous errors are to be avoided. Design involves a compromise between many conflicting requirements, hence interpreting the model results is also an art.

The stages associated with problem definition and the interpretation of the results in design are not discussed in this book; early experience is usually gained in project work. It is on the techniques for use in stages two and three, modelling and the analysis of the model, that emphasis is laid. The models examined are mathematical models involving linear differential equations with constant coefficients. In devising the models and discovering their behaviour, familiarity with the statics of elastic systems and the dynamics of particles, together with the theory of linear differential equations, vector algebra and matrix algebra has been assumed. It has also been assumed that systems of units present no difficulty to readers at this stage, for although SI units have been used in the numerical examples and exercises, it should be obvious that any consistent set of units may be ascribed to the quantities involved.

In an attempt to guide a student through the subject, exercises to reinforce the knowledge imparted and to develop understanding have been inserted in the text at appropriate points. Many of these exercises are modified or revised versions of exercises and examination questions used in a final year honours level course which I gradually developed, and taught, over many years in the University of Nottingham. In assessing the level of the course, readers should bear in mind that vibration studies formed the basis of a very active research school at Nottingham. Students should not too easily become dismayed when tackling the more difficult of these exercises; they are included to exercise the mind – that old Yorkshire saying 'Think on' being the best advice I can give. The hours spent in toil will be amply repaid by occasional exhilaration on finally mastering an intellectually difficult topic.

At the outset, this book presents a general picture of the influence, on a simple oscillatory system, of variations in mass, flexibility and damping; that of the type of excitation follows.

Much of the more advanced analysis depends upon an understanding that, for a conservative system, the principal modes of vibration are orthogonal. Chapter 2 presents this on a physical basis before making the abstractions necessary in general. As an alternative to Newtonian concepts, energy methods in the form of Lagrange's equations and Hamilton's principle are used to derive equations of motion. Upon this foundation is built, in Chapters 3 and 4, the student's knowledge of the behaviour of more complex systems, leading them to a perception of the basis for the important, modern, sophisticated, computer-based techniques used in vibration analysis. These include finite element and finite difference techniques.

Approximate methods of analysis associated with the work of Rayleigh and of Ritz are included in Chapter 5, which examines procedures for finding numerical values of natural frequencies. An attempt has been made to give a clear exposition of Rayleigh's Principle, stress being laid upon the conditions to be met by the assumed vibratory forms, and an attempt has been made to give an explanation of why the Rayleigh–Ritz method yields upper bounds of the lower natural frequencies. So far as I am aware, no proof understandable to other than mathematicians specialising in the eigenvalue problem appears in the literature, and yet engineers are utilising this fact more and more. Also discussed in Chapter 5 are the basic concepts of transfer matrices, matrix iteration and the eigenvalue-seeking techniques of Givens and Householder, together with the Q-R Algorithm.

The Calculus of Finite Differences is applied in Chapter 6 to estimate errors inherent in the piecewise representation of distributed parameter systems. Longitudinal oscillations of bars and transverse oscillations of beams are examples studied. Applications to uniform systems of beams are indicated in relation to turbine blading and mechanical wave filters. I feel that if this powerful calculus were more widely known, it could bring within the scope of computer storage capacity normally available, many systems otherwise too large.

The final two chapters, 7 and 8, expose some of the idiosyncrasies of quickly-rotating systems of shafts and rotors. Although most of this information has been known for nearly 40 years, few engineers have an understanding of it, despite its growing importance in the design of turbo-alternator sets, rotary pumps, and the like. I have attempted to present it in an easily digestible form. I have also suggested possible reasons for one or two confusing results in the research literature.

I wish to record my thanks to the many students with whom I interacted during the development of the course at the University of Nottingham; to Dr B. R. Dudley who kindly endured the course at its mid-stage of development, who gave the constructive criticism of an experienced teacher, and who, from his research and design experience, supplied the frameworks upon which were built a number of the exercises; to Professor D. Dowson, who persuaded me that the world was waiting for a version of my course to be cast into print; to Miss Susan Rook who did an admirable job of typing from my barely decipherable script; to Mr J. P. Ellington, Dr J. N. Fawcett and Dr N. S. Palmer for each reading the whole of the typescript and for making many valuable constructive criticisms; to Mr P. M. Ware for coming to my assistance at the hurried proof-reading stage; and finally to my wife, for tolerating my 'spare time' emulation of a recluse.

As many generations of Nottingham graduates plus, to date, one generation of Canterbury graduates can testify, I develop the course afresh each year, without notes. This book was written in the same manner, as a logical development from beginning to end; not as a blend of vintage notes. In the forms presented, most of the exercises are also new and the answers given are not time-tested. Although great care has been taken to avoid errors, infallibility has never been my claim, so do not be inhibited; notify me of errors, fundamental, careless, or typographical.

H. McCallion

Christchurch,
New Zealand,
August 1972.

Chapter 1

Systems with one degree of freedom

Many of the vibratory phenomena associated with complicated systems may be understood by studying the behaviour of simple systems. The simplest type of system has one degree of freedom, that is, the motion of the system may be described by one independent coordinate. Such a system may be used to assess, for example, the influence of the frequency of applied harmonically varying forces and the influence of damping on the amplitude of vibration. Similarly, the rate of increase of amplitude when the frequency of the applied force coincides with the so-called resonant frequency of the system may be calculated, as may the change of amplitude with time when the frequency of the applied force varies through a resonant frequency.

1.1 Equations of motion

Before going on to these studies, it is an essential preliminary to understand the derivation of the equation of motion for any simple system.

Consider the system shown in Fig. 1.1(a); a body of m mass units, which will be referred to as a mass, rests on a frictionless plane and is attached to a massless spring of stiffness k force units per unit change of length. If the mass were displaced a distance $+\boldsymbol{x}$ from its rest position and released, the sign convention being on Fig. 1.1(a), then, by Newton's third law, the force exerted by the spring on the mass, $\boldsymbol{F} = -k\boldsymbol{x}$, and by the mass on the spring, $\boldsymbol{F}' = k\boldsymbol{x}$, would be as shown in Fig. 1.1(b). For a constant mass system, Newton's second law states that

$$\text{Force} = \text{mass} \times \text{acceleration} \tag{1.1}$$

Both force and acceleration are vector quantities, mass being scalar, therefore a force exerted on a mass causes an acceleration in the direction of the force.

In the case under consideration, the force exerted upon the mass is $-k\boldsymbol{x}$ (see Fig. 1.1(b)). Therefore, by Newton's second law

$$-k\boldsymbol{x} = m\ddot{\boldsymbol{x}} \tag{1.2}$$

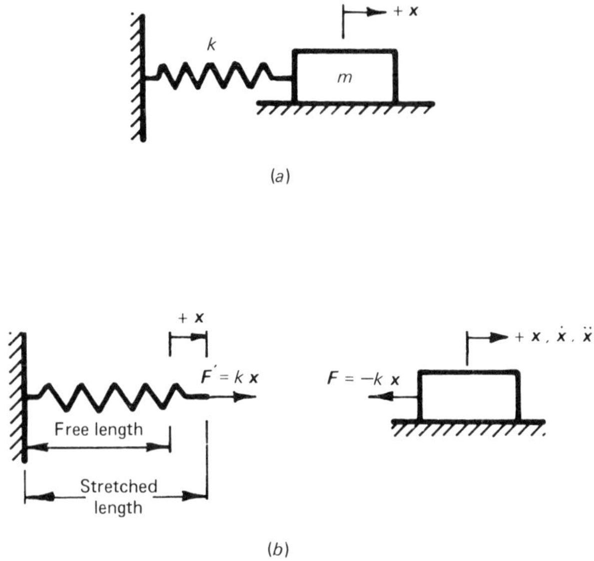

Fig. 1.1(*a*), (*b*)

This is known as the equation of motion for the mass and because the vectors may be expressed in terms of a single unit vector it is usually presented in the scalar form

$$m\ddot{x} + kx = 0 \tag{1.3}$$

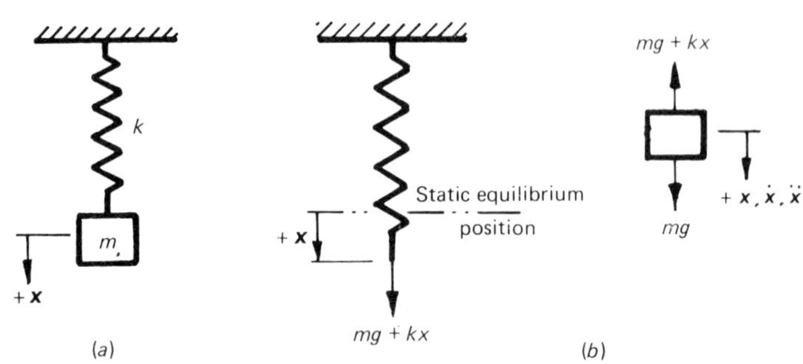

Fig. 1.2(*a*), (*b*)

A slightly more complicated case is illustrated in Fig. 1.2, in which the mass hangs on the lower end of a massless spring, which in turn is attached at its upper end to a rigid support.

At rest the mass will hang in a position referred to as the static equilibrium position, or equilibrium position, in which the upwards spring force exactly counterbalances the downwards force on the mass due to gravitational attraction. If the mass were deflected a distance $+x$ from its static equilibrium position and released, the forces acting on it would be as shown in Fig. 1.2(*b*). Then by Newton's second law we have

$$-mg - kx + mg = m\ddot{x} \tag{1.4}$$

or

$$m\ddot{x} + kx = 0 \tag{1.5}$$

which is the same as the equation (1.3) derived for the first case. So the small vibrations about the equilibrium position are independent of the gravity force,

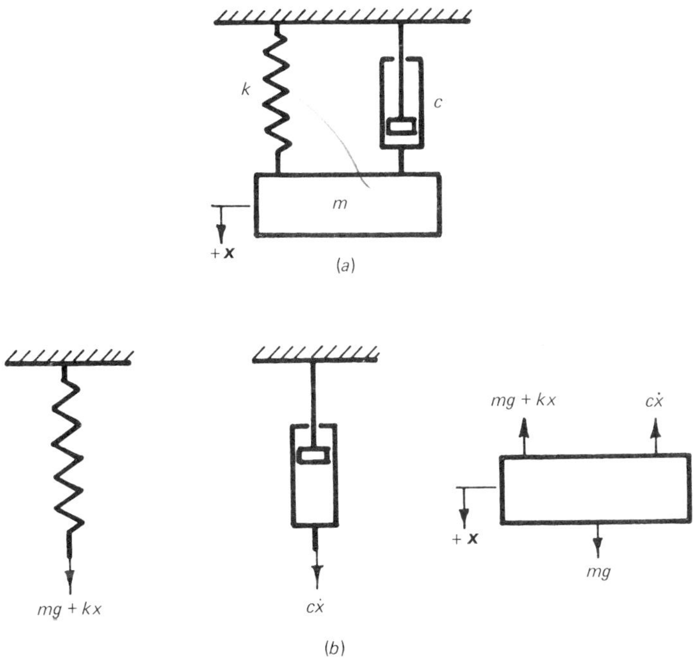

Fig. 1.3(*a*), (*b*)

since it does not appear in the equation of motion of the mass for this type of system. This fact will be used in later sections.

During the vibratory motion of most physical systems the energy associated with the oscillations, potential and kinetic, is gradually converted to other forms of energy, such as heat or sound. This energy dissipation leads to a gradual decrease of the amplitude of a freely vibrating system. It also plays an important role in limiting the amplitude, at resonance, of systems undergoing forced vibratory motion. To make a quantitative study of the influence of energy dissipation it is usual to represent it as a viscous force, proportional to the relevant relative velocity and opposing motion. It is then referred to as viscous damping, and depicted on diagrams as a linear dashpot. For example, in the system shown in Fig. 1.3 the damping force is taken as proportional to the velocity of the mass relative to earth, the constant of proportionality being c.

The equation of motion for the mass may be obtained with the help of Fig. 1.3. That is

$$m\ddot{x} + c\dot{x} + kx = 0 \tag{1.6}$$

The above are each cases of free vibration, in which the mass, once displaced and released, is not acted upon by forces from outside the system; the resulting motion depends upon the physical properties of the system alone. However, in physical systems, engines for example, the mass is often acted upon by another

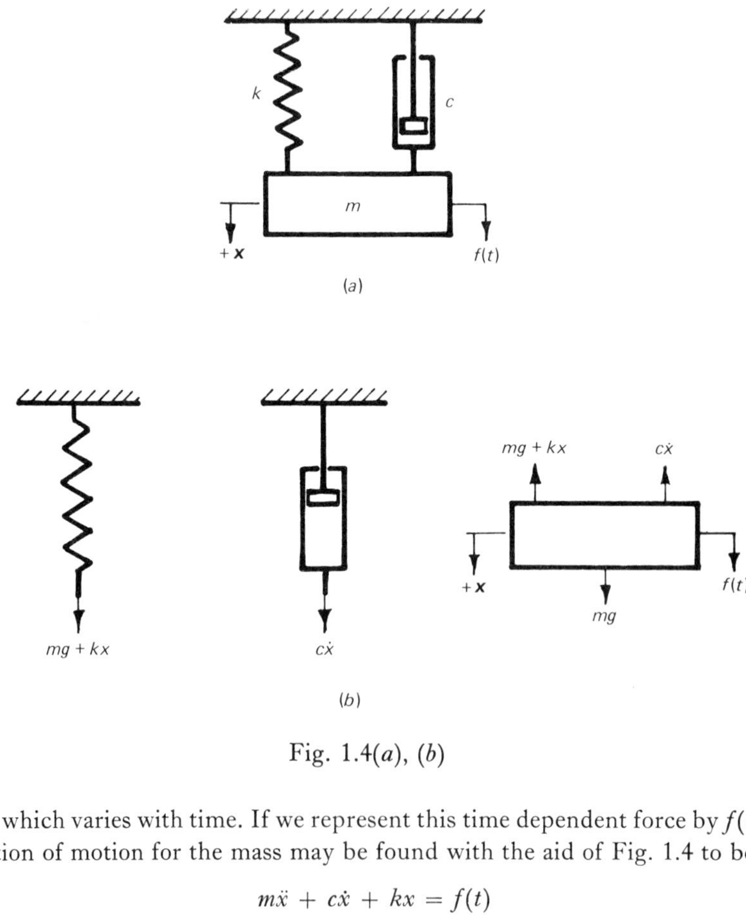

Fig. 1.4(*a*), (*b*)

force which varies with time. If we represent this time dependent force by $f(t)$, the equation of motion for the mass may be found with the aid of Fig. 1.4 to be

$$m\ddot{x} + c\dot{x} + kx = f(t) \tag{1.7}$$

This is the equation of motion for forced vibrations of a system with one degree of freedom.

In some practical situations, for example systems for the vibration isolation of delicate electronic equipment, the vibratory motion of the mass is excited through the isolating spring, or protection isolator, by motion of the support. One such system is shown diagrammatically in Fig. 1.5, from which it is evident that two coordinates, x_1 and x_2, are required to describe the state of the system at any instant. The number of coordinates required defines the number of degrees of freedom the system possesses and, therefore, this is a system with two degrees of freedom. Of these two degrees of freedom, one represents the rigid body motion of the system and the second may be regarded as a vibratory degree of freedom. For the present discussion x_2 is a specified function of time which means that this system may be specified by only one independent coordinate. In effect the rigid body motion is imposed by the outside world, consequently the system is reduced

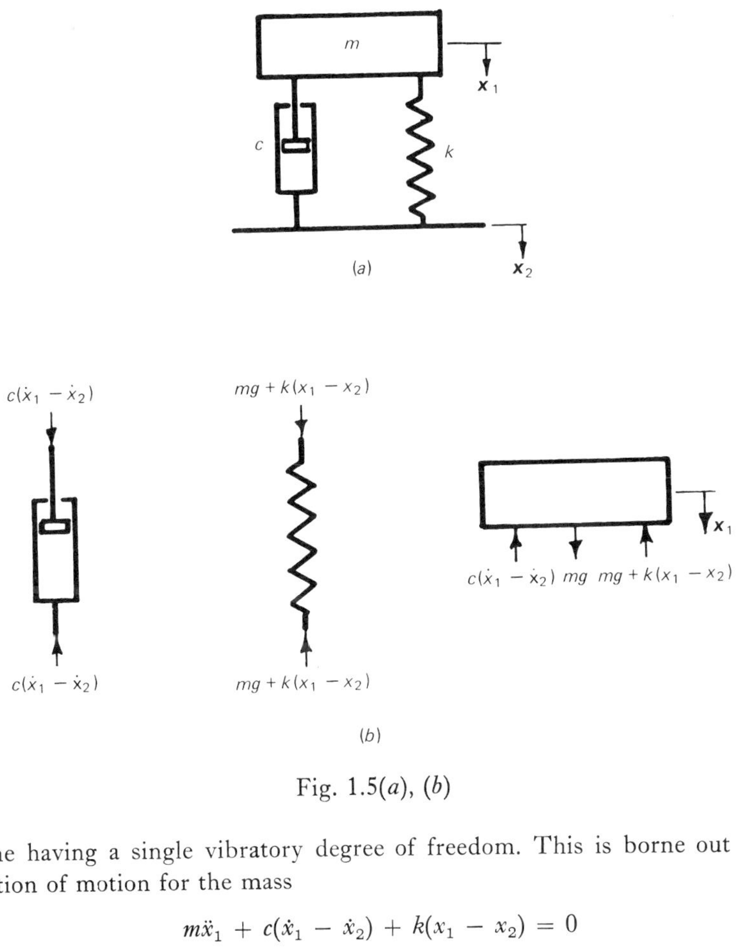

Fig. 1.5(*a*), (*b*)

to one having a single vibratory degree of freedom. This is borne out by the equation of motion for the mass

$$m\ddot{x}_1 + c(\dot{x}_1 - \dot{x}_2) + k(x_1 - x_2) = 0 \tag{1.8}$$

or

$$m\ddot{q} + c\dot{q} + kq = -m\ddot{x}_2$$

where $q = x_1 - x_2$, i.e., the motion of the mass relative to the support. x_2 is the rigid body motion and is a specified function of time. Hence the equation of motion may be written in the form

$$m\ddot{q} + c\dot{q} + kq = f(t) \tag{1.9}$$

which is similar to equation (1.7).

Exercises

Note: *Many of the ideas involved in this and subsequent chapters will be unfamiliar to you. These exercises are intended to help you gain a clear grasp of them. Later in the book, other methods of solution will be presented As your knowledge progresses, you are strongly advised to make up similar exercises and to check the solutions by a number of methods.*

Draw the free body diagrams and set up the equations of motion for small vibrations of the following systems:

(*a*) *Free vibrations*

1. The spring restrained pendulum shown in Fig. 1.6 consists of a rigid massless rod of length L pivoted at B and carrying a concentrated mass m at its free end.

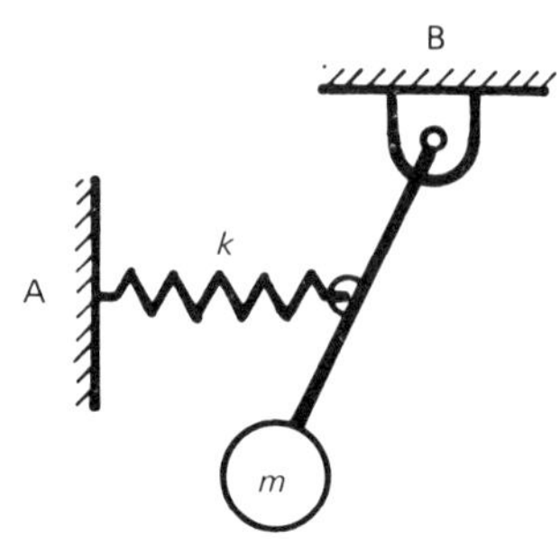

Fig. 1.6

Attached to the rod, at a distance a from the pivot, is a spring of stiffness k the other end of which is fastened to earth at A. In the static equilibrium position the pendulum is vertical.

2. The loaded beam shown in Fig. 1.7: the beam, which may be taken as massless, is encastered at one end and carries a concentrated mass m at its other end.

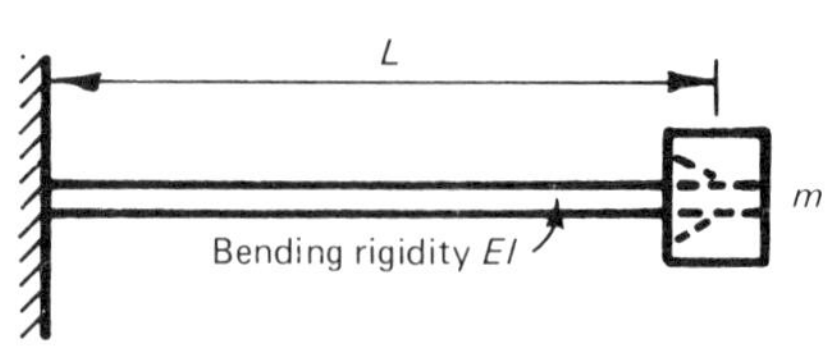

Fig. 1.7

The beam has a length L and a bending rigidity EI. Consider flexural vibrations. (The deflection of a cantilever under the action of a force W is given by $y = WL^3/3EI$.)

3. The damped torsional system shown in Fig. 1.8. The shaft of length L, diameter D, and modulus of rigidity G is firmly clamped at its left hand end and

carries a flywheel of moment of inertia J at its right-hand end. A linear damping torque proportional to its angular velocity acts upon the flywheel, the constant of proportionality being c. Neglect the mass of the shaft.

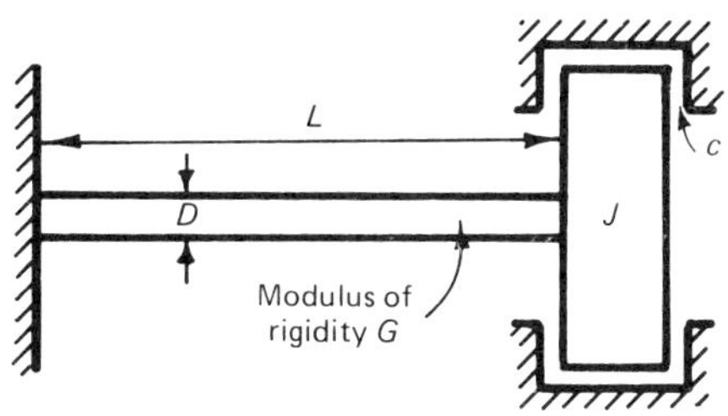

Fig. 1.8

4. The spring restrained mass shown in Fig. 1.9. The mass is constrained to move in frictionless guides, which rotate at a uniform angular velocity ω in the horizontal plane. A spring of stiffness k connects the mass to the centre of rotation.

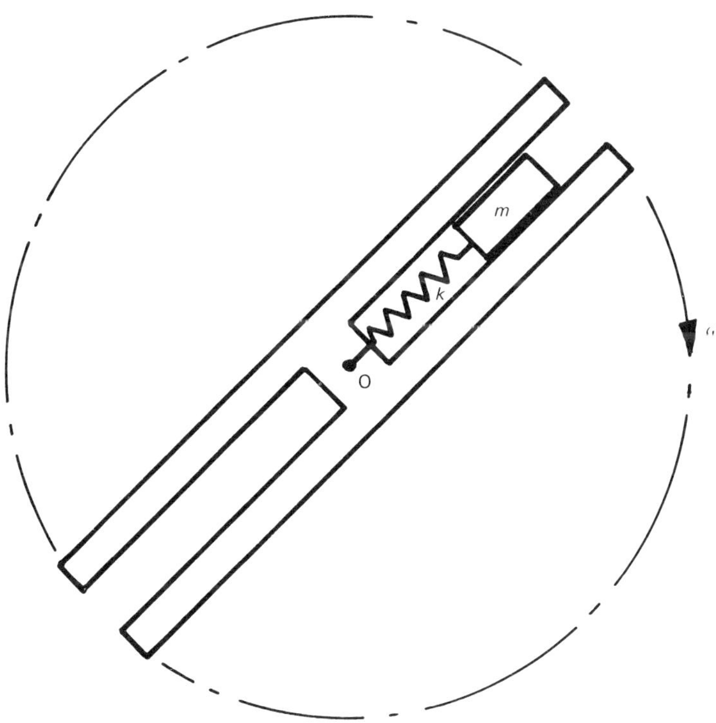

Fig. 1.9

5. The heavy disc and arm shown in Fig. 1.10. The rigid arm SO may be taken as weightless. It is supported at S and may swing as a pendulum. Two cases are to

be considered (*a*) the disc is free to rotate on a pin at O; the pin being fastened to the arm, (*b*) the disc is fastened to the arm by a pin at A in addition to that at O.

You may assume that the disc has mass m and radius R. The distance between S and O is L.

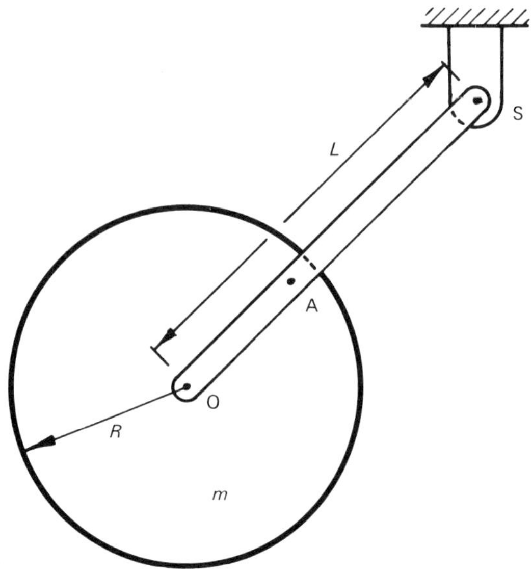

Fig. 1.10

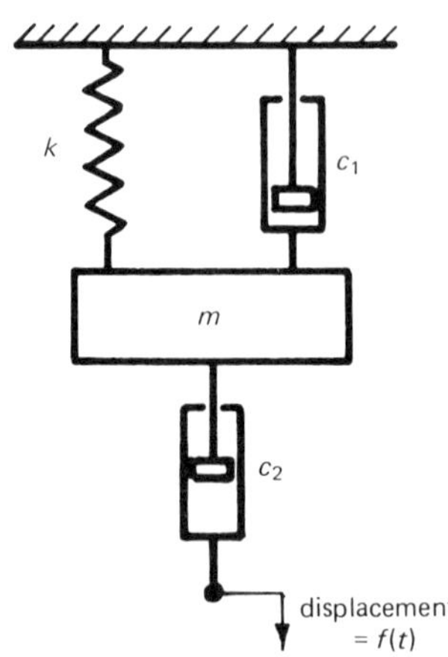

Fig. 1.11

(*b*) *Forced vibrations*

6. The system shown in Fig. 1.6. Derive the equation of motion for the following type of excitation:

(*i*) the horizontal position of support A varies sinusoidally with time, the position of support B remaining fixed.
(*ii*) the horizontal position of support B varies sinusoidally with time, the position of support A remaining fixed.

7. The system shown in Fig. 1.7. A time dependent force $F(t)$ acts in a vertical direction upon the mass.
8. The system shown in Fig. 1.8. A time dependent couple $T(t)$ acts upon the flywheel so as to rotate it about the longitudinal axis of the shaft.
9. The system shown in Fig. 1.11. The mass is connected to earth by means of a spring of stiffness k and a linear dashpot c_1. The system is excited by imposing a time dependent displacement on one element of the linear dashpot c_2 as indicated in Fig. 1.11.

1.2 Solution of the equation of motion

Having considered a systematic method for deriving the equation of motion for a system with one degree of freedom, we now turn our attention to finding its solution in mathematical terms and to interpreting this physically. Again, for convenience, we shall deal separately with the cases of free and of forced vibrations.

1.2.1 Free vibration

The equation of motion for the mass in Fig. 1.3 was shown to be

$$m\ddot{x} + c\dot{x} + kx = 0 \qquad \text{[from equation (1.6)]}$$

This is a second order linear homogeneous differential equation the solution of which can be found easily, as follows. Assume the solution to be of the form

$$x = A\,e^{rt} \tag{1.10}$$

which on substitution in equation (1.6) gives

$$(mr^2 + cr + k)A\,e^{rt} = 0 \tag{1.11}$$

Examination of equation (1.11) indicates that either

$$mr^2 + cr + k = 0 \quad \text{or} \quad A = 0$$

Now if $A = 0$ then $x = 0$ for all values of time. Therefore, unless the system is permanently at rest A is not zero, which means that

$$mr^2 + cr + k = 0 \tag{1.12}$$

for equation (1.11) to be satisfied.

Equation (1.12) is satisfied by the following particular values of r, that is,

$$r = -\frac{c}{2m} \pm \sqrt{\left(\frac{c^2}{4m^2} - \frac{k}{m}\right)} \tag{1.13}$$

which we may write for convenience as

$$r = -\alpha \pm \beta \tag{1.14}$$

The type of motion executed by the mass after an initial disturbance will depend upon the value of β, that is, solely upon the physical parameters of the system. If β is imaginary, that is if

$$\frac{k}{m} > \frac{c^2}{4m^2}$$

the motion is periodic. Under these conditions let $\beta = i\gamma$, the solution of equation (1.11) then has the form

$$x = e^{-\alpha t}\{A_1\, e^{i\gamma t} + A_2\, e^{-i\gamma t}\} \tag{1.15}$$

or

$$x = e^{-\alpha t}\{B_1 \cos \gamma t + B_2 \sin \gamma t\} \tag{1.16}$$

The constants are given by

$$\gamma = \left(\frac{k}{m} - \frac{c^2}{4m^2}\right)^{1/2} \quad \text{and} \quad \alpha = \frac{c}{2m}$$

the values of B_1 and B_2 depending on the initial conditions of the motion. γ is known as the damped natural angular frequency, and $\gamma/2\pi$ the damped natural frequency for the system. When the value of c is zero, $\gamma/2\pi$ is referred to as the undamped natural frequency of the system.

For systems in which $\beta = 0$, that is when

$$\frac{k}{m} = \frac{c^2}{4m^2}$$

the motion is given by

$$x = e^{-\alpha t}(A_1 + A_2 t) \tag{1.17}$$

where

$$\alpha = \frac{c}{2m}$$

and the values of A_1 and A_2 depend upon the initial state of motion of the mass. The motion described by equation (1.17) is non-oscillatory or aperiodic. As c has the lowest value for which aperiodic motion can exist, the damping in this particular case is referred to as 'critical' damping. We shall use the notation c_{crit} to denote the value of the critical damping coefficient for a system:

$$c_{crit} = (4mk)^{1/2} \tag{1.18}$$

for the system under consideration.

Finally when β is real, that is when

$$\frac{k}{m} < \frac{c^2}{4m^2},$$

the motion is again aperiodic and has the form

$$x = e^{-\alpha t}\{B_1 \cosh \beta t + B_2 \sinh \beta t\} \tag{1.19}$$

where

$$\beta = \left(\frac{c^2}{4m^2} - \frac{k}{m}\right)^{1/2}; \qquad \alpha = \frac{c}{2m}$$

and the values of B_1 and B_2 depend upon the initial conditions of the motion.

To illustrate the time dependent behaviour of the system shown in Fig. 1.3, for various amounts of damping let us assume the mass to be 10 kg, the spring stiffness to be 10 kN/m and that the mass is displaced 0·01 m from its static equilibrium position and released from rest. For this system

$$\begin{aligned} c_{crit} &= (4km)^{1/2} \\ &= 632{\cdot}4 \text{ N s/m} \end{aligned}$$

We shall consider four cases with the ratio c/c_{crit}, known as the damping ratio, taking the values zero, 0·2, 1·0, and 1·2. The expressions for these cases are shown in Table 1.1

TABLE 1.1 Response of a simple system to displacement and release

$\frac{c}{c_{crit}}$	x metres
0	$0{\cdot}01 \cos 31{\cdot}62t$
0·2	$\exp(-6{\cdot}324t)\{0{\cdot}01 \cos 30{\cdot}98t + 0{\cdot}002041 \sin 30{\cdot}98t\}$
1·0	$\exp(-31{\cdot}62t)\{0{\cdot}01 + 0{\cdot}3162t\}$
1·2	$\exp(-37{\cdot}94t)\{0{\cdot}01 \cosh 20{\cdot}97t + 0{\cdot}01809 \sinh 20{\cdot}97t\}$

The expressions are shown graphically in Fig. 1.12.

Later in this chapter the unit impulse response of a system will be referred to. Its importance lies in the fact that it enables the response of a system to any time-dependent force to be calculated. The unit impulse response is the motion of the system which results from applying an instantaneous unit impulse to the mass of a quiescent system. An instantaneous unit impulse causes the velocity of the mass to increase immediately from zero to $1/m$ m/s, so the initial conditions used in the evaluation of the constants of integration in equations (1.16) to (1.19) are, at time $t = 0$, $x = 0$, and $\dot{x} = 1/m$ m/s. The impulse response for each of the cases considered in Table 1.1 are given in Table 1.2. They have been evaluated and plotted against time in Fig. 1.13.

In many cases, the amount of damping in a system is difficult to assess without recourse to experiment. For oscillating systems it is often determined by exciting the system and measuring the resulting damped oscillations. The excitation may be a harmonic force applied for a short period of time, an impulsive blow, or a displacement of the system from the equilibrium position and subsequent release.

TABLE 1.2 Response of a simple system to a unit impulse

$\dfrac{c}{c_{crit}}$	x metres
0	$0{\cdot}003162 \sin 31{\cdot}62t$
0·2	$0{\cdot}003228 \exp(-6{\cdot}324t) \sin 30{\cdot}98t$
1·0	$0{\cdot}1t \exp(-31{\cdot}62t)$
1·2	$0{\cdot}004769 \exp(-37{\cdot}94t) \sinh 20{\cdot}97t$

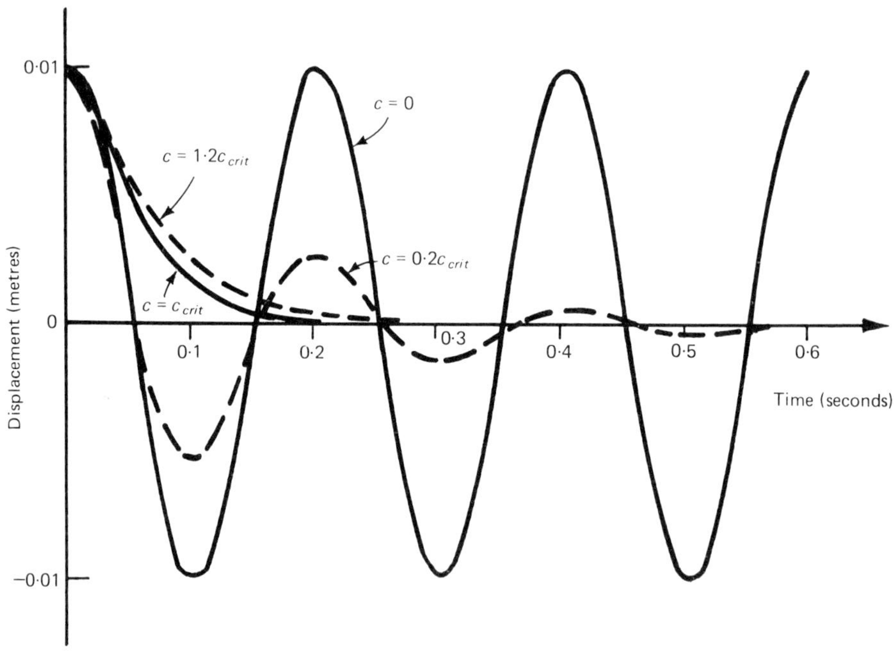

Fig. 1.12 Response of a simple system to a displacement and release

The resulting motion is periodic with a constant period, the period or periodic time being the time required to complete one cycle of the motion. At each end of a time interval equal to the periodic time of the system, therefore, the term of equation (1.16) in braces containing the trigonometrical functions will have the same value. Let the displacement have the value x_1 at time t_1 and the value x_2 at time t_2 where $t_2 - t_1 = \tau$, the periodic time, then

$$\frac{x_1}{x_2} = e^{\alpha\tau} \tag{1.20}$$

or

$$\Lambda = \log_e\left(\frac{x_1}{x_2}\right) = \alpha\tau = \frac{c\tau}{2m} = \frac{2\pi c}{(4km - c^2)^{1/2}} \tag{1.21}$$

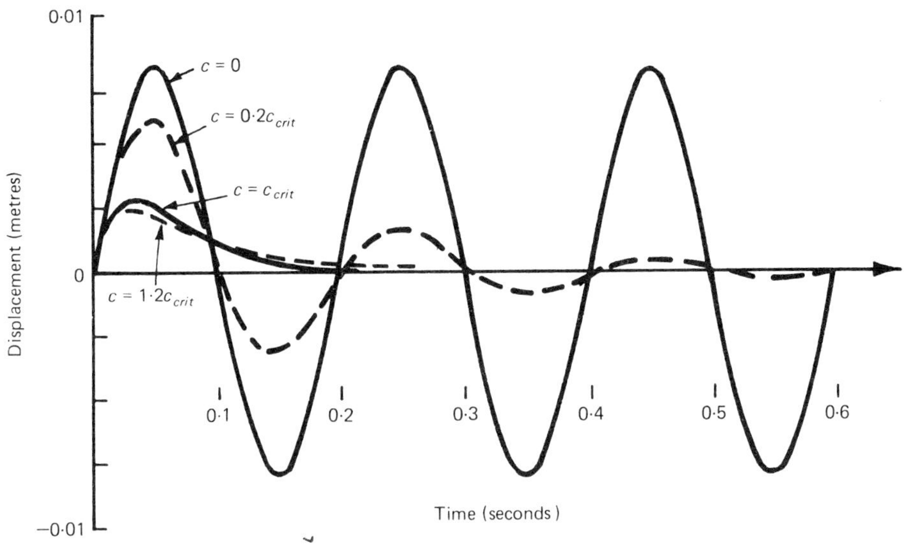

Fig. 1.13 Response of a simple system to a unit impulse

and hence the value of c may be found.

The ratio Λ is referred to as the logarithmic decrement (BS. 3015:1958). Note that its value does not depend upon the amplitude.

Exercises

1. Derive an expression for the motion of the mass shown in Fig. 1.14 subsequent to its release from a position distant x_0 below its static equilibrium position with zero initial velocity. Take $m = 2$ kg, $k = 1$ kN/m, and $c = 0{\cdot}07\ c_{crit}$.

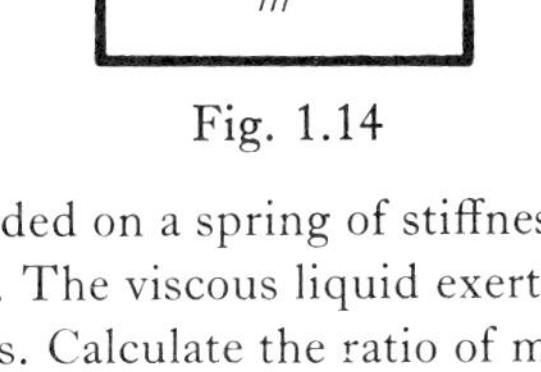

Fig. 1.14

2. A mass of 3 kg is suspended on a spring of stiffness 2 kN/m in a viscous liquid which damps its motion. The viscous liquid exerts a drag on the mass of 15 N when its velocity is 1 m/s. Calculate the ratio of maxima of the first downward displacement from the mean to the fourth downward displacement from the mean, after the mass is lifted and released.
3. If the system described in question 2 is set in motion by suddenly displacing the support point of the spring by 0·02 m draw a graph of the subsequent motion.

4. When the viscous liquid in the system described in question 2 is replaced by another viscous liquid and the mass lifted and released, it is observed that the ratio of the amplitude of the first downward displacement to that of the third downward displacement is three. Find the drag force per unit velocity for the new liquid. What percentage of critical damping does this represent and by what percentage does the damped natural frequency differ from the undamped natural frequency?

1.2.2 Forced vibration

It has been shown, equation (1.7), that the equation of motion for the mass in the system illustrated in Fig. 1.15 is

$$m\ddot{x} + c\dot{x} + kx = F \cos \omega t \tag{1.22}$$

where F is the amplitude of the exciting force and ω is its angular frequency.

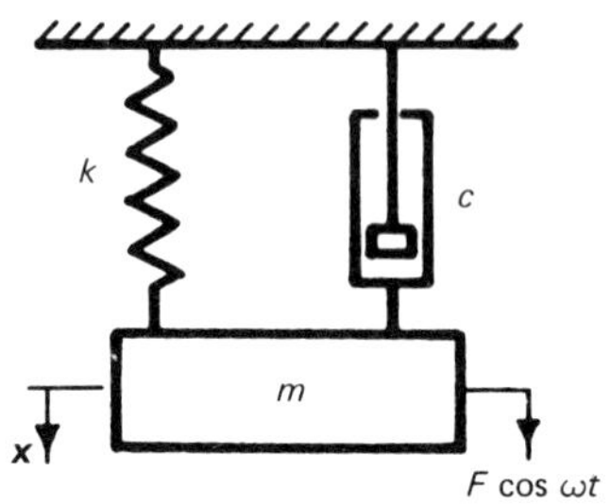

Fig. 1.15

The complete solution of this differential equation consists of a transient part corresponding to free vibrations and given in equations (1.16) to (1.19), together with a constant amplitude or steady state part. For the above equation the steady state solution is

$$x_{ss} = \frac{F\{(k - m\omega^2) \cos \omega t + c\omega \sin \omega t\}}{(k - m\omega^2)^2 + (c\omega)^2} \tag{1.23}$$

Therefore, the complete solution for the case in which

$$\frac{k}{m} > \frac{c^2}{4m^2}$$

is

$$x = e^{-\alpha t}\{B_1 \cos \gamma t + B_2 \sin \gamma t\} + \frac{F\{(k - m\omega^2) \cos \omega t + c\omega \sin \omega t\}}{(k - m\omega^2)^2 + (c\omega)^2} \tag{1.24}$$

where

$$\gamma = \left(\frac{k}{m} - \frac{c^2}{4m^2}\right)^{1/2}$$

$$\alpha = \frac{c}{2m},$$

and B_1 and B_2 depend upon the initial conditions of the motion. As t becomes large, $\exp(-\alpha t)$ becomes very small, leaving only the constant amplitude or steady state term which has the same frequency as that of the exciting force.

The steady state motion of a system excited by a sinusoidally varying force of constant amplitude and frequency may be interpreted in terms of rotating vectors. By this means a better physical understanding of the influence of mass, stiffness, damping, and the exciting force frequency on the amplitude of the resulting motion may be obtained. Therefore, before discussing these influences a vector representation of equation (1.23) will be introduced.

It is evident from equation (1.23) that the steady state solution for the oscillatory displacement has two components; one, of amplitude

$$\frac{F(k - m\omega^2)}{(k - m\omega^2)^2 + (c\omega)^2}$$

in phase with the force, and the second, of amplitude

$$\frac{Fc\omega}{(k - m\omega^2)^2 + (c\omega)^2}$$

lagging behind the force by 90 degrees. Clearly, if we regard the force as the real part of the vector $F \exp(i\omega t)$ in the complex plane, then we may regard the displacement as the real part of the vector $X \exp(-i\phi) \exp(i\omega t)$ where X, the displacement amplitude, is

$$\frac{F}{\{(k - m\omega^2)^2 + (c\omega)^2\}^{1/2}} \tag{1.25}$$

ϕ, the phase angle between the displacement vector and the force vector, is

$$\phi = \tan^{-1}\left\{\frac{c\omega}{k - m\omega^2}\right\} \tag{1.26}$$

and $i = \sqrt{-1}$.

From equation (1.23) we see that

$$\text{Re}\,\{X \exp(-i\phi) \exp(i\omega t)\} = \text{Re}\left\{\frac{F \exp(i\omega t)}{(k - m\omega^2) + ic\omega}\right\} \tag{1.27}$$

As the real parts are equal and both vectors rotate with the same angular velocity, the imaginary parts must also be equal. Hence we may compose them to give

$$X \exp(-i\phi) \exp(i\omega t) = \frac{F \exp(i\omega t)}{k - m\omega^2 + ic\omega} \tag{1.28}$$

In a coordinate system which rotates at a uniform angular velocity ω, equation (1.28) becomes

$$(-m\omega^2 + k + ic\omega)X = F \exp(i\phi) \tag{1.29}$$

Equation (1.29) is expressed graphically in Fig. 1.16, which shows clearly that the force leads the displacement by an angle ϕ. It also illustrates the following well known facts: (*i*) the force in the spring is in phase with the displacement, (*ii*) the force required to overcome the viscous damping force leads the displacement by 90 degrees, and (*iii*) the force required to accelerate the mass leads the displacement by 180 degrees. The graphical construction used in Fig. 1.16 is a convenient method for solving for the ratio X/F and for ϕ given values of m, k, c, and ω. The values used in plotting Fig. 1.17 were derived by this means.

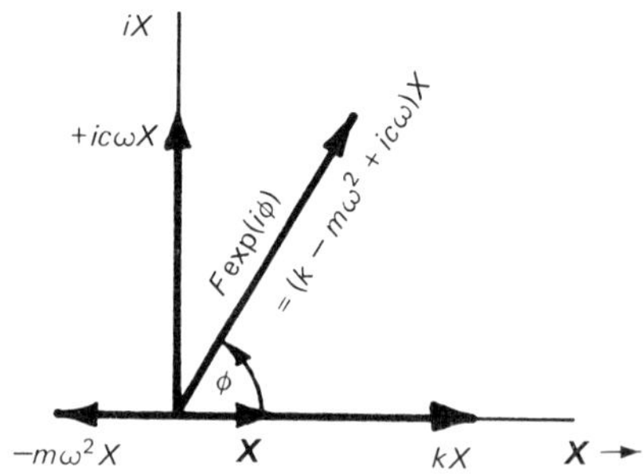

Fig. 1.16 Relationship between the force and displacement vectors in a coordinate system which rotates at a uniform angular velocity ω

Figure 1.17(*a*) shows the variation in the steady state displacement amplitude, for a constant force amplitude, when the frequency of the force varies between zero and 100 rad/s. The mass was taken as 2 kg and the spring stiffness as 5 kN/m. Response curves have been plotted for three values of the damping coefficient c, namely 10, 40, and 80 N s/m. The curve for $c = 10$ N s/m shows that the displacement amplitude rises to a large value in the vicinity of the undamped natural angular frequency of the system, that is at

$$\omega = \left(\frac{k}{m}\right)^{1/2} = 50 \text{ rad/s}.$$

Also at this frequency, Fig. 1.17(*b*) shows that the phase angle between the applied force and the displacement is 90 degrees, the displacement lagging. This arises from the fact that under these conditions, the spring force is exactly equal to the force required to accelerate the mass and, as may be reasoned from Fig. 1.16, the applied force has only the drag force due to viscous damping to overcome. Consequently, the greater the drag force, that is the greater the value of c, the

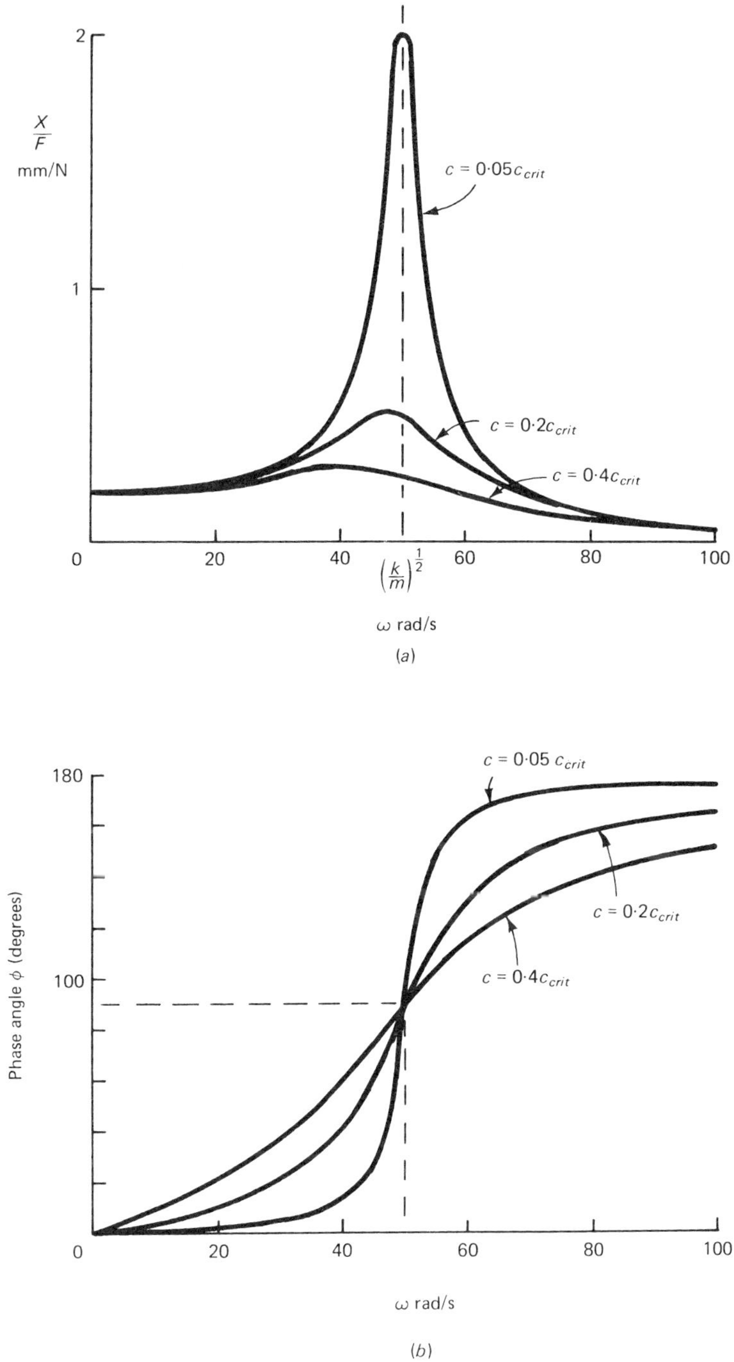

Fig. 1.17(*a*), (*b*): Response of a simple system to sinusoidal excitation (c_{crit} = 200 N s/m)

smaller the displacement amplitude. On careful examination of the response curve for $c = 40$ N s/m it can be seen that the maximum displacement amplitude, for a constant force amplitude, occurs at a frequency slightly lower than the undamped natural frequency. The frequency of the applied force at which the maximum of the response curve occurs is called the resonant frequency. An expression for the resonant frequency of the system shown in Fig. 1.15 excited by a harmonic force of constant amplitude will now be derived.

From equation (1.25) we have

$$\frac{X}{F} = \{(k - m\omega^2)^2 + (c\omega)^2\}^{-1/2}$$

The maximum of this ratio occurs when $\{(k - m\omega^2)^2 + (c\omega)^2\}$ in a minimum. That is, when

$$\frac{\mathrm{d}}{\mathrm{d}\omega}\{(k - m\omega^2)^2 + (c\omega)^2\} = -4\omega m(k - m\omega^2) + 2c^2\omega = 0$$

or when

$$\omega^2 = \frac{k}{m} - \frac{c^2}{2m^2} \tag{1.30}$$

A few other comments on the behaviour indicated in Fig. 1.17 are worthy of note. For the type of system investigated, the displacement amplitude/force amplitude ratio starts at a value of $1/k$, as would be found from statics, increases to a maximum at the resonant frequency, decreases through the damped natural frequency and the undamped natural frequency, and continues to decrease, approaching zero asymptotically, as the frequency increases. Also, regardless of the value of the damping constant, the phase angle between the force and the displacement varies from zero at zero frequency through 90 degrees at the undamped natural frequency to approach 180 degrees asymptotically as the frequency increases.

The damping inherent in many mechanical systems, due to energy losses within the material, friction at joints and in bearings, is very low. In these circumstances the resonant frequency, the damped natural frequency and the undamped natural frequency almost coincide. This condition of low damping prevails in engine installations and has led to a considerable simplification in the vibration analysis of these systems.

When the damping is small the phase shift in passing through a natural frequency is very sharp. This leads to an accurate method for determining the natural frequencies of systems. It is normally carried out by measuring the applied force and displacement with appropriate transducers and feeding the signals from them to an oscilloscope, the displacement being displayed on the screen as ordinate (say) and the force as abscissa. The combined effects of these signals will give what is called a Lissajous' figure. At low frequencies the Lissajous' figure will be a straight line, its slope depending upon the ratio of the amplitudes of the two signals, as shown at (*a*) in Fig. 1.18. The straight line opens out into an ellipse as the frequency is increased towards the undamped natural frequency, the slope and mag-

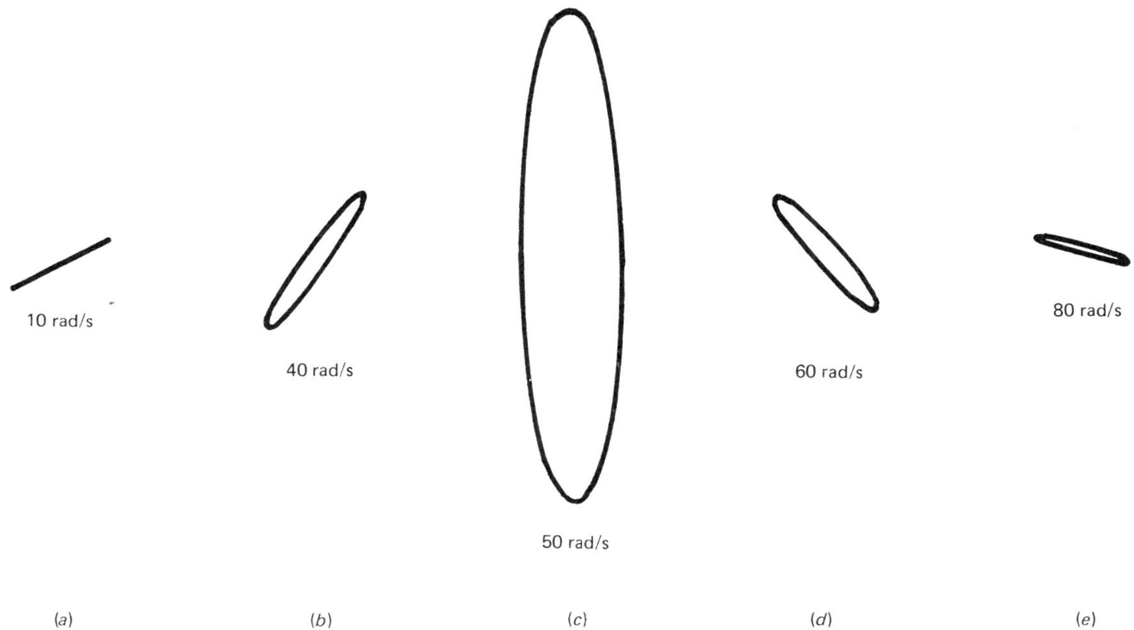

Fig. 1.18(*a*), (*b*), (*c*), (*d*), (*e*): Lissajous' figures for constant force amplitude

nitude of the major axis of the ellipse increasing as the exciting frequency increases. This is illustrated at (*b*) in Fig. 1.18. The major axis of the ellipse is very large and vertical when the oscillations are excited at the undamped natural frequency, Fig. 1.18(*c*). As the frequency continues to be increased the major axis continues rotating but decreases in magnitude. Also the width of the ellipse decreases, until at frequencies well above the natural frequency the ellipse is again reduced to a line which lies almost parallel to the x axis.

When the system damping is very small the frequency at which the major axis of the ellipse is vertical, a phase shift of 90 degrees, may be measured and taken as the natural frequency.

Exercise

Sketch the shape of the Lissajous' figure for the following combinations, marking the direction of motion around the figure. Take $b < a$ in all cases.

$x = a \sin \omega t$ for
(*a*) $y = b \sin \omega t$
(*b*) $y = b \sin (\omega t + 0{\cdot}5\pi)$
(*c*) $y = b \sin (\omega t - 0{\cdot}5\pi)$
(*d*) $y = b \sin 2\omega t$
(*e*) $y = b \sin (2\omega t + 0{\cdot}5\pi)$
(*f*) $y = b \sin 3\omega t$

1.3 Dimensionless parameters

The curves of Fig. 1.17 are particular to a system in which k/m has the value 2500 s^{-2}. For any other value of this ratio it can be shown that the response curves are similar, differing only in scale from those illustrated in Fig. 1.17. When a large number of different systems of the same generic form are to be designed, or when general characteristics of a family of systems are to be investigated it is desirable to plot the results in terms of dimensionless parameters, if possible. In the present case it is possible, as may be shown by transforming the equation of motion, equation (1.22), as follows

$$m\ddot{x} + c\dot{x} + kx = F \cos \omega t \qquad [(1.22)]$$

Expressing it in terms of $\theta = \omega t$ we get

$$m\omega^2 \frac{d^2x}{d\theta^2} + c\omega \frac{dx}{d\theta} + kx = F \cos \theta \qquad (1.31)$$

Dividing through by $m\omega^2$ gives

$$\frac{d^2x}{d\theta^2} + \frac{c}{m\omega}\frac{dx}{d\theta} + \frac{k}{m\omega^2}x = \frac{F}{m\omega^2}\cos \theta \qquad (1.32)$$

If we let p be the undamped natural angular frequency $[=(k/m)^{1/2}]$
c_{crit} be the critical damping

and X_s be the displacement amplitude as ω approaches zero, that is the static deflection,

then equation (1.32) becomes

$$\frac{d^2x}{d\theta^2} + \frac{2c}{c_{crit}}\frac{p}{\omega}\frac{dx}{d\theta} + \frac{p^2}{\omega^2}x = X_s\frac{p^2}{\omega^2}\cos\theta \tag{1.33}$$

from which it is evident that the results will be a function of the dimensionless ratios

$$\frac{2c}{c_{crit}}\ (=\delta) \quad \text{and} \quad \frac{\omega}{p}\ (=n)$$

With these substitutions equation (1.33) becomes

$$n^2\frac{d^2x}{d\theta^2} + n\,\delta\frac{dx}{d\theta} + x = X_s\cos\theta \tag{1.34}$$

The dynamic amplitude, the amplitude of x when n is not equal to zero, may therefore be expressed as

$$X_n = X_s\mathscr{M} \tag{1.35}$$

where $\mathscr{M}$ is known as the dynamic magnifier

$$\mathscr{M} = \frac{1}{\{(1-n^2)^2 + (n\,\delta)^2\}^{1/2}} \tag{1.36}$$

The phase angle between the exciting force and the displacement, in terms of n and δ, is

$$\phi = \tan^{-1}\frac{n\,\delta}{1-n^2} \tag{1.37}$$

Another dimensionless parameter Q is often used to express the magnitude of the damping, especially in electrical circuits. Referred to as the Q-factor, it is the reciprocal of the damping factor δ. Hence a high-Q system is one with very little damping.

Summarising, we have shown that the results given in Fig. 1.17 may be generalised by plotting the non-dimensional quantities $\mathscr{M}$ and the phase angle ϕ against the ratio n for various values of δ.

1.4 Hysteretic damping

So far the system damping considered has been assumed to obey a viscosity law relating to liquids. When the damping arises from the internal friction of a solid, experiments indicate that this law is not obeyed, see for example Kimbal and Lovell (1927).

Viscous damping implies that the damping force depends upon the velocity of strain, but for tests in which the material was strained sinusoidally at constant frequencies, Kimbal and Lovell found the internal friction force to be independent of velocity. It depended upon the amplitude of strain and, approximately, the energy loss per cycle was proportional to the square of the amplitude; the strain was, of course, within the elastic limit throughout.

Damping of this nature is now generally referred to as hysteretic, material, structural, or internal damping. When the energy loss is proportional to the square of the amplitude of strain it is known as linear hysteretic damping. In that case the hysteretic resistance may be represented by a force in phase with the velocity $\dot{x}$ and proportional to $\dot{x}/\omega$, ω being the constant frequency of the sinusoidal straining. Referring again to Fig. 1.15, if the damping in that system were hysteretic the equation of motion corresponding to equation (1.22) would become

$$m\ddot{x} + c_H\dot{x}/\omega + kx = F\cos\omega t$$

or

$$m\ddot{x} + (k + ic_H)x = F\cos\omega t \tag{1.38}$$

where c_H is the hysteretic damping coefficient. Thus the combined effect of the elastic and hysteretic resistance may be represented as a complex stiffness.

Exercise

For the system shown in Fig. 1.15, assume that the damping is linear hysteretic in character and show that the steady state amplitude is then

$$F/\{(k - m\omega^2)^2 + c_H{}^2\}^{1/2}$$

and the steady state phase-angle between the displacement and the force vectors is

$$\phi = \tan^{-1}\left\{\frac{c_H}{k - m\omega^2}\right\}$$

1.5 Exciting force of complex waveform

In many practical situations the exciting force has a more complex waveform than the simple harmonic function. For example, it is well known that reciprocating masses driven by slider-crank mechanisms do not execute simple harmonic motion, but a motion composed of a number of harmonic components. Consequently, the force required to cause the motion is also of complex form. A second example is the torque generated by a reciprocating engine, which is also a complex periodic function of time. Another well known case is that of the force generated by electromagnetic vibrators, which often deviates from a simple harmonic form due to non-linearities in the electrical and magnetic circuits. However, each of these and the many other periodic forcing functions which arise in engineering practice may be represented by the Fourier series. Fortunately, this does not make the vibration analysis of such systems much more complicated as they may generally be considered as linear, and therefore the principle of superposition applies. That is, we may solve for the displacement due to each component of the Fourier series and add them vectorially to obtain the displacement due to the complex waveform of the exciting force. Thus, if we have

$$m\ddot{x} + c\dot{x} + kx = F_1\cos q_1\omega t + F_2\cos q_2\omega t + F_3\sin q_3\omega t \tag{1.39}$$

where F_j are the amplitudes of the individual harmonic components which make up the complex waveform of the exciting force and q_j are positive integers, or in terms of non-dimensional parameters

$$n^2 \frac{d^2x}{d\theta^2} + n\,\delta \frac{dx}{d\theta} + x = X_{s1} \cos q_1\theta + X_{s2} \cos q_2\theta + X_{s3} \sin q_3\theta \quad (1.40)$$

where X_{sj} are the static displacements of the system due to forces F_j, then

$$x = \text{Re}\, \frac{X_{s1}\, e^{iq_1\theta}}{1 - q_1{}^2n^2 + iq_1 n\,\delta} + \text{Re}\, \frac{X_{s2}\, e^{iq_2\theta}}{1 - q_2{}^2n^2 + iq_2 n\,\delta} + \text{Im}\, \frac{X_{s3}\, e^{iq_3\theta}}{1 - q_3{}^2n^2 + iq_3 n\,\delta} \quad (1.41)$$

Writing the vector $1 - q_j{}^2n^2 + iq_j n\,\delta$ as $\exp(i\phi_j)/\mathcal{M}_j$

where $\mathcal{M}_j = \{(1 - q_j{}^2n^2)^2 + (q_j n\,\delta)^2\}^{-1/2}$

and $\phi_j = \tan^{-1} \dfrac{q_j n\,\delta}{1 - q_j{}^2n^2}$

gives $x = \mathcal{M}_1 X_{s1} \cos (q_1\omega t - \phi_1) + \mathcal{M}_2 X_{s2} \cos (q_2\omega t - \phi_2) + \mathcal{M}_3 X_{s3} \sin (q_3\omega t - \phi_3)$ (1.42)

This shows that the motion of the system is the sum, in their correct phase relationships, of the motions arising from each of the harmonics applied separately.

As this is an important point in understanding the vibratory phenomena found, for example, in reciprocating engine installations, a case with $X_{s1} = 0{\cdot}01$ m, $X_{s2} = 0{\cdot}005$ m, $X_{s3} = 0{\cdot}002$ m, $q_1 = 1$, $q_2 = 2$, $q_3 = 3$, and $\delta = 0.1$ has been calculated and the results given in Fig. 1.19 and 1.20. Figure 1.19(*a*) shows the amplitude variations with n which result from the application of each of the harmonic components separately. Figures 1.19(*b*), 1.19(*c*), 1.20(*a*), and 1.20(*b*) show the combined displacement waveform at the values of n given. The waveform of the exciting force is shown in Fig. 1.20(*c*). The natural frequency of the system is $n = 1$ but due to the second harmonic component in the exciting force a high displacement amplitude arises at $n = 0{\cdot}5$. When dealing with rotating or reciprocating machinery the value of ω associated with $n = 0{\cdot}5$ is often called the second-order critical speed and two cycles of vibration occur during one revolution of the crankshaft or rotor shaft. Similarly the value of ω associated with $n = \frac{1}{3}$ is often called the third-order critical speed. In other words, the number of complete cycles of the vibration per revolution of the reference shaft of the machine is referred to as the 'order' of the vibration. In Figs. 1.19 and 1.20 the horizontal dotted lines indicate the magnitude that the peak-to-peak displacement would assume if it resulted from the simple sum of the peak amplitudes. It would, in each case, be an overestimate. Also shown in Fig. 1.19(*b*) is the amplitude which would arise from the third-order vibration alone. This would considerably underestimate the peak-to-peak movement. These vibratory movements induce stresses, in

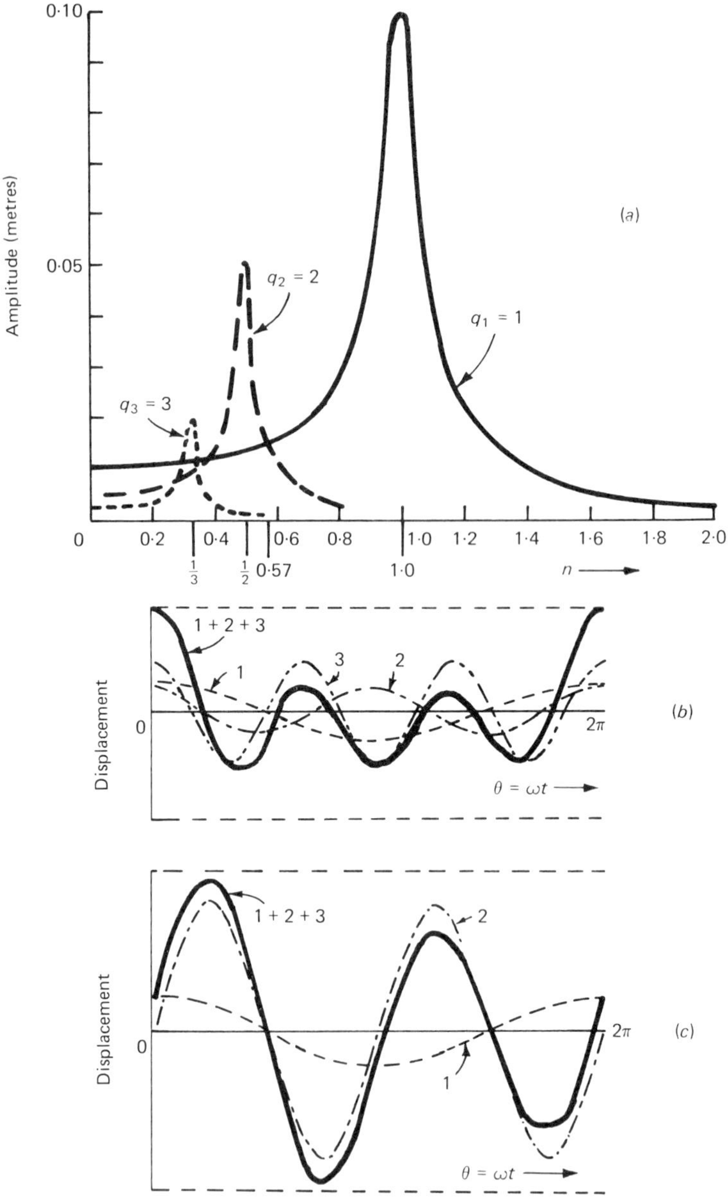

Fig. 1.19 Response of a system to a complex cyclic exciting force (*i*):

(*a*) Amplitude due to each component of force separately

(*b*) Variation of displacement with time at $n = \frac{1}{3}$

(*c*) Variation of displacement with time at $n = \frac{1}{2}$

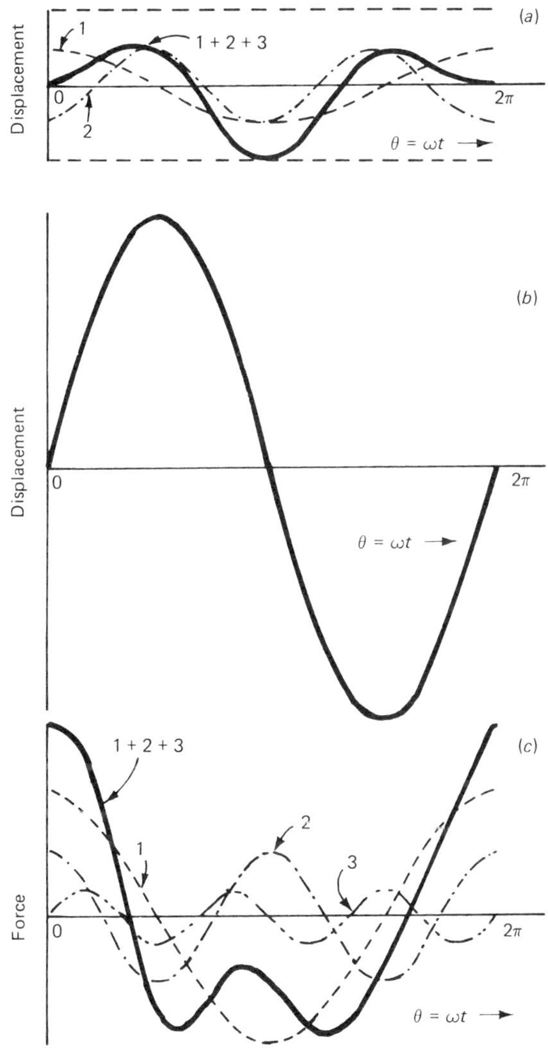

Fig. 1.20 Response of a system to a complex cyclic exciting force (*ii*):

(*a*) Variation of displacement with time at $n = 0{\cdot}57$

(*b*) Variation of displacement with time at $n = 1{\cdot}0$

(*c*) Exciting force and its components

the elastic members (support springs, crankshafts, etc.) which, if too high, can lead to fatigue failure. Therefore, at a critical speed, it is important to check that the peak-to-peak stresses, including those arising from the flanks of the response curves associated with the other harmonics, will not lead to premature failure of the component.

1.6 Force transmitted to the foundation

The displacement amplitude and its dynamic magnification are very important factors to consider in the design of a spring-mounted machine, but at least of equal importance in many cases is the force transmitted to its foundations. This force causes disturbances to be propagated through the ground, which could interfere with the proper functioning of delicate instruments or machines, or, if large, cause damage to nearby buildings. The force transmitted to the foundations may be found easily from the vector sum of the spring and damping forces. Again we may use a non-dimensional form

$$F_n = F_s \mathcal{T} \qquad (1.43)$$

where F_n is the amplitude of the force transmitted to the foundations when the ratio ω/p has the value n.

F_s is the amplitude of the force acting on the mass as this would be transmitted to the foundations unchanged in the static case and $\mathcal{T}$ is a non-dimensional factor which we shall call the transmissibility.

Exercise

Show that the amplitude of the force transmitted to the support of the system, F_n, shown in Fig. 1.15, when the exciting force on the mass has an angular frequency ω and an amplitude F_s, is

$$F_n = F_s \left\{ \frac{k^2 + (c\omega)^2}{(k - m\omega^2)^2 + (c\omega)^2} \right\}^{1/2}$$

if c is less than c_{crit}.

Let $f_n = F_s \mathcal{T} \sin(\omega t - \phi)$ and consider the case of $c < c_{crit}$, then show that in non-dimensional terms

$$\mathcal{T} = \left\{ \frac{1 + (n\,\delta)^2}{(1 - n^2)^2 + (n\,\delta)^2} \right\}^{1/2}$$

and

$$\phi = \tan^{-1} \frac{n^3\,\delta}{1 - n^2 + n^2\,\delta^2}$$

Plot $\mathcal{T}$ and ϕ as functions of n when $\delta = 0.2$.

1.7 Excitation by a non-periodic force

Periodic forces of any general waveform may be represented by Fourier's series as a superposition of harmonic components of various frequencies. The response of a linear system is then found by superposition of the harmonic response to each of the exciting forces. When the exciting force is non-periodic, such as that arising from an earthquake or that due to the blast from an explosion, another method for calculating the response of the system is required. A standard method used is based upon the following physical reasoning. We know the response of a system to a unit impulse. This was discussed in section 1.2, and there is no difficulty in representing any varying force as a series of impulses, as shown in Fig. 1.21. Therefore, the total response of the system to a non-periodic force may be obtained approximately by superimposing the responses to the discrete impulses

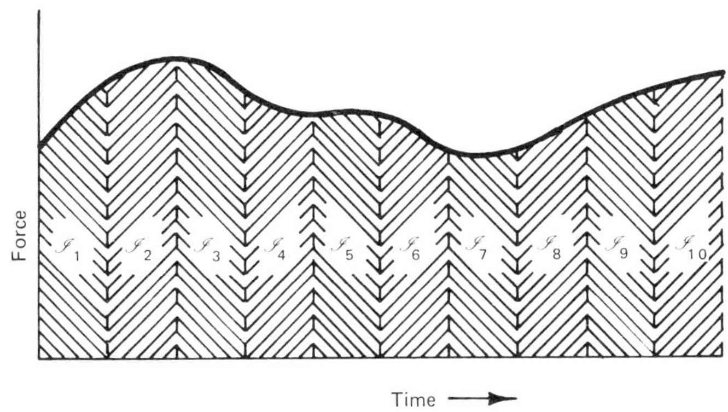

Fig. 1.21 The representation of a varying force as a series of impulses

$\mathscr{I}(\tau_k)$ which approximately represent the applied force. Consider the two impulses shown in Fig. 1.22. The impulse response of a system, $i(t - \tau)$, is the displacement of an initially quiescent system at time $(t - \tau)$ after being struck by a unit impulse at time τ. Therefore, the response of the system we are considering is

(*i*) between $t = 0$ and $t = \tau_1$, $x = 0$

(*ii*) between $t = \tau_1$ and $t = \tau_2$, $x = \mathscr{I}(\tau_1)i(t - \tau_1)$

(*iii*) from $t = \tau_2$ onwards, $x = \mathscr{I}(\tau_1)i(t - \tau_1) + \mathscr{I}(\tau_2)i(t - \tau_2)$

For a complicated variation of force with time the response would be

$$x(t) = \sum_k \mathscr{I}(\tau_k)i(t - \tau_k) \quad \text{for } t > \tau_k \tag{1.44}$$

When $(t - \tau_k)$ is negative $i(t - \tau_k)$ is zero.

Obviously the approximate response may be made more exact by refining the steps in the time dimension used to represent the applied force, until in the limit it becomes

$$x(t) = \int_{t_0}^{t} f(\tau)i(t - \tau)\,\mathrm{d}\tau; \tag{1.45}$$

the system being quiescent at time t_0. If the system is not initially quiescent, but has an initial displacement $x(0)$ and velocity $\dot{x}(0)$ these are easily allowed for by superposition of the appropriate transients.

The integral, equation (1.45), is known in mathematical texts as the convolution integral or Duhamel's integral. Whereas we have obtained it by physical reasoning, it is not difficult to show that it satisfies the equation of motion for a particular system. As an example we shall prove that this is so for the system in Fig. 1.15.

Recalling the equation of motion, equation (1.22), and generalising the applied force to be $f(t)$, any function of time, gives

$$m\ddot{x} + c\dot{x} + kx = f(t) \tag{1.46}$$

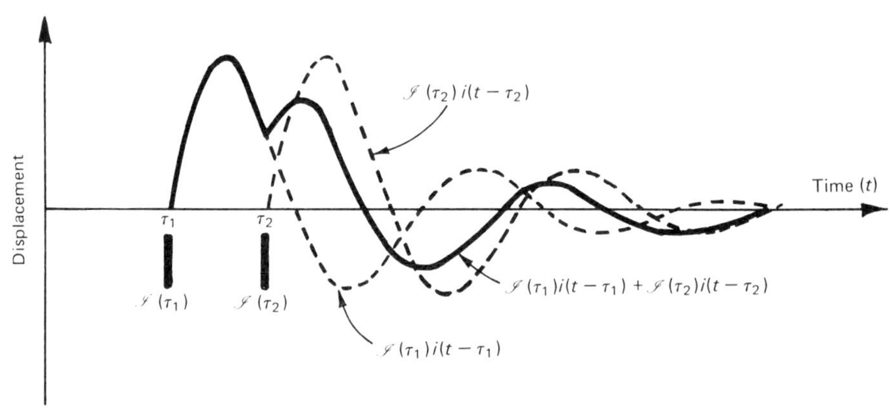

Fig. 1.22 Response of an initially quiescent system to two impulses of the same value

We shall now show that the complete solution of this equation may be expressed in the form

$$x = \int_{t_0}^{t} f(\tau)x_1(t - \tau)\,\mathrm{d}\tau \tag{1.47}$$

where $x_1(\xi)$ is the solution of the equation

$$m\ddot{x}_1 + c\dot{x}_1 + kx_1 = 0 \tag{1.48}$$

subject to the initial conditions $x_1(0) = 0$ and $\dot{x}_1(0) = 1/m$. That is, x_1 is the response of the system to a unit impulse at $\xi = 0$.

To show that this is true we differentiate equation (1.47) with respect to t and substitute for $\ddot{x}$, $\dot{x}$, and x in equation (1.46). This involves the differentiation of a finite definite integral and in mathematical texts, see for example Sokolnikoff and and Redheffer (1958), it is shown that

$$\frac{d}{dt}\int_a^b f(x, t)\,dx = \int_a^b \frac{\partial}{\partial t} f(x, t)\,dx + f(b, t)\frac{db}{dt} - f(a, t)\frac{da}{dt} \tag{1.49}$$

Using this relationship and taking into account the initial conditions on x_1, gives in our case

$$\dot{x} = \int_{t_0}^{t} f(\tau)\dot{x}_1(t - \tau)\,d\tau \tag{1.50}$$

$$\ddot{x} = \int_{t_0}^{t} f(\tau)\ddot{x}_1(t - \tau)\,d\tau + \frac{f(t)}{m} \tag{1.51}$$

Substituting for $\ddot{x}$, $\dot{x}$, and x in equation (1.46) gives

$$\int_{t_0}^{t} f(\tau)\{m\ddot{x}_1(\xi) + c\dot{x}_1(\xi) + kx_1(\xi)\}\,d\tau = 0 \tag{1.52}$$

Since the term in braces is identically zero, the value of x given by equation (1.47) satisfies the complete equation (1.46); and from equations (1.47) and (1.50) we have $x(t_0) = 0$ and $\dot{x}(t_0) = 0$.

1.8 The variation of amplitude with time

Previously we have not been concerned with the manner in which the steady state amplitude has been achieved. This is reasonable because for most practically important systems the steady amplitude is achieved quickly and the rate at which it is approached is of little interest. However, many mechanical systems have one or more resonant frequencies, or critical speeds, below their normal operating speeds and it is common knowledge that a system may be run through critical speeds up to its normal operating speed, without deleterious effects. Thus the time variation of amplitude may be of major interest when running through a critical speed but unfortunately great difficulty is experienced in finding a mathematical expression for it (for attempts see bibliography). Therefore, we shall demonstrate the approach by examining an easier case, the build-up of amplitude at a constant exciting-force frequency equal to the natural frequency of the system.

Consider the case in which, starting from rest, an undamped system is subjected to a force of instantaneous magnitude $F \cos pt$, where p is the natural frequency of the one degree of freedom system being excited. By equation (1.47) the response is

$$x = \int_0^t \frac{F}{pm} \cos p\tau \sin p(t - \tau)\,d\tau \tag{1.53}$$

therefore

$$x = X_s \frac{pt}{2} \sin pt \tag{1.54}$$

Thus, when excited at its natural frequency, the amplitude of an undamped system grows linearly with time.

Exercises

1. Show that when an exciting force $F \sin \omega t$, ω not equal to the system natural frequency, is applied to an initially quiescent system comprised of a mass m attached to a rigid support by means of a spring of stiffness k, the amplitude of oscillation of the mass builds up to a value of $F/(k - m\omega^2)$. Derive an expression for the time to reach this steady state condition.
2. Repeat the task set in question 1 for a system in which, in addition to the spring, viscous damping acts between the mass and the support.
3. For the system examined in question 2 take $m = 10$ kg, $k = 10$ kN m^{-1}, and $c = 126{\cdot}5$ N s m^{-1} and find, graphically, its response to a single triangular pulse of force. The pulse shape is a linear variation from zero to 20 N in 4 s, then an instantaneous drop to zero. The system is initially quiescent.

References

KIMBAL, A. L. and LOVELL, D. E.
'Internal friction in solids', *Phys. Rev.*, 2nd Ser., **30**, 1927, 948–59.

SOKOLNIKOFF, I. S. and REDHEFFER, R. M.
Mathematics of physics and modern engineering, McGraw-Hill, 1958, pp. 261–2.

Bibliography

ELLINGTON, J. P. and McCALLION, H.
'On running a machine through its resonant frequency', *J. Roy. Aeronaut. Soc.*, **60**, 1956, 549.

LEWIS, F. M.
'Vibration during acceleration through a critical speed', *Trans. Am. Soc. Mech. Engrs.*, **54**, 1932, 253.

HOTHER-LUSHINGTON, S. and JOHNSON, D. C.
'The acceleration of a single-degree-of-freedom system through its resonant frequency', *J. Roy. Aeronaut. Soc.*, **62**, 1958, 752.

MARPLES, V.
'Transition of a rotating shaft through a critical speed', *Proc. Inst. Mech. Engrs.*, **180**, part 3I, 1965–6, 8.

Chapter 2

Systems with more than one degree of freedom

These systems also have more than one natural frequency; we are now ready to study their vibratory behaviour. For undamped systems an important principle will be demonstrated, namely that the principal modes of vibration are orthogonal. Use will be made of this fact in chapter 5.

Under conditions of forced vibration these systems also have more than one resonant frequency. Sometimes, in practice, a resonant frequency of the primary system, for example an engine alternator set, falls within its operating range. It will be shown that this dangerous condition may be removed by adding a subsystem known as a vibration absorber. The influence of damping will also be discussed.

To set up the equations of motion by Newtonian methods often requires considerable thought. As a step towards reducing it, Lagrange's form of the equations of motion will be introduced. Later a routine method, suitable for use with modern automatic digital computers, for setting them up in matrix form is given.

Physical understanding of the forced vibratory behaviour of more complex systems can often be aided by examining the variation with frequency of the dynamic stiffnesses of its subsystems. This will be illustrated in one of the final sections.

2.1 Orthogonality of principal modes

2.1.1 Planar systems

The mass m shown in Fig. 2.1(a) is constrained to vibrate only in the plane containing the centre lines of the springs of stiffnesses k_1 and k_2. The equations of motion of the system in the x, y coordinate system are

$$\begin{aligned} m\ddot{x} + k_1 x &= 0 \\ m\ddot{y} + k_2 y &= 0 \end{aligned} \tag{2.1}$$

These equations are independent, thus their solutions are

$$\begin{aligned} x &= A_1 \sin \omega_1 t + B_1 \cos \omega_1 t \\ y &= A_2 \sin \omega_2 t + B_2 \cos \omega_2 t \end{aligned} \tag{2.2}$$

where $\omega_1{}^2 = k_1/m \quad \text{and} \quad \omega_2{}^2 = k_2/m$

That is, the system may vibrate in the x-direction at the natural angular frequency ω_1, the first mode (assuming $k_1 < k_2$), and it may vibrate in the y-direction at a higher frequency ω_2, the second mode. When the modes of vibration are independent, as found here, they are called principal modes. As the coordinates x and y are at 90 degrees to one another, the motion in one mode is orthogonal to that in the other mode.

Exercise

If for the system shown in Fig. 2.1(*a*) the mass is 2 kg and the spring stiffnesses are $k_1 = 50$ N/m and $k_2 = 200$ N/m, draw the locus of the mass when released from rest at $x_0 = y_0 = 50$ mm, the static equilibrium position being $x = y = 0$.

The above result should be independent of the coordinate system chosen. Let us show that this is so by setting up the equations of motion in the X, Y coordinate

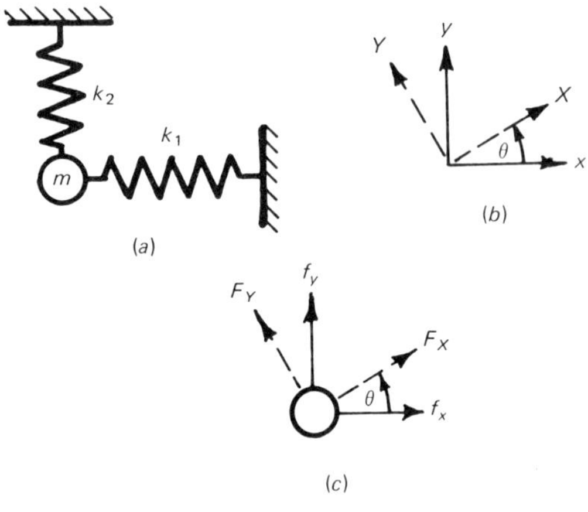

Fig. 2.1(*a*), (*b*), (*c*)

system shown in Fig. 2.1(*b*). In matrix notation the relationship between x, y and X, Y is

$$\begin{bmatrix} x \\ y \end{bmatrix} = \begin{bmatrix} \cos\theta & -\sin\theta \\ \sin\theta & \cos\theta \end{bmatrix} \begin{bmatrix} X \\ Y \end{bmatrix} \tag{2.3}$$

and that between the forces on the mass in the two coordinates systems is

$$\begin{bmatrix} F_X \\ F_Y \end{bmatrix} = \begin{bmatrix} \cos\theta & \sin\theta \\ -\sin\theta & \cos\theta \end{bmatrix} \begin{bmatrix} f_x \\ f_y \end{bmatrix} \tag{2.4}$$

The equations of motion in matrix notation become

$$\begin{bmatrix} f_x \\ f_y \end{bmatrix} = \begin{bmatrix} -k_1 & 0 \\ 0 & -k_2 \end{bmatrix} \begin{bmatrix} x \\ y \end{bmatrix} = \begin{bmatrix} m\mathrm{D}^2 & 0 \\ 0 & m\mathrm{D}^2 \end{bmatrix} \begin{bmatrix} x \\ y \end{bmatrix} \tag{2.5}$$

where
$$D \equiv \frac{\mathrm{d}}{\mathrm{d}t}$$

Substituting equation (2.5) in (2.4) making use of (2.3) gives

$$\begin{bmatrix} F_X \\ F_Y \end{bmatrix} = \begin{bmatrix} \cos\theta & \sin\theta \\ -\sin\theta & \cos\theta \end{bmatrix} \begin{bmatrix} -k_1 & 0 \\ 0 & -k_2 \end{bmatrix} \begin{bmatrix} \cos\theta & -\sin\theta \\ \sin\theta & \cos\theta \end{bmatrix} \begin{bmatrix} X \\ Y \end{bmatrix}$$
$$= \begin{bmatrix} \cos\theta & \sin\theta \\ -\sin\theta & \cos\theta \end{bmatrix} \begin{bmatrix} m\mathrm{D}^2 & 0 \\ 0 & m\mathrm{D}^2 \end{bmatrix} \begin{bmatrix} \cos\theta & -\sin\theta \\ \sin\theta & \cos\theta \end{bmatrix} \begin{bmatrix} X \\ Y \end{bmatrix} \tag{2.6}$$

Therefore in the X, Y coordinate system the equation of motion is

$$\begin{bmatrix} \cos\theta & \sin\theta \\ -\sin\theta & \cos\theta \end{bmatrix} \begin{bmatrix} k_1 + m\mathrm{D}^2 & 0 \\ 0 & k_2 + m\mathrm{D}^2 \end{bmatrix} \begin{bmatrix} \cos\theta & -\sin\theta \\ \sin\theta & \cos\theta \end{bmatrix} \begin{bmatrix} X \\ Y \end{bmatrix} = 0 \tag{2.7}$$

As the system is linear the solution is of the form

$$\begin{bmatrix} X \\ Y \end{bmatrix} = \begin{bmatrix} A_1 & B_1 \\ A_2 & B_2 \end{bmatrix} \begin{bmatrix} \sin\omega t \\ \cos\omega t \end{bmatrix} \tag{2.8}$$

Substituting this in equation (2.7) gives

$$\begin{bmatrix} \cos\theta & \sin\theta \\ -\sin\theta & \cos\theta \end{bmatrix} \begin{bmatrix} k_1 - m\omega^2 & 0 \\ 0 & k_2 - m\omega^2 \end{bmatrix} \begin{bmatrix} \cos\theta & -\sin\theta \\ \sin\theta & \cos\theta \end{bmatrix} \begin{bmatrix} A_1 & B_1 \\ A_2 & B_2 \end{bmatrix} \begin{bmatrix} \sin\omega t \\ \cos\omega t \end{bmatrix} = 0 \tag{2.9}$$

The value on the right-hand side is independent of time, also all of A_1, A_2, B_1, and B_2 cannot be zero or there would be no vibration. The case of no vibration is referred to as a trivial solution to equation (2.9). Therefore, for the whole expression to be equal to zero the determinant of the coefficients of the A's and B's must be zero. That is

$$\left| \begin{bmatrix} \cos\theta & \sin\theta \\ -\sin\theta & \cos\theta \end{bmatrix} \begin{bmatrix} k_1 - m\omega^2 & 0 \\ 0 & k_2 - m\omega^2 \end{bmatrix} \begin{bmatrix} \cos\theta & -\sin\theta \\ \sin\theta & \cos\theta \end{bmatrix} \right| = 0 \tag{2.10}$$

This determinant is often called the eliminant [Aitken, 1946*a*]. It is also called the frequency equation because it is only satisfied by special values of the frequency ω. These special values are called characteristic values, all other values being known as ordinary values. The characteristic values of ω^2 are generally known as the eigenvalues of the matrix.

Now the determinant of the product of a number of matrices is equal to the product of the determinants of the matrices [Aitken, 1946*b*], giving

$$\left| \begin{bmatrix} \cos\theta & \sin\theta \\ -\sin\theta & \cos\theta \end{bmatrix} \right| \times \left| \begin{bmatrix} k_1 - m\omega^2 & 0 \\ 0 & k_2 - m\omega^2 \end{bmatrix} \right| \times \left| \begin{bmatrix} \cos\theta & -\sin\theta \\ \sin\theta & \cos\theta \end{bmatrix} \right| = 0 \tag{2.11}$$

Only the determinant in the centre is a function of ω, therefore ω must satisfy the equation

$$\begin{vmatrix} k_1 - m\omega^2 & 0 \\ 0 & k_2 - m\omega^2 \end{vmatrix} = 0 \qquad (2.12)$$

that is $\omega^2 = k_1/m$ or k_2/m as before. Hence we may conclude that the natural frequencies or eigen frequencies, and consequently eigenvalues, are independent of the coordinate system chosen. Let us now check that the modes of vibration are

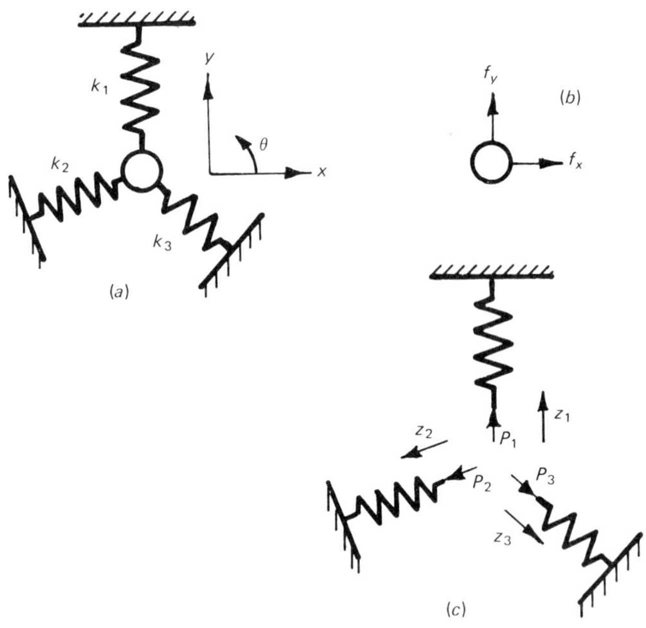

Fig. 2.2(*a*), (*b*), (*c*)

also independent of the coordinate system. Substituting $\omega^2 = k_1/m$ into equation (2.9) and multiplying the matrices gives

$$0x = 0 \qquad [2.13(a)]$$

$$(k_2 - k_1)y = 0 \qquad [2.13(b)]$$

From equation [2.13(*a*)], the amplitude of x is indeterminate, it may take any value, and from equation [2.13(*b*)] the amplitude of y must be zero. A similar substitution for the other eigenvalue gives the amplitude of x as zero with that of y being indeterminate. These are identical with the mode shapes in the x, y coordinate system.

We are now in a position to examine the more complex system shown in Fig. 2.2. In this case the mass m is free to vibrate in the plane containing the centre lines of the three springs of stiffness k_1, k_2, and k_3. Referring to Fig. 2.2*c*, the axial thrust P_i required to shorten the *i*'th spring by an amount z_i is given by

$$\begin{bmatrix} P_1 \\ P_2 \\ P_3 \end{bmatrix} = \begin{bmatrix} k_1 & 0 & 0 \\ 0 & k_2 & 0 \\ 0 & 0 & k_3 \end{bmatrix} \begin{bmatrix} z_1 \\ z_2 \\ z_3 \end{bmatrix} \tag{2.14}$$

Also from Figs. 2.2(*b*) and (*c*) it is obvious that

$$\begin{bmatrix} f_x \\ f_y \end{bmatrix} + \begin{bmatrix} \cos\theta_1 & \cos\theta_2 & \cos\theta_3 \\ \sin\theta_1 & \sin\theta_2 & \sin\theta_3 \end{bmatrix} \begin{bmatrix} P_1 \\ P_2 \\ P_3 \end{bmatrix} = 0 \tag{2.15}$$

and from (*a*) and (*c*) of that figure it is obvious that

$$\begin{bmatrix} z_1 \\ z_2 \\ z_3 \end{bmatrix} = \begin{bmatrix} \cos\theta_1 & \sin\theta_1 \\ \cos\theta_2 & \sin\theta_2 \\ \cos\theta_3 & \sin\theta_3 \end{bmatrix} \begin{bmatrix} x \\ y \end{bmatrix} \tag{2.16}$$

Using equations (2.14), (2.15), and (2.16) together with Newton's laws gives the equations of motion for the mass as

$$\begin{bmatrix} f_x \\ f_y \end{bmatrix} = - \begin{bmatrix} \cos\theta_1 & \cos\theta_2 & \cos\theta_3 \\ \sin\theta_1 & \sin\theta_2 & \sin\theta_3 \end{bmatrix} \begin{bmatrix} k_1 & 0 & 0 \\ 0 & k_2 & 0 \\ 0 & 0 & k_3 \end{bmatrix} \begin{bmatrix} \cos\theta_1 & \sin\theta_1 \\ \cos\theta_2 & \sin\theta_2 \\ \cos\theta_3 & \sin\theta_3 \end{bmatrix} \begin{bmatrix} x \\ y \end{bmatrix}$$

$$= \begin{bmatrix} mD^2 & 0 \\ 0 & mD^2 \end{bmatrix} \begin{bmatrix} x \\ y \end{bmatrix} \tag{2.17}$$

As the solution is of the form

$$x = A_1 \sin(\omega t + \phi_1); \qquad y = A_2 \sin(\omega t + \phi_2) \tag{2.18}$$

multiplying out the matrices and substituting gives

$$- \begin{bmatrix} \sum_i k_i \cos^2\theta_i & \sum_i k_i \sin\theta_i \cos\theta_i \\ \sum_i k_i \sin\theta_i \cos\theta_i & \sum_i k_i \sin^2\theta_i \end{bmatrix} \begin{bmatrix} x \\ y \end{bmatrix} = -m\omega^2 \begin{bmatrix} 1 & 0 \\ 0 & 1 \end{bmatrix} \begin{bmatrix} x \\ y \end{bmatrix} \tag{2.19}$$

There are two degrees of freedom and, therefore, two principal modes of vibration. If these principal modes are orthogonal it should be possible to rotate the coordinate system, by an angle α, to a new coordinate system X, Y in which the vibrations on the coordinate axes are independent of one another. By the same considerations as in the two-spring case, we can show that the resulting equations are

$$\begin{bmatrix} \cos\alpha & \sin\alpha \\ -\sin\alpha & \cos\alpha \end{bmatrix} \begin{bmatrix} \sum_i k_i \cos^2\theta_i & \sum_i k_i \sin\theta_i \cos\theta_i \\ \sum_i k_i \sin\theta_i \cos\theta_i & \sum_i k_i \sin^2\theta_i \end{bmatrix} \begin{bmatrix} \cos\alpha & -\sin\alpha \\ \sin\alpha & \cos\alpha \end{bmatrix} \begin{bmatrix} X \\ Y \end{bmatrix}$$

$$= m\omega^2 \begin{bmatrix} 1 & 0 \\ 0 & 1 \end{bmatrix} \begin{bmatrix} X \\ Y \end{bmatrix} \tag{2.20}$$

That is

$$\begin{bmatrix} \sum_i \{k_i[\cos^2\theta_i \cos^2\alpha + \sin^2\theta_i \sin^2\alpha & \sum_i \{k_i[(\sin^2\theta_i - \cos^2\theta_i)\sin\alpha\cos\alpha \\ \quad + 2\sin\theta_i\cos\theta_i\sin\alpha\cos\alpha]\} & \quad + \sin\theta_i\cos\theta_i(\cos^2\alpha - \sin^2\alpha)]\} \\ \sum_i \{k_i[(\sin^2\theta_i - \cos^2\theta_i)\sin\alpha\cos\alpha & \sum_i \{k_i[\cos^2\theta_i \sin^2\alpha + \sin^2\theta_i \cos^2\alpha \\ \quad + \sin\theta_i\cos\theta_i(\cos^2\alpha - \sin^2\alpha)]\} & \quad - 2\sin\theta_i\cos\theta_i\sin\alpha\cos\alpha]\} \end{bmatrix} \begin{bmatrix} X \\ Y \end{bmatrix}$$

$$= m\omega^2 \begin{bmatrix} 1 & 0 \\ 0 & 1 \end{bmatrix} \begin{bmatrix} X \\ Y \end{bmatrix} \quad (2.21)$$

The matrix on the left-hand side of this equation may be made diagonal by a suitable choice of α. That is, by making

$$\sum_i k_i(-\cos 2\theta_i \sin 2\alpha + \sin 2\theta_i \cos 2\alpha) = 0 \quad (2.22)$$

or

$$\tan 2\alpha = \frac{\sum_i k_i \sin 2\theta_i}{\sum_i k_i \cos 2\theta_i} \quad (2.23)$$

Obviously the lowest two values of 2α are separated by 180 degrees. Hence there are two values of α separated by 90 degrees, as in the previous two-spring case, and the principal modes are again orthogonal.

Observe that the selection of the values of α to bring the coordinate axes and the directions of the principal modes in line could have been made on physical considerations alone. From the statics of elastic systems, it is well known that planar systems, such as the one under consideration, have two principal axes of stiffness, which are at 90 degrees to each other. By definition, a force directed along a principal axis will cause a deflection along that axis only. There is no cross coupling. Hence the following uncoupled frequency equations could have been written down from the above considerations.

$$-2m\omega^2 + \sum_i k_i\{1 + \cos 2(\theta_i \pm \alpha)\} = 0 \quad (2.24)$$

where α is the smallest value which satisfies

$$\tan 2\alpha = \frac{\sum_i k_i \sin 2\theta_i}{\sum_i k_i \cos 2\theta_i}$$

2.1.2 Non-planar systems

In the previous section the coordinate axes, and the principal modes of vibration, were at 90 degrees to one another. Geometrically they were orthogonal. When in the physical space the coordinate axes of the two degrees of freedom are not at 90 degrees, as in the system shown in Fig. 2.3, the principal modes are still said to be orthogonal, but now it is in an abstract space. We shall lead up to this by considering the system of Fig. 2.3(*a*). The frequency equation is

$$\begin{vmatrix} -m\omega^2 + k_1 + k_2 & -k_2 \\ -k_2 & -m\omega^2 + k_2 \end{vmatrix} = 0 \qquad (2.25)$$

giving the solutions

$$\begin{aligned} \omega_{1,2}^2 &= \{p_1^2 + 2p_2^2 \pm \sqrt{(p_1^4 + 4p_2^4)}\}/2 \\ &= p_2^2\{\alpha + 1 \pm \beta\} \end{aligned} \qquad (2.26)$$

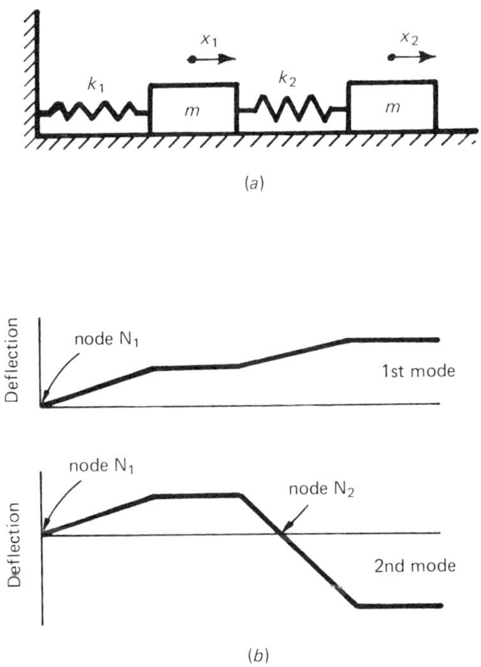

Fig. 2.3(*a*), (*b*)

where $$\alpha = p_1^2/2p_2^2; \qquad \beta^2 = \alpha^2 + 1$$

$$p_1^2 = \frac{k_1}{m} \quad \text{and} \quad p_2^2 = \frac{k_2}{m}$$

Substituting for ω in one of the equations of motion gives x_1 in terms of x_2 as follows

$$-p_2^2 x_1 + (-\omega^2 + p_2^2)x_2 = 0$$

or $$x_1 = -(\alpha \pm \beta)x_2 \qquad (2.27)$$

We are considering an undamped vibration, therefore we may arbitrarily fix the value of the amplitude of the mass at the free end for each of the frequencies. Let ${}_1A_1$ be the amplitude of x_1 in the first mode of vibration, ${}_1A_2$ be the amplitude of

x_2 in the first mode of vibration and ${}_2A_1$ and ${}_2A_2$ be the amplitude of x_1 and x_2 respectively in the second mode of vibration. Then in the first mode the relation between the amplitudes becomes

$$ {}_1A_1 = -(\alpha - \beta)\, {}_1A_2 \tag{2.28} $$

and in the second mode

$$ {}_2A_1 = -(\alpha + \beta)\, {}_2A_2 \tag{2.29} $$

Figure 2.3(*b*) shows, diagrammatically, the deflection of individual points in the system for each of the two modes. That is, they illustrate the mode shapes. A point which is permanently at rest or of zero amplitude relative to the reference system is called a node. The first mode of this system has one node, N_1, and the second mode two, N_1 and N_2.

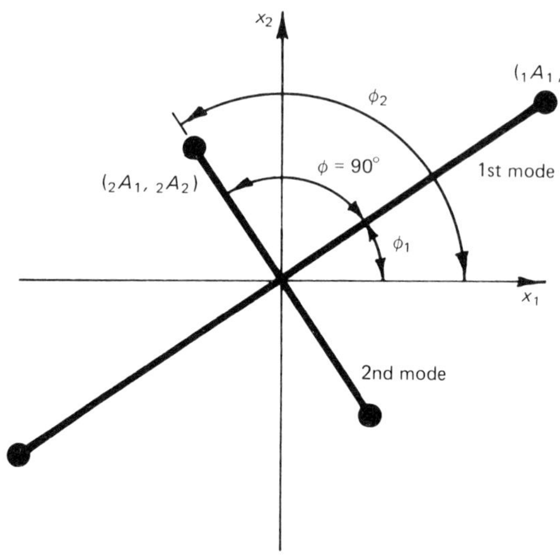

Fig. 2.4

If an abstract two-dimensional space is set up with coordinates x_1 and x_2 as shown in Fig. 2.4 then the principal modes of vibration occur on the straight lines given by equation (2.27). The angle ϕ_1 of the first mode is given by

$$ \tan\phi_1 = \frac{{}_1A_2}{{}_1A_1} = \frac{-1}{\alpha - \beta} $$

and the angle ϕ_2 of the second mode by

$$ \tan\phi_2 = \frac{{}_2A_2}{{}_2A_1} = \frac{-1}{\alpha + \beta} $$

Now,

$$ \tan\phi_1 \tan\phi_2 = \frac{1}{\alpha^2 - \beta^2} = -1 \tag{2.30} $$

Therefore in this abstract space the modes are at 90 degrees to one another; they are orthogonal, that is

$$\tan \phi = \tan (\phi_2 - \phi_1) = \frac{\tan \phi_2 - \tan \phi_1}{1 + \tan \phi_2 \tan \phi_1}$$

When the masses are equal, as in this example, the orthogonality condition may be written as

$$_1A_1\,{}_2A_1 + {}_1A_2\,{}_2A_2 = 0$$

for a system with two degrees of freedom. For a system with three degrees of freedom a similar geometrical interpretation to that given above may be found for the orthogonality condition.

If the masses are not equal, the coordinates representing the instantaneous state of the system in the abstract space, which we shall call state space, will depend upon the masses. That is, instead of representing the state by the coordinates x_1 and x_2 we represent them by the coordinates $m_1^{1/2}x_1$ ($=s_1$) and $m_2^{1/2}x_2$ ($=s_2$).

We shall now carry out the analysis for the system shown in Fig. 2.3 with the mass at 1 being m_1 and that at 2 being m_2. In the state space s_1, s_2 we get from the equations of motion

$$\begin{bmatrix} (-\omega^2 + p_{11}{}^2 + p_{21}{}^2)m_1{}^{-1/2} & -p_{21}{}^2 m_2{}^{-1/2} \\ -p_{22}{}^2 m_1{}^{-1/2} & (-\omega^2 + p_{22}{}^2)m_2{}^{-1/2} \end{bmatrix} \begin{bmatrix} s_1 \\ s_2 \end{bmatrix} = 0 \tag{2.31}$$

where

$$p_{ij}{}^2 = k_i/m_j$$

The frequency equation then becomes

$$\omega^4 - (p_{11}{}^2 + p_{21}{}^2 + p_{22}{}^2)\omega^2 + p_{11}{}^2 p_{22}{}^2 = 0 \tag{2.32}$$

or

$$\omega^4 - 2\alpha\omega^2 + \beta = 0$$

therefore,

$$\omega_{1,2}^2 = \alpha \pm \sqrt{(\alpha^2 - \beta)} \tag{2.33}$$

For the principal modes to be orthogonal in this space we must have

$$\frac{_1s_1}{_1s_2} \cdot \frac{_2s_1}{_2s_2} = -1 \tag{2.34}$$

That this is true may be established by substituting $\omega_1{}^2$ for ω^2 in the first line of equation (2.31) and $\omega_2{}^2$ in the second line to form the ratios required. The orthogonality condition for this case, when the masses are unequal, is then

$$_1s_1\,{}_2s_1 + {}_1s_2\,{}_2s_2 = 0 \tag{2.35}$$

or

$$m_1\,{}_1x_1\,{}_2x_1 + m_2\,{}_1x_2\,{}_2x_2 = 0 \tag{2.36}$$

The state space concept may be generalised to n dimensional space for a system with n degrees of freedom, but before embarking upon such a generalisation we shall examine the forms of the expressions for the kinetic and potential energies of our simple two degrees of freedom system.

By definition the instantaneous value of the kinetic energy is

$$T = \tfrac{1}{2}\{m_1\dot{x}_1^2 + m_2\dot{x}_2^2\} \tag{2.37}$$

But, as in general the free motion of an undamped system is comprised of motion in each of its principal modes, that is the solutions

$$x_1 = {}_1x_1 + {}_2x_1 = {}_1A_1 \sin(\omega_1 t + \psi_1) + {}_2A_1 \sin(\omega_2 t + \psi_2)$$
$$x_2 = {}_1x_2 + {}_2x_2 = {}_1A_2 \sin(\omega_1 t + \psi_1) + {}_2A_2 \sin(\omega_2 t + \psi_2)$$

satisfy the equations of motion for the present case,

$$\begin{aligned} T &= \tfrac{1}{2}\{m_1({}_1\dot{x}_1 + {}_2\dot{x}_1)^2 + m_2({}_1\dot{x}_2 + {}_2\dot{x}_2)^2\} \\ &= \tfrac{1}{2}\{\omega_1^2(m_1\,{}_1x_1^2 + m_2\,{}_1x_2^2) + \omega_2^2(m_1\,{}_2x_1^2 + m_2\,{}_2x_2^2) \\ &\qquad + 2\omega_1\omega_2(m_1\,{}_1x_1\,{}_2x_1 + m_2\,{}_1x_2\,{}_2x_2)\} \end{aligned}$$

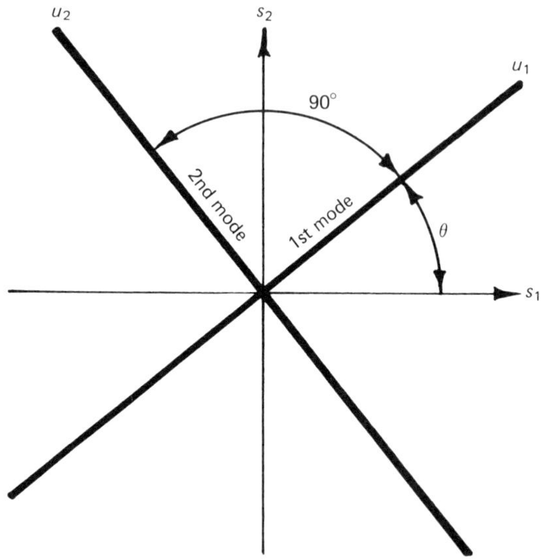

Fig. 2.5

But as the modes are orthogonal, the last term in parentheses is zero giving

$$T = \tfrac{1}{2}\{\omega_1^2({}_1s_1^2 + {}_1s_2^2) + \omega_2^2({}_2s_1^2 + {}_2s_2^2)\} \tag{2.38}$$

We may rotate the coordinate system from the s_1, s_2 system to u_1, u_2 as shown in Fig. 2.5. The coordinates u_1, u_2 are known as principal coordinates.

Thus
$$\left.\begin{aligned} {}_1s_1 &= u_1\cos\theta; \qquad {}_1s_2 = u_1\sin\theta \\ {}_2s_1 &= u_2\cos\left(\theta + \frac{\pi}{2}\right); \quad {}_2s_2 = u_2\sin\left(\theta + \frac{\pi}{2}\right) \\ \text{where} \quad \cot\theta &= \frac{p_{21}^2 m_2^{-1/2}}{(-\omega_1^2 + p_{11}^2 + p_{21}^2)m_1^{-1/2}} \end{aligned}\right\} \tag{2.39}$$

In the u_1, u_2 coordinate system the expression for kinetic energy becomes

$$T = \tfrac{1}{2}\{\omega_1{}^2 u_1{}^2 + \omega_2{}^2 u_2{}^2\} \tag{2.40}$$

or

$$T = \tfrac{1}{2}(\dot{u}_1{}^2 + \dot{u}_1{}^2)$$

Turning now to the instantaneous value of the potential energy of the system we have

$$V = \tfrac{1}{2}\{k_1 x_1{}^2 + k_2(x_2 - x_1)^2\} \tag{2.41}$$

In terms of principal coordinates this becomes

$$\begin{aligned} V &= \tfrac{1}{2}u_1{}^2\{p_{11}{}^2 \cos^2\theta + (p_{22}\sin\theta - p_{21}\cos\theta)^2\} \\ &\qquad + \tfrac{1}{2}u_2{}^2\{p_{11}{}^2 \sin^2\theta + (p_{22}\cos\theta + p_{21}\sin\theta)^2\} \\ &= \tfrac{1}{2}\lambda_1 u_1{}^2 + \tfrac{1}{2}\lambda_2 u_2{}^2 \end{aligned} \tag{2.42}$$

Exercise

In expression (2.42) show that $\lambda_1 = \omega_1{}^2$ and $\lambda_2 = \omega_2{}^2$ where ω_1 and ω_2 are the natural angular frequencies in the first and second modes of vibration respectively.

We have now seen that, in the state space used to describe the instantaneous position of the system, the principal modes are orthogonal. Resulting from this fact the total kinetic energy of the system is the simple sum of the kinetic energy in each principal mode. When expressed in principal coordinates there is no cross-product term. Similarly, we have shown that the potential energy of the system is the simple sum of the potential energy in each principal mode.

The amplitude of motion of the i'th mode, U_i, of an undamped system may be ascribed arbitrarily. The modes are then called normal modes. That is, a normal mode is a principal mode to which has been assigned a magnitude.

In normalising some workers make the amplitude of the mass point with the greatest amplitude equal to unity for each mode of vibration; some make the kinetic energy in passing through the static equilibrium position, for each mode, equal to unity; whilst others adopt the mathematical definition for normalising the vector $\mathrm{m}^{1/2}\mathrm{x}$ for each mode which gives $\sum_i m_i\, {}_jx_i\, {}_jx_i = 1$, making the length of the resultant vector in the s space unity.

Exercises

1. A rolled-steel beam of unequal-angle section acts as a horizontal cantilever carrying a mass of 3 kg at its free end. Experimentally it was found that a force of 4 N, applied vertically at the free end of the beam, produced deflections of 2×10^{-4} m in the vertical direction and 4×10^{-5} m in the horizontal direction, up and to the right being taken as positive. Similarly, it was found that a force of 4 N applied in the horizontal direction at the free end produced a deflection of 3×10^{-4} m in the horizontal direction and 4×10^{-5} m in the vertical direction.

   Neglecting the mass of the beam, find the frequencies of the principal oscillations, the ratio of their amplitudes and the angle between their directions.

2. Find the natural frequencies and associated mode shapes for torsional vibrations of the system shown in Fig. 2.6. The stiffnesses of the shafts are $k_1 = 12 \times 10^5$ N m/rad and $k_2 = 3 \times 10^5$ N m/rad. The moments of inertia of the flywheels are $I_1 = 80$ kg m^2 and $I_2 = 40$ kg m^2.

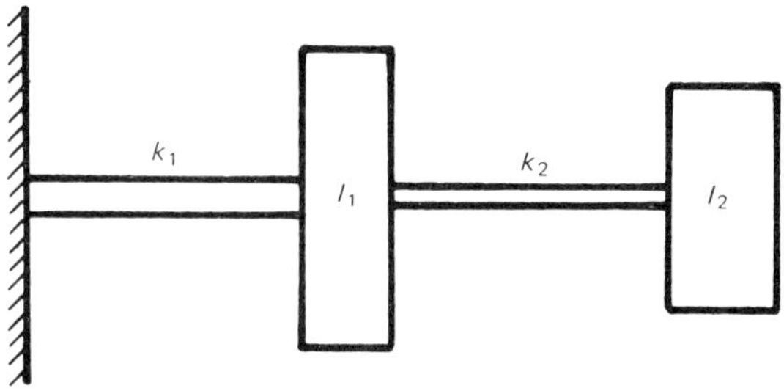

Fig. 2.6

Show that the principal modes are orthogonal and then, in terms of principal coordinates, derive expressions for the potential and kinetic energies at any instant.

2.2 Static and dynamic coupling

In general for the free vibration of an undamped system the equations of motion may be written in the form

$$\{\mathrm{m}D^2 + \mathrm{k}\}\mathrm{x} = 0 \tag{2.43}$$

where
$$D \equiv \frac{\mathrm{d}}{\mathrm{d}t};$$

m is known as the mass matrix and k as the stiffness matrix. If, as for the previously considered case in the x or s coordinate system, the stiffness matrix is not a diagonal matrix, then the system is said to have static coupling between the coordinates. The static coupling forces, those associated with the off-diagonal elements of the stiffness matrix, are due to the parts of the system which store energy as potential energy, that is energy due to displacement. The same values of force would act at a coordinate point if the displacements were static displacements.

When the mass matrix of the system has off-diagonal elements dynamic coupling forces are said to be present. These arise from the changes in velocity of the masses. The energy associated with dynamic coupling forces is the kinetic energy of the system.

As an example of a system with dynamic coupling we shall consider the double pendulum shown in Fig. 2.7. Each pendulum is considered to consist of a light rigid rod to the end of which is attached a concentrated mass. Pendulum 1 is attached, at its upper end, to a fixed point by a frictionless support and pendulum 2 is attached, at its upper end by a frictionless support, to the centre of mass m_1.

There are two degrees of freedom. We shall take the defining coordinates as θ_1

and θ_2. However, in writing down the equations of motion it is convenient to express them initially in the x, y coordinate system, and transform them later.

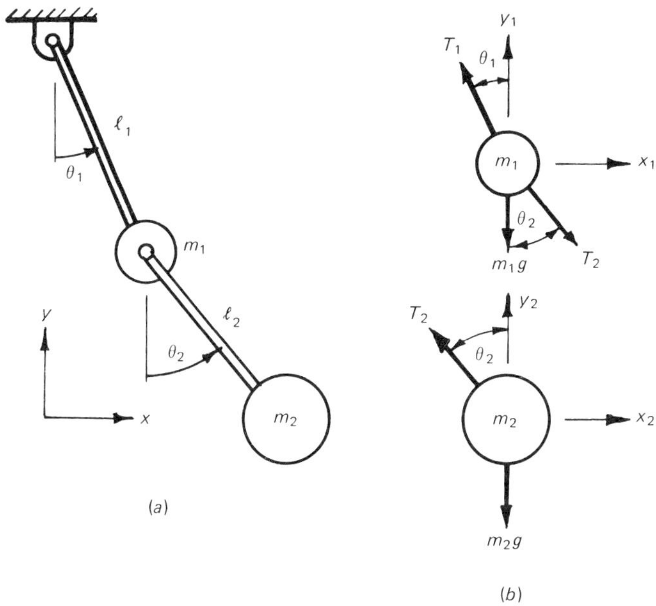

Fig. 2.7(*a*), (*b*)

Referring to the free body diagrams of Fig. 2.7(*b*) and applying Newton's laws to each mass in turn we find

(*i*) for mass m_1

(*a*) resolving vertically

$$T_1 \cos \theta_1 - m_1 g - T_2 \cos \theta_2 = m_1 \ddot{y}_1 \tag{2.44}$$

(*b*) resolving horizontally

$$T_2 \sin \theta_2 - T_1 \sin \theta_1 = m_1 \ddot{x}_1 \tag{2.45}$$

(*ii*) for mass m_2

(*a*) resolving vertically

$$T_2 \cos \theta_2 - m_2 g = m_2 \ddot{y}_2 \tag{2.46}$$

(*b*) resolving horizontally

$$-T_2 \sin \theta_2 = m_2 \ddot{x}_2 \tag{2.47}$$

We shall concern ourselves only with small amplitude vibrations, giving the geometrical relationships between x_i, y_i and θ_i as

$$y_1 = y_2 = 0; \qquad x_1 = \ell_1 \theta_1; \qquad x_2 = \ell_1 \theta_1 + \ell_2 \theta_2$$

$\sin \theta = \theta$ and $\cos \theta = 1$.

When substituted in the equations of motion these became

$$T_2 = m_2 g; \qquad T_1 = (m_1 + m_2)g \tag{2.48}$$

$$\begin{bmatrix} (m_1 + m_2)(\ell_1 \mathrm{D}^2 + g) & m_2 \ell_2 \mathrm{D}^2 \\ m_2 \ell_1 \mathrm{D}^2 & m_2(\ell_2 \mathrm{D}^2 + g) \end{bmatrix} \begin{bmatrix} \theta_1 \\ \theta_2 \end{bmatrix} = 0 \tag{2.49}$$

From this the mass matrix m is

$$\begin{bmatrix} (m_1 + m_2)\ell_1 & m_2 \ell_2 \\ m_2 \ell_1 & m_2 \ell_2 \end{bmatrix} \tag{2.50}$$

and the stiffness matrix k is

$$\begin{bmatrix} (m_1 + m_2)g & 0 \\ 0 & m_2 g \end{bmatrix} \tag{2.51}$$

Therefore, from our previous definitions the system is a dynamically coupled system. The natural frequencies and mode shapes may be found by the methods given previously.

Exercises

Derive the stiffness and mass matrices for small amplitude oscillations of each of the following systems, illustrated in Fig. 2.8. State whether or not static and/or dynamic coupling forces will exist within each system when it is vibrating.

(*a*) Two masses m_1 and m_2, supported on springs of stiffness k_1 and k_2 as shown, are free to vibrate in the vertical direction only.

(*b*) A bar, of inertia I_0 about the support point 0, is pivoted at 0 and supported at a point A, distance ℓ from 0, by the spring of stiffness k_1. Also attached to A is the spring of stiffness k_2 which supports the mass m. The mass m may vibrate in a vertical direction only.

(*c*) A mass m_1 is supported on frictionless rollers and restrained by spring k_1. Pivoted to the centre of mass m_1 is the end of a light rigid rod of length ℓ to the end of which is fastened a mass m_2.

(*d*) The uppermost mass m_1 of a double pendulum has an additional restraint in the form of a spring of stiffness k. ℓ_1 is vertical when stationary.

2.3 Further consideration on the orthogonality of principal modes

For systems with both dynamic and static coupling the space in which the modes are orthogonal is more complex than those considered so far. When considering harmonic oscillations the equations of motion reduce to the matrix form

$$\mathrm{kx} - \omega^2 \mathrm{mx} = 0 \tag{2.52}$$

If we premultiply equation (2.52) by x^T we get

$$\mathrm{x}^T \mathrm{kx} - \omega^2 \mathrm{x}^T \mathrm{mx} = 0 \tag{2.53}$$

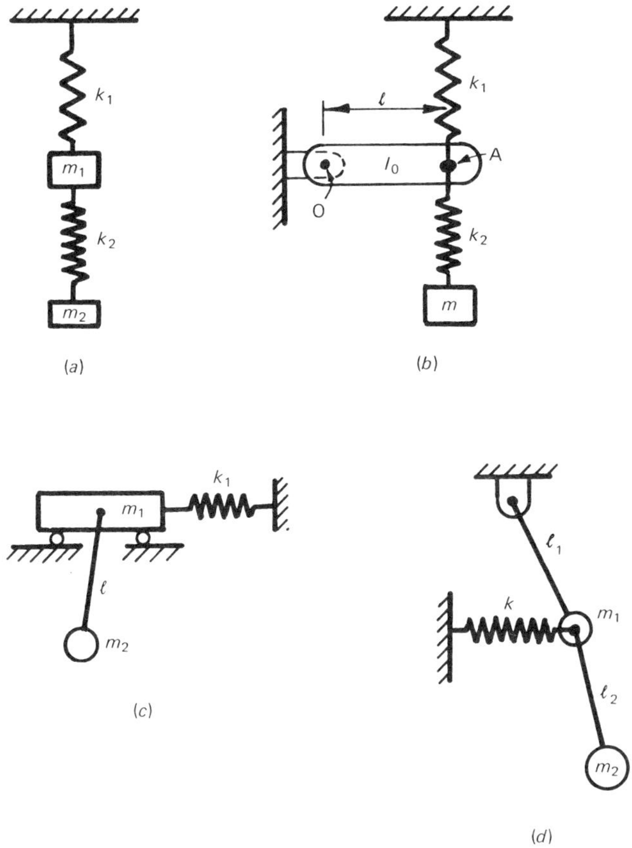

Fig. 2.8(*a*), (*b*), (*c*), (*d*)

that is
$$2(V_{max} - T_{max}) = 0 \tag{2.54}$$

It is possible to factorise m into $\mu^T\mu$.

Thus
$$x^T\mu^T(\mu^T)^{-1}k\mu^{-1}\mu x - \omega^2 x^T\mu^T I\mu x = 0 \tag{2.55}$$

or
$$s^T(\mu^T)^{-1}k\mu^{-1}s - \omega^2 s^T I s = 0 \tag{2.56}$$

or
$$s^T K s - \omega^2 s^T I s = 0 \tag{2.57}$$

The equations of motion in this space are

$$Ks - \omega^2 Is = 0 \tag{2.58}$$

Therefore, the values of ω^2 required are the eigenvalues of K, which was the case discussed in section 2.1. That is, the principal modes of systems with both dynamic and static coupling are orthogonal in the *s* space defined above. When the coordinate system defining the state space is rotated to coincide with the principal coordinates K will have the form

$$\begin{bmatrix} \lambda_1 & 0 & 0 & \dots & 0 \\ 0 & \lambda_2 & 0 & \dots & 0 \\ 0 & 0 & \lambda_3 & \dots & 0 \\ \vdots & & & & \vdots \\ \cdot & \cdot & \cdot & \dots & \lambda_n \end{bmatrix} \tag{2.59}$$

That is, it will be a diagonal matrix of the eigenvalues λ_i where

$$\lambda_i = \omega_i^2$$

Exercise

Prove in general for a conservative system that its principal modes are orthogonal.
[**Hint:** compare the virtual work due to a virtual displacement in the p'th mode whilst vibrating in the q'th mode with that due to a virtual displacement in the q'th mode whilst vibrating in the p'th mode.]

2.4 Systems with rigid body degrees of freedom

Many engineering systems are not tied rigidly in all directions relative to earth. They have one or more rigid body degrees of freedom. The shafting of an engine driven installation must be capable of rotating in its bearings as a rigid body. The vibratory motion, which at present is our main interest, is superimposed on this rigid body motion. Similarly, an aeroplane when flying, has six rigid body degrees of freedom upon which may be superimposed the vibratory motion of its structure.

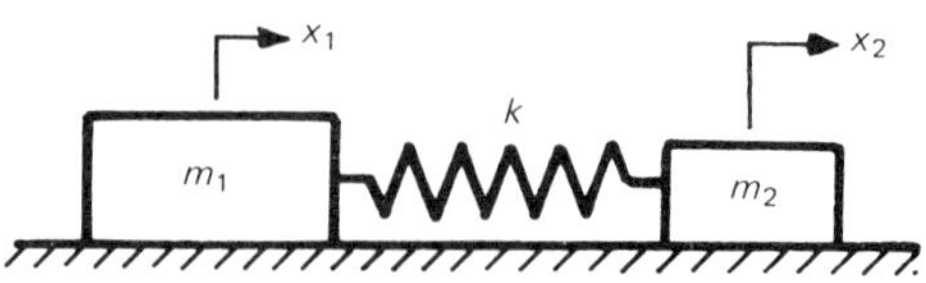

Fig. 2.9

We now wish to know whether or not these rigid body degrees of freedom require us to modify the method used in previous sections for finding the system natural frequencies. To take a specific case, the equations of motion for the undamped system shown in Fig. 2.9 are

$$\left\{\begin{bmatrix} m_1 & 0 \\ 0 & m_2 \end{bmatrix} D^2 + \begin{bmatrix} k & -k \\ -k & k \end{bmatrix}\right\} \begin{bmatrix} x_1 \\ x_2 \end{bmatrix} = 0 \tag{2.60}$$

giving the frequency equation

$$\omega^2\{m_1 m_2 \omega^2 - (m_1 + m_2)k\} = 0 \tag{2.61}$$

The first natural frequency ω_1 is given by

$$\omega_1{}^2 = 0$$

and the second by

$$\omega_2{}^2 = \left(\frac{m_1 + m_2}{m_1 \quad m_2}\right) k \tag{2.62}$$

The solution $\omega_1{}^2 = 0$ indicates that rigid body displacement is possible. This may, of course, be due to a static displacement or to a uniform velocity as is clearly evident when the equations of motion are solved for the motion. That is,

$$x_1 = A + Bt + C \sin \omega_2 t + D \cos \omega_2 t \tag{2.63}$$

$$x_2 = A + Bt + \left(1 - \frac{m_1 \omega_2{}^2}{k}\right)(C \sin \omega_2 t + D \cos \omega_2 t) \tag{2.64}$$

where A, B, C, and D are constants of integration. The method used previously for finding natural frequencies does not, therefore, require modification, the solution $\omega_1{}^2 = 0$ being taken to indicate the possibility of rigid body motion.

Exercises

1. Show that the rigid body mode and the vibratory mode of the undamped system illustrated in Fig. 2.9 are orthogonal.
2. Find the natural frequencies and normal modes for the undamped system shown in Fig. 2.10. Take $m_1 = 1$ kg, $m_2 = 2$ kg, $m_3 = 3$ kg, $k_1 = 2\ \text{kN m}^{-1}$, and $k_2 = 4\ \text{kN m}^{-1}$. Take the normalisation condition to be the kinetic energy in each mode equals 1 J. Check that the normal modes are orthogonal.

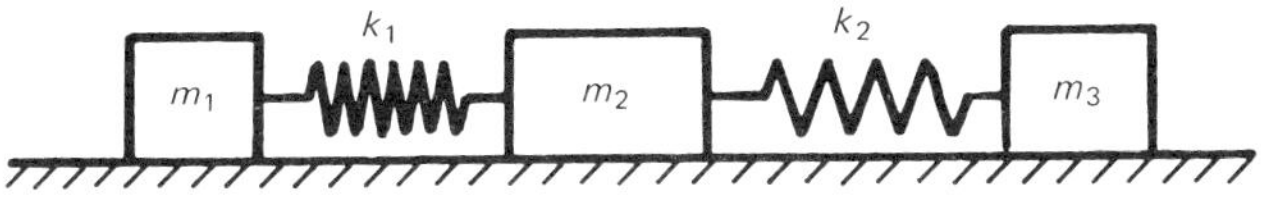

Fig. 2.10

2.5 Geared systems

A system comprised of one or more engines driving one or more power absorbing devices, such as say propellors and auxiliaries in a marine propulsion system, is referred to as a branched system, for obvious reasons. Physically the junctions between the branches and the main path along which the power flows usually consist of gear boxes. Care has to be taken in setting up the equations of motion for branched or geared systems, especially in adhering strictly to a sign convention.

We shall illustrate a process for setting up the equations of motion by considering the simple geared system of Fig. 2.11(a). It is convenient to describe the instantaneous position of the system in terms of the three coordinates θ_1, θ_2, and θ_3

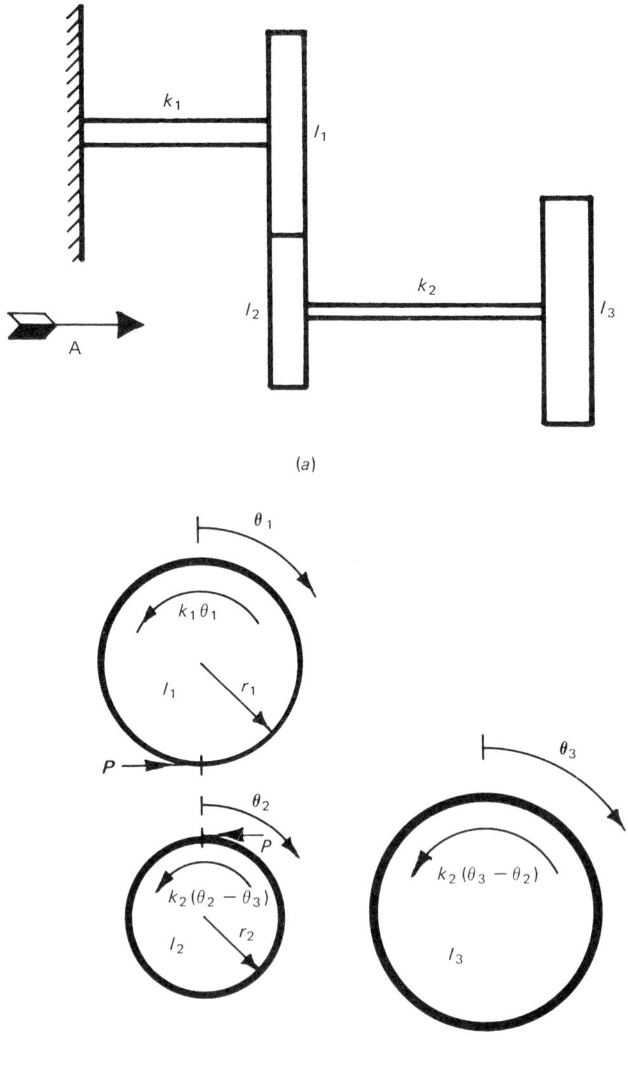

Fig. 2.11(*a*); (*b*) View in direction of arrow A

as shown in Fig. 2.11(*b*). If the gears I_1 and I_2 are rigid and their teeth remain in contact there will be a geometrical relationship between θ_1 and θ_2. Therefore, in addition to satisfying Newton's laws for each inertia we must satisfy the geometrical constraints which we shall call compatibility conditions.

The compatibility condition is

$$\theta_1 = -n\theta_2 \tag{2.65}$$

where $$n = r_2/r_1$$

and by Newton's laws we have for I_1

$$-k_1\theta_1 - Pr_1 = I_1\ddot{\theta}_1 \qquad [2.66(a)]$$

where P is the magnitude of the tangential forces acting on the contact surfaces of the gear teeth. Also

for I_2
$$-k_2(\theta_2 - \theta_3) - Pr_2 = I_2\ddot{\theta}_2 \qquad [2.66(b)]$$

for I_3
$$-k_2(\theta_3 - \theta_2) = I_3\ddot{\theta}_3 \qquad [2.66(c)]$$

Eliminating P between equations (2.66*a*) and (2.66*b*) and substituting from equation (2.65) into the resulting expression gives

$$(n^2I_1 + I_2)\ddot{\theta}_2 + (n^2k_1 + k_2)\theta_2 - k_2\theta_3 = 0 \qquad (2.67)$$

The natural frequencies and mode shapes may be found from equations (2.66*c*) and (2.67).

Exercises

1. Using the following values, $k_1 = 1000$ N m/rad, $k_2 = 1500$ N m/rad, $k_3 = 3000$ N m/rad, $k_4 = 4000$ N m/rad, $I_1 = 2$ kg m^2, $I_2 = 8$ kg m^2, and $r_2/r_1 = 2$, find the natural frequencies and mode shapes for free torsional vibrations of the system shown in Fig. 2.12. You may assume the gears to be light and rigid.
Check that the principal modes are orthogonal.

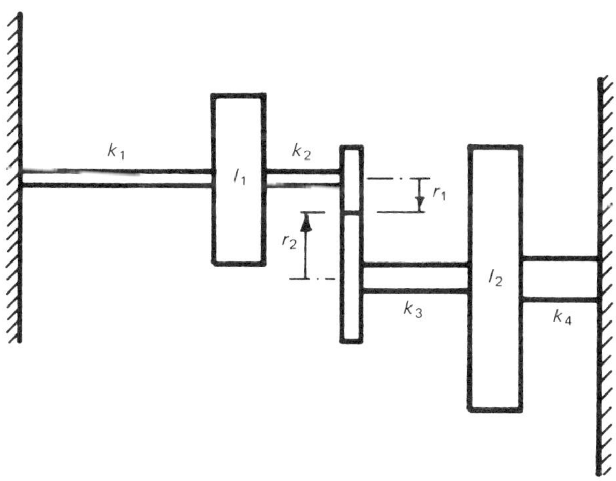

Fig. 2.12

2.6 Forced vibration

2.6.1 An undamped system

We shall now study the response of the system shown in Fig. 2.13 when acted upon by the harmonically varying forces f_1 and f_2. The masses m_1 and m_2 may oscillate in the vertical direction only. The equations of motion are

$$\begin{bmatrix} m_1D^2 & 0 \\ 0 & m_2D^2 \end{bmatrix}\begin{bmatrix} x_1 \\ x_2 \end{bmatrix} + \begin{bmatrix} k_1 + k_2 & -k_2 \\ -k_2 & k_2 \end{bmatrix}\begin{bmatrix} x_1 \\ x_2 \end{bmatrix} = \begin{bmatrix} f_1 \\ f_2 \end{bmatrix}$$

or

$$\{\mathrm{m}D^2 + \mathrm{k}\}\mathrm{x} = \mathrm{f} \tag{2.68}$$

where m is the mass matrix of the system
k is the stiffness matrix of the system
x is a vector of displacements
and f is the vector of applied forces

If the forces have the form

$$f_1 = F_1 \cos \omega t; \qquad f_2 = F_2 \cos (\omega t + \phi) \tag{2.69}$$

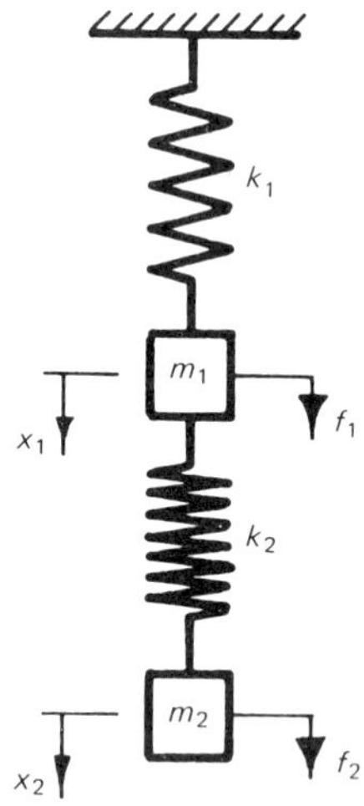

Fig. 2.13

that is they have the same frequency but differ in amplitude and phase, then the steady state solution to equation (2.68) may be written as

$$\{-\omega^2\mathrm{m} + \mathrm{k}\}\mathrm{x} = \mathrm{f}$$

or

$$\mathrm{Yx} = \mathrm{f} \tag{2.70}$$

Y relates force amplitudes F and displacement amplitude X. The equation (2.70) may, therefore, be looked upon as a generalised force versus displacement relationship with Y as the generalised stiffness matrix. We shall call Y the dynamic stiffness matrix. Equation (2.70) may be solved for the displacements giving

$$\mathrm{x} = \mathrm{Y}^{-1}\mathrm{f} \tag{2.71}$$

Y^{-1} is the dynamic flexibility matrix. In most of the recent mechanical literature it has been named the receptance matrix. For a comprehensive discussion on receptances and their use, see Bishop and Johnson (1960). For the receptance matrix we shall adopt the symbol Z, z_{ij} being the elements.

The elements of the receptance matrix under consideration may easily be worked out, giving

$$\left.\begin{aligned} z_{11} &= (-m_2\omega^2 + k_2)/\Delta \\ z_{12} &= z_{21} = k_2/\Delta \\ z_{22} &= (-m_1\omega^2 + k_1 + k_2)/\Delta \end{aligned}\right\} \qquad (2.72)$$

where $$\Delta = (-m_1\omega^2 + k_1 + k_2)(-m_2\omega^2 + k_2) - k_2^2$$

It will be noted that, for this system, the natural frequencies are given by the frequency equation

$$\Delta = 0 \qquad (2.73)$$

The receptances, as the elements are called, are, therefore, infinite at each of the natural frequencies. Their values are plotted for a range of frequencies in Fig. 2.14 taking the following values for the masses and stiffnesses:

$m_1 = 2$ Mg, $m_2 = 200$ kg, $k_1 = 5$ MN m^{-1}, and $k_2 = 500$ kN m^{-1}

As the systems we are considering are linear, no loss of generality is imposed by considering this system excited by one force at a time. If $F_2 = 0$ then the amplitude of x_1 $(=X_1)$ is given by $|z_{11}|F_1$. Therefore, for a constant force amplitude, the displacement amplitude X_1 increases from its near static value of F_1/k_1 at near zero frequency to infinity at the first natural frequency. That is, there is a resonant frequency at the first natural frequency. X_1 then decreases as frequency increases until it reaches zero at $\omega^2 = k_2/m_2$, a subsystem natural frequency. As shown on Fig. 2.15, with further increase in frequency, X_1 increases, again becoming infinite at the second natural frequency of the system. After this it decreases, approaching zero asymptotically at very high frequencies. The frequency of the exciting force, such as $\omega^2 = k_2/m_2$, at which the amplitude of motion of a specified point increases with any small change in the frequency is called the anti-resonant frequency.

The variation in X_2 with frequency is also shown in Fig. 2.15. The curves clearly show that the system has two resonant frequencies, one coinciding with each of the natural frequencies. Note that one of these is greater than, and one less than, the natural frequency of the single degree of freedom system comprised of m_1 and k_1 and of that comprised of m_2 and k_2. An important practical use is made of the fact that the amplitude X_1 is zero at the natural frequency of the subsystem comprised of k_2 and m_2. If, for example, we had a main system represented by the spring k_1 and the mass m_1 which experienced an exciting force under normal running conditions of frequency ω $\{=(k_1/m_1)^{1/2}\}$, the amplitude of the mass would be very large. Additionally, if it were not possible to eliminate the exciting force or change the natural frequency of the system by modifying k_1 or m_1, then it would not be possible to operate the system unless another means were found for changing the value of the system natural frequency. Such a method is suggested by the curves of Fig. 2.15. For if an additional spring-mass system were attached to the main system to give a combined system of the form illustrated in Fig. 2.13, having

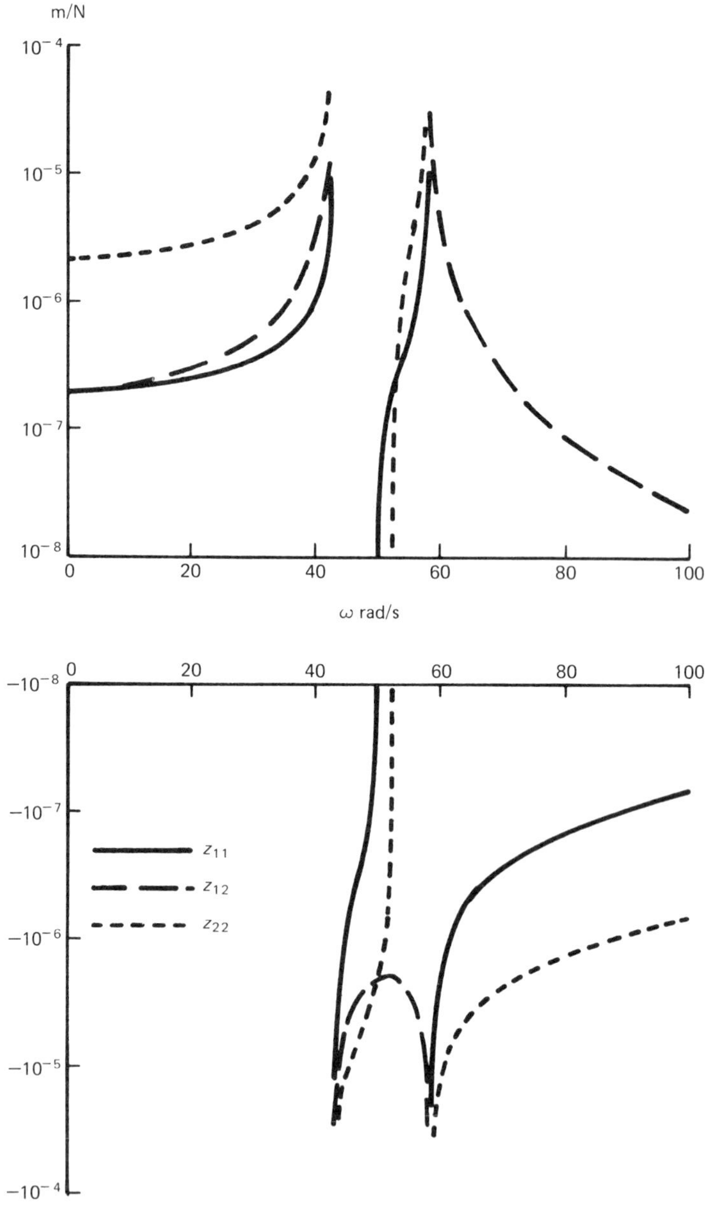

Fig. 2.14 Receptances for system shown in Fig. 2.13 (m_1 = 2 Mg, m_2 = 200 kg, k_1 = 5 MN/m, and k_2 = 500 kN/m)

physical parameters $k_2/m_2 = k_1/m_1$, then the dangerously high oscillations of mass m_1 would be eliminated. At the frequency of the exciting force $\omega^2 = k_1/m_1$, the force acting on the mass m_1 due to the extension of the spring k_2 exactly balances the exciting force. The hardware of an added system k_2, m_2 is called an undamped

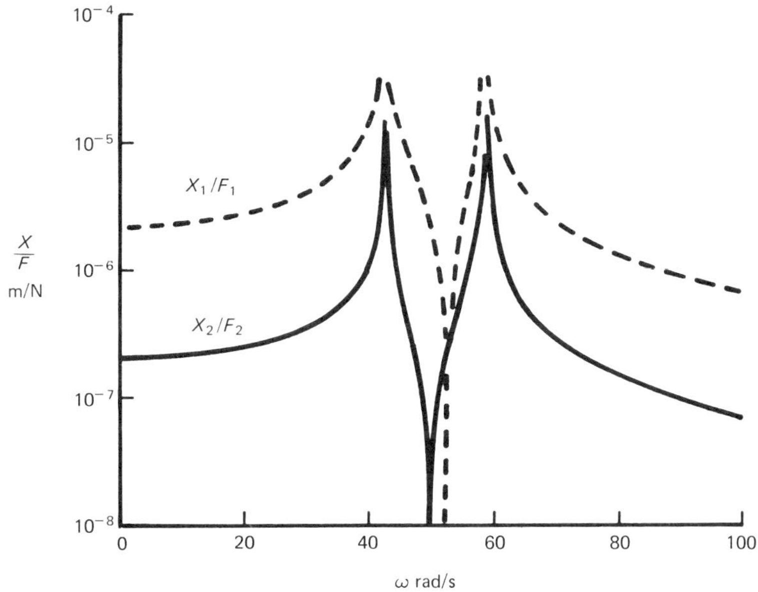

Fig. 2.15 Responses of system shown in Fig. 2.13 ($m_1 = 2$ Mg, $m_2 = 200$ kg, $k_1 = 5$ MN/m, and $k_2 = 500$ kN/m)

vibration absorber. (B.S. 3015:1958 Glossary of terms used in vibration and shock testing.)

Vibration absorbers are often used to move the natural frequencies of reciprocating engine installations to more desirable values.

2.6.2 A damped system

The dynamic response of the undamped system, shown in Fig. 2.15, has one main undesirable feature from the practical point of view. That is, although the natural frequencies of the combined system do not coincide with that of the single degree of freedom system (m_1 and k_1), they lie each side of it. Consequently, if the running speed of the system, and therefore the exciting force frequency, deviated a little from its normal value large amplitudes could result. In practical systems damping is present naturally, due to materials properties, and if insufficient to avoid dangerous amplitudes more is introduced by including hardware devices known as dampers.

For systems with more than one degree of freedom the inclusion of damping terms in the equations of motion complicates the analysis, as will be seen in the following example. Consider the system with viscous damping shown in Fig. 2.16. The equations of motion are

$$m_1\ddot{x}_1 + (c_1 + c_2)\dot{x}_1 + (k_1 + k_2)x_1 - c_2\dot{x}_2 - k_2x_2 = F\cos\omega t$$

and

$$m_2\ddot{x}_2 + c_2\dot{x}_2 + k_2x_2 - c_2\dot{x}_1 - k_2x_1 = 0 \tag{2.74}$$

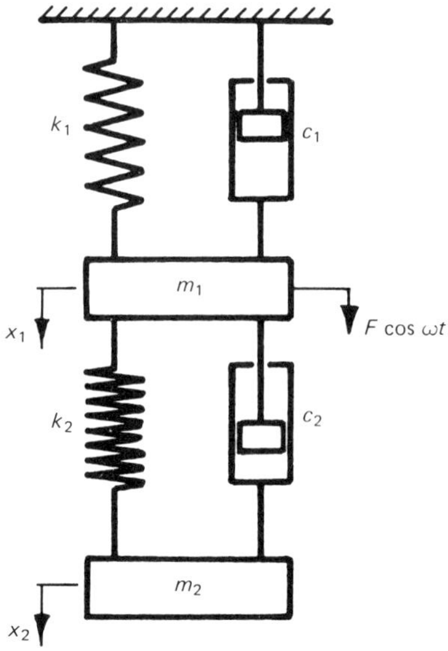

Fig. 2.16

The steady state solution for the motion of mass m_1, that is x_1, is given by

$$x_1 = \mathrm{Re}\left\{\frac{(-m_2\omega^2 + ic_2\omega + k_2)F\,e^{i\omega t}}{m_1m_2\omega^4 - (m_1k_2 + m_2k_1 + m_2k_2 + c_1c_2)\omega^2 + k_1k_2 + i\omega[-(m_1c_2 + m_2c_1 + m_2c_2)\omega^2 + k_1c_2 + k_2c_1]}\right\} \tag{2.75}$$

We are interested in the ratio of the amplitude of x_1 to the amplitude of the force, F, which is given by

$$\frac{X_1}{F} = \left\{\frac{(k_2 - m_2\omega^2)^2 + (c_2\omega)^2}{[m_1m_2\omega^4 - (m_1k_2 + m_2k_1 + m_2k_2 + c_1c_2)\omega^2 + k_1k_2]^2 + \omega^2[k_1c_2 + k_2c_1 - (m_1c_2 + m_2c_1 + m_2c_2)\omega^2]^2}\right\}^{1/2} \tag{2.76}$$

The variation in this ratio with frequency is illustrated, for the particular case of $m_1 = 2$ Mg, $m_2 = 200$ kg, $k_1 = 5$ MN m^{-1}, $k_2 = 500$ kN m^{-1}, and $c_1 = c_2 = 1$ kN s m^{-1}, in Fig. 2.17. When compared with the response of the undamped system, it may be seen that the amplitude at the two resonant frequencies has remained finite for the damped system, but this has been obtained at the expense of a finite amplitude occurring at the frequency $\omega = 50$ rad/s, where originally we had zero amplitude. It is also interesting to note that the damping has caused the

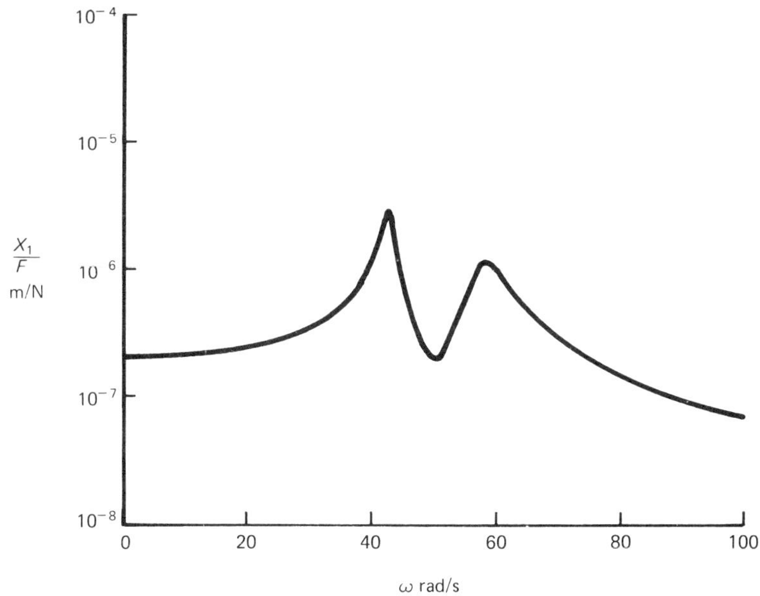

Fig. 2.17 Response of system shown in Fig. 2.16 ($m_1 = 2$ Mg, $m_2 = 200$ kg, $k_1 = 5$ MN/m, $k_2 = 500$ kN/m, and $c_1 = c_2 = 1$ kN s/m)

lower resonant frequency to increase and the higher resonant frequency to decrease. With increasing damping this trend would continue until they coalesced. This is because m_2 and k_2 are of much lower magnitudes than m_1 and k_1. Therefore, the damping would have a greater effect on the subsystem than on the main system, causing the motion of m_2 to be locked to that of m_1 before it was great enough to prevent dynamic magnification of the oscillatory motion of the system as a whole.

Exercises

1. It is normal to assume that the gas forces in the cylinder of a reciprocating engine do not cause vibration of the engine, as a whole, on its mountings. Show that this assumption is not strictly correct if the flexibility of the members in the load path between the piston and the cylinder head is allowed for. Use the system shown schematically in Fig. 2.18.
2. Examine the influence of the size of the mass m_2 on the resonance curve of X_1/F for the system shown in Fig. 2.16. Take $m_1 = 2$ Mg, $k_1 = 5$ MN/m and

assume a range of values of c_1 ($=c_2$), the ratio k_2/m_2 being constant and equal to k_1/m_1.

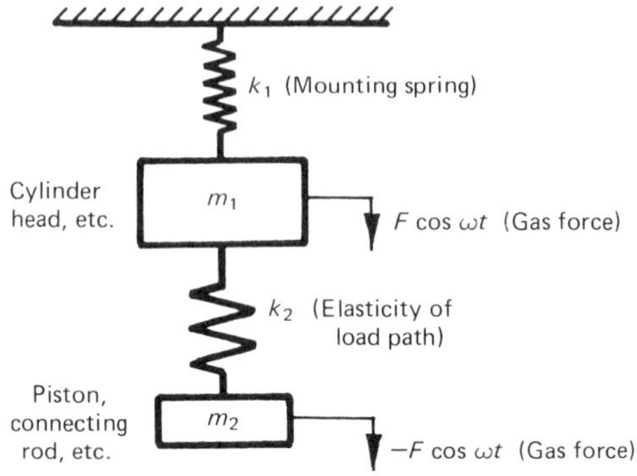

Fig. 2.18

2.7 Lagrange's equations of motion

The examples and exercises presented so far are based upon systems which are simple compared with many that arise in practice. Even so, we have seen that the Newtonian formulation of the equations of motion, involving as it does vector quantities (displacement, velocity, acceleration, and force), entails extreme care to ensure strict adherence to a consistent sign convention for all these vector quantities. From previous experience you will already be familiar with the possibility of formulating problems in dynamics in terms of the scalar quantities potential and kinetic energy. It is natural, therefore, to seek a formulation of the equations of motion in terms of energy functions; Lagrange was the first to achieve this, and the resulting equations have been named after him. They have the following form:

$$\frac{\mathrm{d}}{\mathrm{d}t}\left(\frac{\partial T}{\partial \dot{q}_k}\right) - \frac{\partial T}{\partial q_k} + \frac{\partial V}{\partial q_k} = Q_k \tag{2.77}$$

where T is the kinetic energy
V is the potential energy
t is time
q_k is termed a generalised coordinate
Q_k is called a generalised force

We shall discuss the characteristics of generalised coordinates and forces.

2.7.1 Generalised coordinates

By definition the position of a system with one degree of freedom is determined by the value of one variable, that is it may be specified completely by one coordinate. Similarly, the configuration of a system with two degrees of freedom may be specified completely by two independent coordinates. By independent we mean that they may be varied arbitrarily and independently without violating the constraints of the system. A logical extension is that a system with n degrees of freedom requires n independent coordinates to specify its configuration completely. It is then known as a holonomic system and the independent coordinates are known as generalised coordinates; we shall denote them by the symbols $q_1, q_2, q_3, \ldots, q_k, \ldots, q_n$.

Too much emphasis cannot be laid upon the fact that by definition the generalised coordinates must be independent, that is, if we let the generalised coordinate q_k increase by a small amount δq_k, which does not violate the constraints, it must be possible to keep the other generalised coordinates constant. By 'does not violate the constraints' we mean that it does not change the length of a rigid bar or deflect a rigid support or the like. As an example, consider a bar moving in the x, y plane. The bar can be replaced by a dynamically equivalent system comprised of two masses m_1 and m_2 a fixed distance ℓ apart. The position of the bar may be specified completely by the coordinates (x_1, y_1, x_2, y_2) of the centre of each mass or by the coordinates (r, θ, ϕ) shown in Fig. 2.19. It is easily seen that the first set of coordinates are not independent, for if x_1 is given a small increment δx_1 it is impossible for all of y_1, x_2, and y_2 to remain unaltered without violating a constraint, that the distance between the masses is constant. The r, θ, and ϕ coordinates are obviously independent, therefore, they form a set of generalised coordinates. There are three generalised coordinates and, by inspection, it can be seen that there are also three degrees of freedom two linear and one rotational.

In deriving the Lagrangean form of the equations of motion we shall require a relationship between the above two coordinate systems, that is between the cartesian system and the generalised coordinate system. From Fig. 2.19 it can be seen that

$$\left.\begin{aligned} x_1 &= r\cos\theta \\ y_1 &= r\sin\theta \\ x_2 &= r\cos\theta + \ell\cos\phi \\ y_2 &= r\sin\theta + \ell\sin\phi \end{aligned}\right\} \tag{2.78}$$

We shall express these in the following abbreviated form

$$\left.\begin{aligned} x_i &= x_i(r, \theta, \phi) \\ y_i &= y_i(r, \theta, \phi) \end{aligned}\right\} \tag{2.79}$$

That is the cartesian coordinates x_i, y_i are each a function of the three generalised coordinates r, θ, ϕ. Expressions (2.79) may be generalised; for a system with s mass points and n degrees of freedom the rectangular cartesian coordinates of the k'th mass point will be

$$\left.\begin{aligned} x_k &= x_k(q_1, q_2, q_3, \ldots, q_n) \\ y_k &= y_k(q_1, q_2, q_3, \ldots, q_n) \\ z_k &= z_k(q_1, q_2, q_3, \ldots, q_n) \end{aligned}\right\} k = 1, 2, 3, \ldots, s.$$

or in vector notation

$$\boldsymbol{r}_k = \boldsymbol{r}_k(q_1, q_2, q_3, \ldots, q_n) \tag{2.80}$$

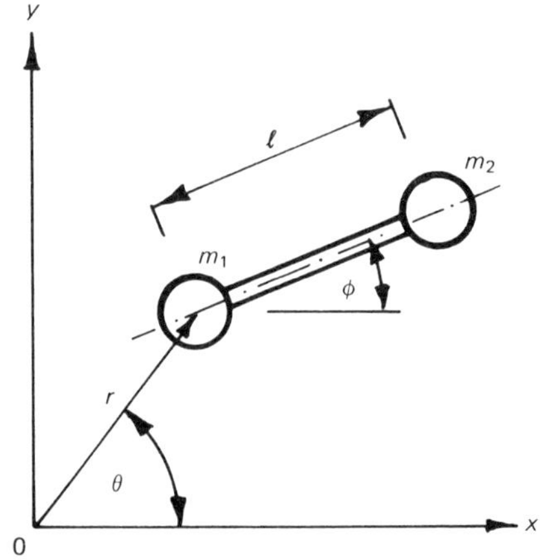

Fig. 2.19

2.7.2 Transformation to Lagrangean form

We shall now show, for a holonomic system, that the Newtonian form of the equations of motion may be transformed to the Lagrangean form.

In vector notation, with s mass points and n degrees of freedom, the principle of virtual work gives for the system

$$\sum_k \{\boldsymbol{F}_k + \boldsymbol{P}_k - m_k\ddot{\boldsymbol{r}}_k\}\cdot\delta\boldsymbol{r}_k = 0 \tag{2.81}$$

where $\boldsymbol{F}_k$ is the resultant external force on the k'th mass point, $\boldsymbol{P}_k$ is the force due to the internal frictionless constraints and $\boldsymbol{r}_k$ is the position vector of the k'th mass m_k. The total work due to the internal frictionless constraints is zero, that is

$$\sum_k \boldsymbol{P}_k\cdot\delta\boldsymbol{r}_k = 0 \tag{2.82}$$

because $\boldsymbol{P}_k$ is perpendicular to $\delta\boldsymbol{r}_k$. We are left with the task of transforming

$$\sum_k (\boldsymbol{F}_k - m_k\ddot{\boldsymbol{r}}_k)\cdot\delta\boldsymbol{r}_k = 0 \tag{2.83}$$

By the total differential theorem

$$\delta\boldsymbol{r}_k = \sum_i \frac{\partial \boldsymbol{r}_k}{\partial q_i}\,\delta q_i \tag{2.84}$$

therefore,

$$\dot{\boldsymbol{r}}_k = \sum_i \frac{\partial \boldsymbol{r}_k}{\partial q_i}\,\dot{q}_i$$

therefore,

$$\frac{\partial \dot{\boldsymbol{r}}_k}{\partial \dot{q}_i} = \frac{\partial \boldsymbol{r}_k}{\partial q_i} \tag{2.85}$$

Substituting for $\delta \boldsymbol{r}_k$ in equation (2.83) gives

$$\sum_k \left(\boldsymbol{F}_k \cdot \sum_i \frac{\partial \boldsymbol{r}_k}{\partial q_i} - m_k \ddot{\boldsymbol{r}}_k \cdot \sum_i \frac{\partial \boldsymbol{r}_k}{\partial q_i} \right) \delta q_i = 0 \tag{2.86}$$

As the q_i are independent, the work due to each virtual displacement alone is zero giving

$$\sum_k \left(\boldsymbol{F}_k \cdot \frac{\partial \boldsymbol{r}_k}{\partial q_i} - m_k \ddot{\boldsymbol{r}}_k \cdot \frac{\partial \boldsymbol{r}_k}{\partial q_i} \right) = 0 \tag{2.87}$$

We wish to relate the second term in the brackets to the kinetic energy which is

$$T = \sum_k \tfrac{1}{2} m_k \dot{\boldsymbol{r}}_k \cdot \dot{\boldsymbol{r}}_k \tag{2.88}$$

That is, knowing the form of the answer, our aim is to show that

$$\begin{aligned} \ddot{\boldsymbol{r}}_k \cdot \frac{\partial \boldsymbol{r}_k}{\partial q_i} &= \frac{\mathrm{d}}{\mathrm{d}t} \left\{ \frac{1}{2} \frac{\partial}{\partial \dot{q}_i} (\dot{\boldsymbol{r}}_k \cdot \dot{\boldsymbol{r}}_k) \right\} - \frac{1}{2} \frac{\partial}{\partial q_i} (\dot{\boldsymbol{r}}_k \cdot \dot{\boldsymbol{r}}_k) \\ &= \frac{\mathrm{d}}{\mathrm{d}t} \left\{ \dot{\boldsymbol{r}}_k \cdot \frac{\partial \dot{\boldsymbol{r}}_k}{\partial \dot{q}_i} \right\} - \dot{\boldsymbol{r}}_k \cdot \frac{\partial \dot{\boldsymbol{r}}_k}{\partial q_i} \\ &= \ddot{\boldsymbol{r}}_k \cdot \frac{\partial \dot{\boldsymbol{r}}_k}{\partial \dot{q}_i} + \dot{\boldsymbol{r}}_k \cdot \frac{\mathrm{d}}{\mathrm{d}t} \left(\frac{\partial \dot{\boldsymbol{r}}_k}{\partial \dot{q}_i} \right) - \dot{\boldsymbol{r}}_k \cdot \frac{\partial \dot{\boldsymbol{r}}_k}{\partial q_i} \end{aligned} \tag{2.89}$$

We must show that

$$\frac{\mathrm{d}}{\mathrm{d}t} \left(\frac{\partial \dot{\boldsymbol{r}}_k}{\partial \dot{q}_i} \right) = \frac{\partial \dot{\boldsymbol{r}}_k}{\partial q_i}$$

Now

$$\frac{\mathrm{d}}{\mathrm{d}t} \left(\frac{\partial \dot{\boldsymbol{r}}_k}{\partial \dot{q}_i} \right) = \frac{\mathrm{d}}{\mathrm{d}t} \left(\frac{\partial \boldsymbol{r}_k}{\partial q_i} \right) \qquad \text{by equation (2.85)}$$

$$\begin{aligned} &= \sum_j \frac{\partial^2 \boldsymbol{r}_k}{\partial q_j \, \partial q_i} \dot{q}_j \\ &= \frac{\partial}{\partial q_i} \sum_j \frac{\partial \boldsymbol{r}_k}{\partial q_j} \dot{q}_j \\ &= \frac{\partial \dot{\boldsymbol{r}}_k}{\partial q_i} \end{aligned} \tag{2.90}$$

Therefore, equation (2.87) becomes

$$\frac{d}{dt}\left\{\frac{\partial}{\partial \dot{q}_i}\sum_k \tfrac{1}{2}m_k\dot{\boldsymbol{r}}_k\cdot\dot{\boldsymbol{r}}_k\right\} - \frac{\partial}{\partial q_i}\sum_k \tfrac{1}{2}m_k\dot{\boldsymbol{r}}_k\cdot\dot{\boldsymbol{r}}_k = \sum_k \boldsymbol{F}_k\cdot\frac{\partial \boldsymbol{r}_k}{\partial q_i}$$

or

$$\frac{d}{dt}\left(\frac{\partial T}{\partial \dot{q}_i}\right) - \frac{\partial T}{\partial q_i} = P_i \tag{2.91}$$

where P_i are known as the generalised forces.

The configuration of a system may be defined by equation (2.80), (*a*) when it is free, and (*b*) when the external constraints on it are fixed. The systems we shall investigate fall within these categories.

If the external constraints on the system move $\boldsymbol{r}_k$ may be an explicit function of time t, that is

$$\boldsymbol{r}_k = \boldsymbol{r}_k(q_1, q_2, q_3, \ldots, q_n, t)$$

So long as the system is holonomic equations (2.91) are the equations of motion, see for example Easthope (1964).

2.7.3 Generalised forces

The generalised forces are expressed in units such that the product $P_i q_i$ has the units of work.

The generalised force P_i can conveniently be considered to consist of three parts; 1. the part R_i due to conservative forces, such as gravity forces and elastic spring restoring forces; 2. the part S_i due to viscous friction; and 3. the part Q_i including all other forces acting on the system, such as pulsating disturbing forces and forces due to Coulomb friction.

We shall now develop expressions for R_i and S_i.

1. Conservative forces, R_i

If the work done on a system by external forces, during a virtual displacement, depends only upon the initial and final coordinates of the system, being independent of the path taken between them, the forces are said to be conservative. If the forces on a system are all conservative, and the system is originally and finally at rest, the work done by the system in a virtual displacement plus the change in its potential energy is zero. By definition the potential energy V may be expressed as

$$V = V(q_1, q_2, q_3, \ldots, q_n) \tag{2.92}$$

then

$$\delta V = \sum_i \frac{\partial V}{\partial q_i}\,\delta q_i \tag{2.93}$$

and the work done by the system must be

$$\delta W = \sum_i R_i\,\delta q_i \tag{2.94}$$

according to the definition of generalised force.

As

$$\delta W + \delta V = 0 \tag{2.95}$$

and the generalised coordinates are independent,

$$R_i = -\frac{\partial V}{\partial q_i} \tag{2.96}$$

2. Dissipative forces, S_i

When viscous friction is present each mass point is acted upon by forces proportional to its component velocities, which retard the motion. The virtual work associated with these forces will be

$$\delta W = -\sum_k (c_{xk}\dot{x}_k\ \delta x_k + c_{yk}\dot{y}_k\ \delta y_k + c_{zk}\dot{z}_k\ \delta z_k) \tag{2.97}$$

where c_{xk}, c_{yk}, and c_{zk} are constants.
This transforms to

$$\delta W = -\sum_k \sum_i \frac{1}{2}\frac{\partial}{\partial \dot{q}_i}\{c_{xk}\dot{x}_k{}^2 + c_{yk}\dot{y}_k{}^2 + c_{zk}\dot{z}_k{}^2\}\ \delta q_i \tag{2.98}$$

We may now define a dissipation function F where

$$F = \frac{1}{2}\sum_k (c_{xk}\dot{x}_k{}^2 + c_{yk}\dot{y}_k{}^2 + c_{zk}\dot{z}_k{}^2)$$

$$= \tfrac{1}{2}c_{11}\dot{q}_1{}^2 + \tfrac{1}{2}c_{22}\dot{q}_2{}^2 + \cdots + c_{12}\dot{q}_1\dot{q}_2 + c_{23}\dot{q}_2\dot{q}_3 + \cdots \tag{2.99}$$

such that

$$\delta W = -\sum_i \frac{\partial F}{\partial \dot{q}_i}\ \delta q_i = \sum_i S_i\ \delta q_i \tag{2.100}$$

giving the generalised force due to viscous friction as

$$S_i = -\frac{\partial F}{\partial \dot{q}_i} \tag{2.101}$$

The formulation of the dissipation function is due to Lord Rayleigh (1894). Lagrange's equations of motion may be written in the form

$$\frac{\mathrm{d}}{\mathrm{d}t}\left(\frac{\partial T}{\partial \dot{q}_i}\right) - \frac{\partial T}{\partial q_i} = R_i + S_i + Q_i$$

or

$$\frac{\mathrm{d}}{\mathrm{d}t}\left(\frac{\partial T}{\partial \dot{q}_i}\right) - \frac{\partial T}{\partial q_i} + \frac{\partial V}{\partial q_i} + \frac{\partial F}{\partial \dot{q}_i} = Q_i \tag{2.102}$$

Applications

To help establish familiarity with the use of Lagrange's equations we shall rework the examples relating to Figs. 2.7, 2.11, and 2.13.

Referring to Fig. 2.7, take θ_1 and θ_2 as the generalised coordinates, then

$$\left.\begin{aligned} T &= \tfrac{1}{2}m_1\ell_1^2\dot{\theta}_1^2 + \tfrac{1}{2}m_2\{(\ell_1\dot{\theta}_1)^2 + (\ell_2\dot{\theta}_2)^2 + 2\ell_1\ell_2\dot{\theta}_1\dot{\theta}_2\cos(\theta_2 - \theta_1)\} \\ V &= m_1 g\ell_1(1 - \cos\theta_1) + m_2 g\{\ell_1(1 - \cos\theta_1) + \ell_2(1 - \cos\theta_2)\} \end{aligned}\right\} \quad (2.103)$$

For small amplitude vibrations these become

$$\left.\begin{aligned} T &= \tfrac{1}{2}m_1\ell_1^2\dot{\theta}_1^2 + \tfrac{1}{2}m_2(\ell_1\dot{\theta}_1 + \ell_2\dot{\theta}_2)^2 \\ V &= \tfrac{1}{2}m_1 g\ell_1\theta_1^2 + \tfrac{1}{2}m_2 g(\ell_1\theta_1^2 + \ell_2\theta_2^2) \end{aligned}\right\} \quad (2.104)$$

Considering θ_1 first, $Q_1 = 0$, therefore,

$$\frac{\mathrm{d}}{\mathrm{d}t}\left(\frac{\partial T}{\partial \dot{\theta}_1}\right) - \frac{\partial T}{\partial \theta_1} + \frac{\partial V}{\partial \theta_1} = 0$$

Hence $$m_1\ell_1^2\ddot{\theta}_1 + m_2\ell_1(\ell_1\ddot{\theta}_1 + \ell_2\ddot{\theta}_2) + (m_1 + m_2)g\ell_1\theta_1 = 0 \quad (2.105)$$

Similarly for θ_2 we get

$$m_2\ell_2(\ell_1\ddot{\theta}_1 + \ell_2\ddot{\theta}_2) + m_2 g\ell_2\theta_2 = 0 \quad (2.106)$$

These agree with expression (2.49).

Referring to Fig. 2.11, take θ_2 and θ_3 as the generalised coordinates, then

$$\left.\begin{aligned} T &= \tfrac{1}{2}I_1(n\dot{\theta}_2)^2 + \tfrac{1}{2}I_2\dot{\theta}_2^2 + \tfrac{1}{2}I_3\dot{\theta}_3^2 \\ V &= \tfrac{1}{2}k_1(n\theta_2)^2 + \tfrac{1}{2}k_2(\theta_3 - \theta_2)^2 \end{aligned}\right\} \quad (2.107)$$

There are no external exciting forces, that is $Q_i = 0$,

hence $$\frac{\mathrm{d}}{\mathrm{d}t}\left(\frac{\partial T}{\partial \dot{\theta}_i}\right) - \frac{\partial T}{\partial \theta_i} + \frac{\partial V}{\partial \theta_i} = 0$$

giving $$(I_1 n^2 + I_2)\ddot{\theta}_2 + (k_1 n^2 + k_2)\theta_2 - k_2\theta_3 = 0 \quad (2.108)$$

and $$I_3\ddot{\theta}_3 + k_2(\theta_3 - \theta_2) = 0 \quad (2.109)$$

which agree with expressions (2.67) and (2.66*c*).

Referring to Fig. 2.13, take the displacements x_1 and x_2, from the static equilibrium position, as the generalised coordinates. Note, the generalised exciting forces Q_1 and Q_2 are f_1 and f_2 respectively. Obviously

$$T = \tfrac{1}{2}m_1\dot{x}_1^2 + \tfrac{1}{2}m_2\dot{x}_2^2 \quad (2.110)$$

but the masses are under the influence of two conservative forces which arise from potential energy functions having different forms, that is the elastic forces and the gravitational forces. The combined potential energy function for cases like these are sometimes not obvious to students, hence we shall consider each term in turn.

$$\begin{aligned} V &= \overset{[1]}{\{-m_1 g x_1 - m_2 g x_2\}} + \overset{[2]}{\{(m_1 g + m_2 g)x_1 + m_2 g(x_2 - x_1)\}} \\ &\qquad + \overset{[3]}{\{\tfrac{1}{2}k_1 x_1^2 + \tfrac{1}{2}k_2(x_2 - x_1)^2\}} \\ &= \tfrac{1}{2}k_1 x_1^2 + \tfrac{1}{2}k_2(x_2 - x_1)^2 \end{aligned} \quad (2.111)$$

The term in brackets [1] is due to the change in the P.E. of the masses; that in brackets [2] is the change in P.E. stored in the springs due to the initial force in each of them at the static equilibrium position; that in brackets [3] is the change in P.E. stored in the springs if initially they had been unloaded. The equations of motion which result

$$\left.\begin{aligned} m_1\ddot{x}_1 + (k_1 + k_2)x_1 - k_2x_2 &= f_1 \\ \text{and} \qquad m_2\ddot{x}_2 + k_2x_2 + k_2x_1 &= f_2 \end{aligned}\right\} \tag{2.112}$$

obviously agree with the result derived previously.

Exercise

Using Lagrange's equations, determine the equations of motion for the systems shown in Fig. 2.8 and described on page 44.

2.8 More complex systems

Using Lagrangean or Newtonian methods, you now know how to set up the equations of motion for any system if its mass and elasticity may be regarded as concentrated in separate elements, that is a lumped-parameter system. We shall extend this knowledge to gain physical understanding and to reduce to a routine the derivation of equations of motion for more complex systems. Three topics will be introduced: the first enables us to visualise the relationships between the natural frequencies of two subsystems and the natural frequencies of the combined system; in the second we shall consider the dynamics of a rigid body as a subsystem; finally we shall be concerned with building up the overall system dynamic stiffness matrix from those of the component subsystems.

2.8.1 Changes due to the addition of a subsystem

As natural frequencies depend solely upon system parameters one would expect those of an undamped system with two or more degrees of freedom to be derivable from the dynamic response of its subsystems. This is so, and for complex systems sometimes the natural frequencies of a complete system are predicted from the measured responses of the subsystems. A classical problem solved by this means was the accurate prediction of aircraft engine-propellor vibrations from the experimental response measurements of the engine and of the shaft plus propellor as subsystems.

To illustrate this approach let us consider the system shown in Fig. 2.20(*a*) and its subsystems as defined in Fig. 2.20(*b*). In joining the subsystems we introduce the constraints of compatability, that is ${}_ax_1 = {}_bx_1$, and equilibrium, that is ${}_aP_1 + {}_bP_1 = P$. Let ${}_aY_{11}$ be the direct dynamic stiffness at coordinate x_1 for subsystem a, then by definition

$$ {}_aP_1\, e^{i\omega t} = {}_aY_{11}\, {}_ax_1 = {}_aY_{11}\, {}_aX_1\, e^{i\omega t} \tag{2.113}$$

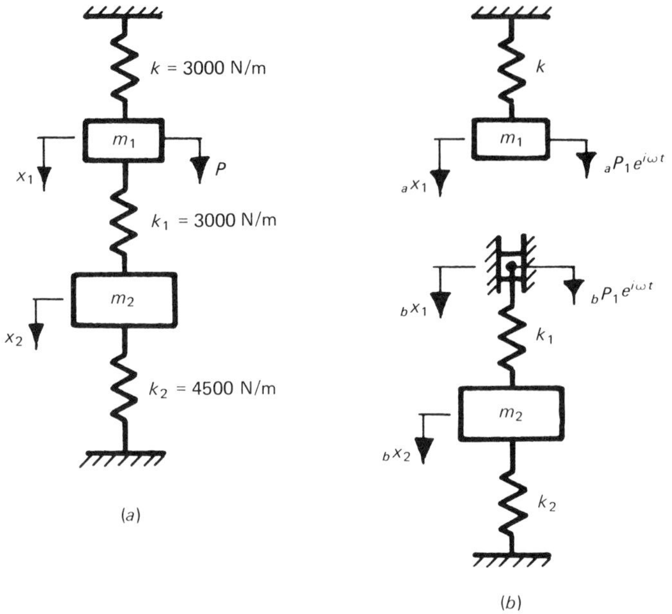

Fig. 2.20(a) and (b); ($m_1 = 1$ kg; $m_2 = 2$ kg)

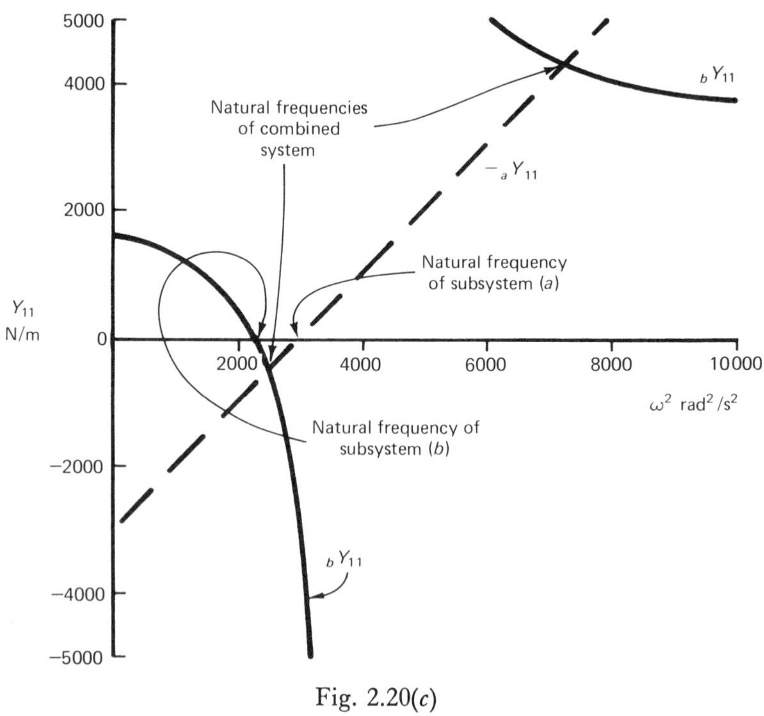

Fig. 2.20(c)

where ${}_aX_1$ is the displacement amplitude. Similarly, for subsystem b, let ${}_bY_{11}$ be the direct dynamic stiffness at coordinate x_1, then

$$ {}_bP_1\,e^{i\omega t} = {}_bY_{11}\,{}_bx_1 = {}_bY_{11}\,{}_bX_1\,e^{i\omega t} \tag{2.114} $$

For the complete system, applying the compatability condition together with the equilibrium condition gives

$$ {}_aP_1 + {}_bP_1 = ({}_aY_{11} + {}_bY_{11})X_1 = P \tag{2.115} $$

Free vibrations may occur at thc natural frequencies when $P = 0$; X_1 is not equal to zero or there would be no vibrations, therefore,

$$ -{}_aY_{11} = {}_bY_{11} \tag{2.116} $$

Thus we may find the natural frequencies by plotting $-{}_aY_{11}$ against ω^2 and ${}_bY_{11}$ against ω^2; where they intersect locates the desired frequency values. This is illustrated in Fig. 2.20(c).

Exercises

1. Show graphically that the addition of a mass at coordinate x_1 to the system given in Fig. 2.21 introduces an extra natural frequency. Discuss the influence of the magnitude of the mass on the positioning of the natural frequencies on the ω^2 axis.

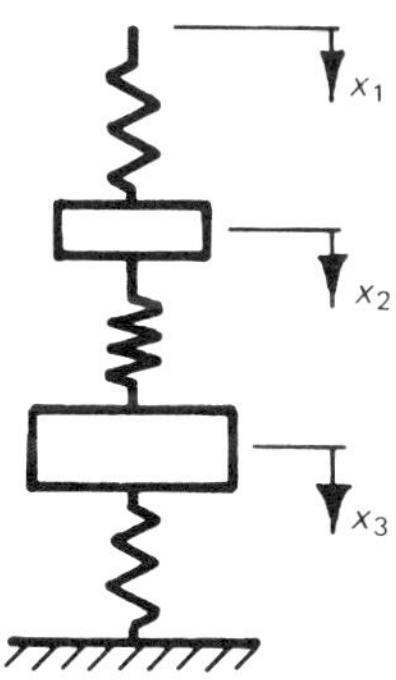

Fig. 2.21

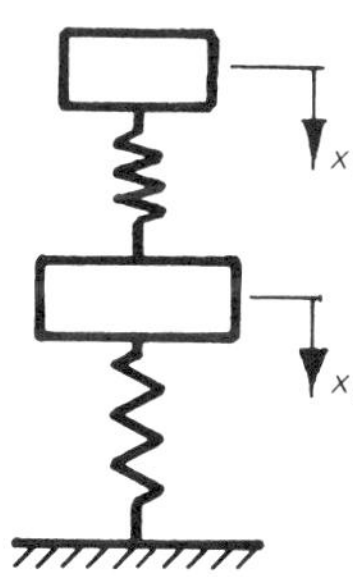

Fig. 2.22

2. For the subsystem shown in Fig. 2.22 show graphically that the addition of a spring at coordinate x_1, the other end of which is attached to a fixed point, does not introduce extra natural frequencies but that it raises those of the original subsystem.

2.8.2 Rigid bodies as system components

Engines on their mountings, the bodies of road and rail vehicles on their suspensions and ships buoyantly supported in water are a few examples of systems which oscillate not as point masses but as rigid bodies. Before the frequencies and modes of their oscillations may be calculated the equations of motion must be set up and solved.

A rigid body has six degrees of freedom (three rotational, three translational), consequently it is cumbersome to derive the equations in terms of cartesian coordinates. Vector algebra is ideal for this task and will be used.

The kinetic energy of a system of particles of mass m_i, instantaneously positioned at $\boldsymbol{r}_i$ from the origin 0 is given by

$$\mathrm{KE} = \frac{1}{2}\sum_i m_i \dot{\boldsymbol{r}}_i \cdot \dot{\boldsymbol{r}}_i \tag{2.117}$$

If the particles are rigidly connected together and their mass centre is at R, then we may work in the coordinate system $\boldsymbol{R}$, and $\boldsymbol{\rho}_i$ where

$$\left.\begin{aligned} \boldsymbol{r}_i &= \boldsymbol{R} + \boldsymbol{\rho}_i \\ \text{and} \qquad \dot{\boldsymbol{r}}_i &= \dot{\boldsymbol{R}} + \dot{\boldsymbol{\rho}}_i = \dot{\boldsymbol{R}} + \boldsymbol{\omega} \times \boldsymbol{\rho}_i \end{aligned}\right\} \tag{2.118}$$

where $\boldsymbol{\omega}$ is the angular velocity of the particle system about the centre of mass. Substitution for $\dot{\boldsymbol{r}}_i$ in equation (2.117), noting that by definition of mass centre

$$\dot{\boldsymbol{R}} \cdot \sum_i m_i \dot{\boldsymbol{\rho}}_i = 0,$$

gives

$$\begin{aligned} \mathrm{KE} &= \frac{1}{2}\sum_i m_i \dot{\boldsymbol{R}} \cdot \dot{\boldsymbol{R}} + \frac{1}{2}\sum_i m_i (\boldsymbol{\omega} \times \boldsymbol{\rho}_i) \cdot (\boldsymbol{\omega} \times \boldsymbol{\rho}_i) \\ &= \mathrm{KE_T} + \mathrm{KE_R} \end{aligned} \tag{2.119}$$

That is, in this coordinate system the kinetic energy of a system of rigidly connected particles is the sum of two independent components, that due to motion of the centre of mass and that due to motion about the centre of mass.

In the case of a rigid body of total mass M we have

$$\mathrm{KE_T} = \tfrac{1}{2} M \dot{\boldsymbol{R}}^2 \tag{2.120}$$

and

$$\mathrm{KE_R} = \tfrac{1}{2} \int_v (\boldsymbol{\omega} \times \boldsymbol{\rho})^2 \,\mathrm{d}m \tag{2.121}$$

where v indicates that the integration is to be performed over the complete volume of the body.

Expression (2.121) is not a convenient form for use hence it must be manipulated. By the rules of vector algebra

$$\mathrm{KE_R} = \frac{1}{2}\int_v [\omega^2 \rho^2 - (\boldsymbol{\omega} \cdot \boldsymbol{\rho})^2] \,\mathrm{d}m \tag{2.122}$$

Expressing the vectors $\boldsymbol{\omega}$ and $\boldsymbol{\rho}$ in terms of cartesian components, whose origin is at the mass centre, that is

$$\left.\begin{aligned} \boldsymbol{\omega} &= \omega_x \boldsymbol{i} + \omega_y \boldsymbol{j} + \omega_z \boldsymbol{k} \\ \text{and} \qquad \boldsymbol{\rho} &= x\boldsymbol{i} + y\boldsymbol{j} + z\boldsymbol{k} \end{aligned}\right\} \tag{2.123}$$

we may write equation (2.122) in the form

$$\mathrm{KE_R} = \frac{1}{2}\int_v [\omega_x\ \omega_y\ \omega_z]\begin{bmatrix} y^2+z^2 & -xy & -xz \\ -yx & z^2+x^2 & -yz \\ -zx & -zy & x^2+y^2 \end{bmatrix}\begin{bmatrix}\omega_x \\ \omega_y \\ \omega_z\end{bmatrix} \mathrm{d}m \quad (2.124)$$

which upon integration of the terms in the square matrix becomes

$$\begin{aligned}\mathrm{KE_R} &= \tfrac{1}{2}[\omega_x\ \omega_y\ \omega_z]\begin{bmatrix} A & -F & -E \\ -F & B & -D \\ -E & -D & C \end{bmatrix}\begin{bmatrix}\omega_x \\ \omega_y \\ \omega_z\end{bmatrix} \\ &= \tfrac{1}{2}\omega^{\mathrm{T}}\mathscr{I}\omega \end{aligned} \quad (2.125)$$

The terms A, B, and C are called moments of inertia of the body; F, E, and D are its products of inertia. When the coordinate axes are such that the products of inertia are each zero the moments of inertia are the principal moments of inertia of the body and the coordinate axes are its principal axes.

If you were to perform a coordinate rotation or a transformation to another system of coordinates you would find that the square array of coefficients in equation (2.125) transforms in such a way that the rotational kinetic energy remains invariant. Entities with such properties are called tensors; this is the inertia tensor of a rigid body. In the shorthand notation of tensor algebra equation (2.125) is written as

$$\mathrm{KE_R} = \tfrac{1}{2}\mathscr{I}_{\alpha\beta}\omega^{\alpha}\omega^{\beta} \quad (2.126)$$

where the repeated affix indicates summation over all permissible values of the affix. Each element of the array is called a component of the tensor. We shall return to the use of tensor concepts in the next section.

Armed with Lagrange's equations of motion our problem was solved when equation (2.125) was achieved, but for unrestricted motion about the centre of mass in the fixed coordinate system used here the values of the components of the inertia tensor will vary with time. Hence in this chapter we shall restrict the motion to that of small amplitude oscillations about the coordinate axes. In chapter 8 this restriction will be removed.

With the above restriction the equation of motion may be written in the form

$$\begin{bmatrix} P_x \\ P_y \\ P_z \\ G_x \\ G_y \\ G_z \end{bmatrix} = \left[\begin{array}{ccc|ccc} M & 0 & 0 & & & \\ 0 & M & 0 & & 0 & \\ 0 & 0 & M & & & \\ \hline & & & A & -F & -E \\ & 0 & & -F & B & -D \\ & & & -E & -D & C \end{array}\right]\begin{bmatrix} \ddot{x} \\ \ddot{y} \\ \ddot{z} \\ \ddot{\theta}_x \\ \ddot{\theta}_y \\ \ddot{\theta}_z \end{bmatrix} \quad (2.127)$$

P_i and G_i are forces and torques applied on the respective axes.

When possible one sets up the coordinate system in the rigid body along the principal axes of inertia. In that case the equations of motion become

$$\left.\begin{aligned} M\ddot{x} &= P_x; & M\ddot{y} &= P_y; & M\ddot{z} &= P_z \\ I_{xx}\ddot{\theta}_x &= G_x; & I_{yy}\ddot{\theta}_y &= G_y; & I_{zz}\ddot{\theta}_z &= G_z \end{aligned}\right\} \qquad (2.128)$$

In terms of radii of gyration $I_{xx} = Mk_x^2$; $I_{yy} = Mk_y^2$, and $I_{zz} = Mk_z^2$.

Exercises

1. Figure 2.23(a) shows a rigid body supported from a rigid anchorage 0 by a flexible rope OA. The body is displaced and allowed to oscillate. Figure 2.23(b) shows its position at time t. Write down the equations of motion for the body in terms of the quantities shown in the figure.

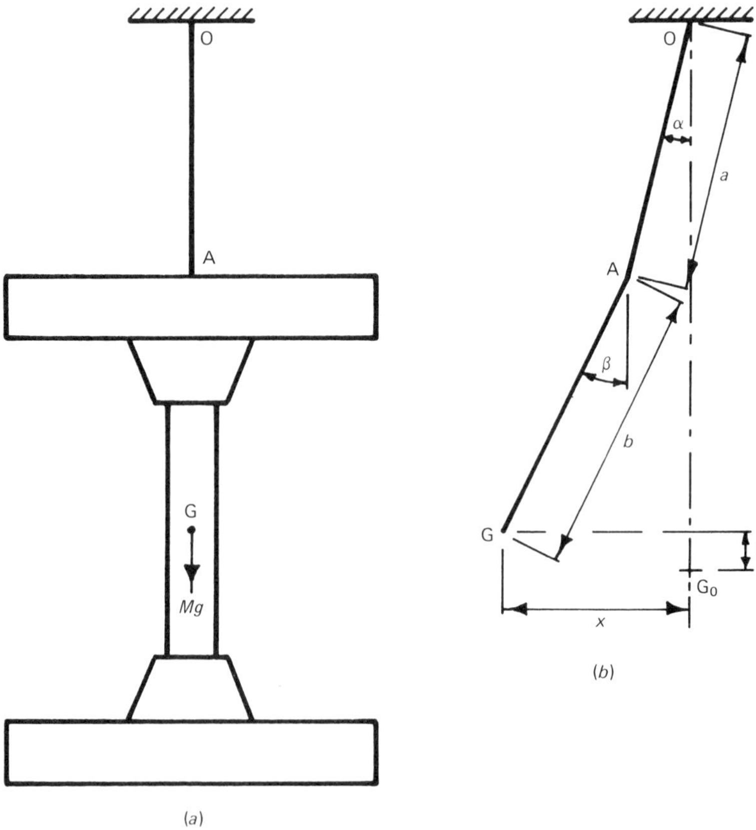

Fig. 2.23 Mass $= M$; radius of gyration about G $= k$

If now the oscillations are restricted so that the angles α and β may be considered small, modify your equations and obtain an equation involving only the frequency and the relevant length dimensions.

Given $a = 0{\cdot}60$ m, $b = 0{\cdot}75$ m, $k = 0{\cdot}90$ m calculate the two natural frequencies.

2. Figure 2.24 shows a vibration experiment in which a rectangular block of mass 2000 kg and radius of gyration 0·46 m about its centre of mass is flexibly suspended on springs which are so soft that their restoring forces for small displacements of the block may be neglected.

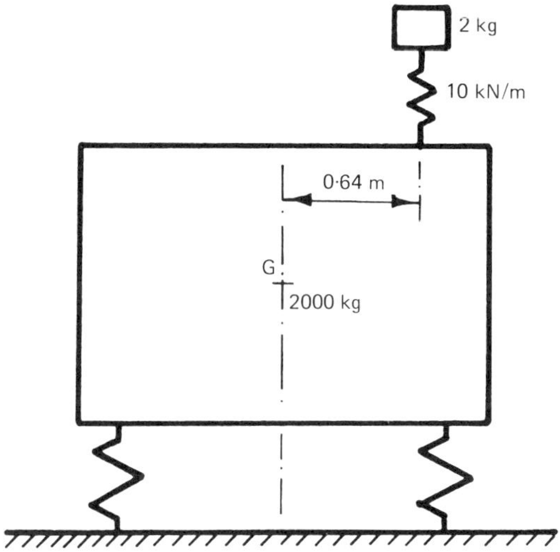

Fig. 2.24

A coil spring of stiffness 10 kN/m stands on the block 0·64 m from the central plane and supports a mass of 2 kg. Show that a natural frequency of the system under these conditions is higher by about 0·15 per cent than it would be if the block were supported rigidly.

3. Derive an expression for the angular momentum of a rigid body in motion about its centre of mass. Using Newtonian concepts show that, for small oscillations about coordinate axes through the centre of mass, the couples to cause given angular accelerations are related as in equation (2.127).

2.8.3 On building up the system dynamic stiffness matrix

High speed digital computers are used extensively for finding the natural frequencies and mode shapes of complex systems; the numerical procedures are discussed in chapter 5, but before one can set a computer to find these numerical values it must, in most instances, be presented with the complete dynamic stiffness matrix for the system. The preferred process for building up this matrix should be routine and general. Kron, in the 1930's [see Kron, (1939)], invented a general and powerful system for building up the oscillatory characteristics of complex electrical circuits and machines. Later, see Kron (1963), he showed how it could be applied to mechanical systems. Many people find Kron's work difficult to understand; it is hoped that what follows will be found a more easily understood interpretation of it.

The system shown in Fig. 2.25(*a*) may be broken up into the elements shown in Fig. 2.25(*b*). For element 1, the end forces and displacements are related by

$$\begin{bmatrix} {}_1p_1 \\ {}_1p_2 \end{bmatrix} = \begin{bmatrix} k_1 & -k_1 \\ -k_1 & k_1 \end{bmatrix} \begin{bmatrix} {}_1y_1 \\ {}_1y_2 \end{bmatrix}$$

or $${}_1\mathrm{p} = {}_1\mathrm{Y}\,{}_1\mathrm{y} \tag{2.129}$$

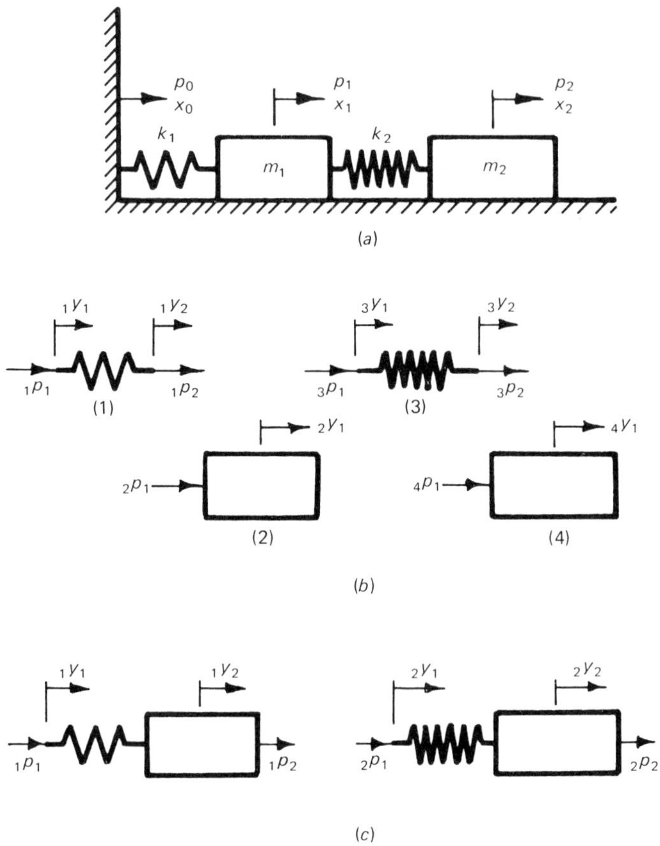

Fig. 2.25

Similarly for element 2

$${}_2p_1 = m_1\mathrm{D}^2\,{}_2y_1$$

or $${}_2\mathrm{p} = {}_2\mathrm{Y}\,{}_2\mathrm{y} \tag{2.130}$$

where $$\mathrm{D} \equiv \frac{\mathrm{d}}{\mathrm{d}t}$$

Now we call ${}_1y_1$, ${}_1y_2$, ${}_2y_1$, etc., the local displacement system. Similarly, the local force system is ${}_1p_1$, ${}_1p_2$, ${}_2p_1$, etc. The relationships between the local forces and local displacements are easily written down, giving the element dynamic stiffness matrix. Also, in local coordinates the complete system dynamic stiffness matrix is easily defined thus,

$$\begin{bmatrix} {}_1p_1 \\ {}_1p_2 \\ {}_2p_1 \\ {}_3p_1 \\ {}_3p_2 \\ {}_4p_1 \end{bmatrix} = \begin{bmatrix} k_1 & -k_1 & 0 & 0 & 0 & 0 \\ -k_1 & k_1 & 0 & 0 & 0 & 0 \\ 0 & 0 & m_1\mathrm{D}^2 & 0 & 0 & 0 \\ 0 & 0 & 0 & k_2 & -k_2 & 0 \\ 0 & 0 & 0 & -k_2 & k_2 & 0 \\ 0 & 0 & 0 & 0 & 0 & m_2\mathrm{D}^2 \end{bmatrix} \begin{bmatrix} {}_1y_1 \\ {}_1y_2 \\ {}_2y_1 \\ {}_3y_1 \\ {}_3y_2 \\ {}_4y_1 \end{bmatrix} \tag{2.131}$$

or

$$_{\mathrm{L}}\mathrm{p} = {}_{\mathrm{L}}\mathrm{Y}\,{}_{\mathrm{L}}\mathrm{y},$$

the subscript L denoting local.

As it stands, equation (2.131) is of little use for the determination of natural frequencies and mode shapes because it does not contain the information relating to the manner in which the elements are connected in the complete system. Obviously there is a connection between the local displacements ${}_{\mathrm{L}}\mathrm{y}$, and the global displacements ${}_{\mathrm{G}}\mathrm{y}$ (i.e. x_0, x_1, and x_2 in this example); these are the compatibility conditions for the system.

Thus

$$\begin{bmatrix} {}_1y_1 \\ {}_1y_2 \\ {}_2y_1 \\ {}_3y_1 \\ {}_3y_2 \\ {}_4y_1 \end{bmatrix} = \begin{bmatrix} 1 & 0 & 0 \\ 0 & 1 & 0 \\ 0 & 1 & 0 \\ 0 & 1 & 0 \\ 0 & 0 & 1 \\ 0 & 0 & 1 \end{bmatrix} \begin{bmatrix} x_0 \\ x_1 \\ x_2 \end{bmatrix} \tag{2.132}$$

or

$$_{\mathrm{L}}\mathrm{y} = \mathrm{C}\,{}_{\mathrm{G}}\mathrm{y}$$

There is also a connection between the local forces ${}_{\mathrm{L}}\mathrm{p}$ and the global forces ${}_{\mathrm{G}}\mathrm{p}$ (i.e., p_0, p_1, and p_2 in this example), these are the equilibrium conditions for the system. Thus

$$\begin{bmatrix} p_0 \\ p_1 \\ p_2 \end{bmatrix} = \begin{bmatrix} 1 & 0 & 0 & 0 & 0 & 0 \\ 0 & 1 & 1 & 1 & 0 & 0 \\ 0 & 0 & 0 & 0 & 1 & 1 \end{bmatrix} \begin{bmatrix} {}_1p_1 \\ {}_1p_2 \\ {}_2p_1 \\ {}_3p_1 \\ {}_3p_2 \\ {}_4p_1 \end{bmatrix} \tag{2.133}$$

or

$$_{\mathrm{G}}\mathrm{p} = \mathrm{C}^{\mathrm{T}}\,{}_{\mathrm{L}}\mathrm{p}$$

Note that the connection matrix in this case is the transpose of that given by the compatability conditions.

Substituting from equations (2.132) and (2.133) into equation (2.131) gives

$$_{G}p = C^{T}\,{}_{L}Y\,C\,{}_{G}y$$

or

$$_{G}p = {}_{G}Y\,{}_{G}y \tag{2.134}$$

Exercise

1. Check that equation (2.134) gives the correct set of equations of motion.
2. Set up the equations of motion for the systems shown in Fig. 2.8 by the above technique.

In finding ${}_{G}Y$ from ${}_{L}Y$ we have merely transformed from a local coordinate system to a global coordinate system. Regardless of the coordinates chosen, the work done for the complete system in a virtual displacement should be the same. That is

$$_{G}y^{T}\,{}_{G}p = {}_{L}y^{T}\,{}_{L}p \tag{2.135}$$

Now

$$_{L}p = {}_{L}Y\,{}_{L}y$$

and

$$_{L}y = C\,{}_{G}y$$

therefore,

$$_{L}y^{T}\,{}_{L}p = {}_{G}y^{T}\,C^{T}\,{}_{L}Y\,C\,{}_{G}y$$

or

$$_{G}p = C^{T}\,{}_{L}Y\,C\,{}_{G}y = {}_{G}Y\,{}_{G}y \tag{2.136}$$

as before. The dynamic stiffness matrix for a complete system may, therefore, be found for any general system from those of its elements, and its connection matrix C.

In using the above technique we are not restricted to the use of primary elements, we may start from the dynamic stiffness of matrices of subsystems as follows. The complete system of Fig. 2.25(*a*) may be split into the subsystem shown in Fig. 2.25(*c*). The process of building up the global dynamic stiffness matrix then becomes

$$\begin{bmatrix} {}_{1}p_{1} \\ {}_{1}p_{2} \\ {}_{2}p_{1} \\ {}_{2}p_{2} \end{bmatrix} = \begin{bmatrix} k_1 & -k_1 & 0 & 0 \\ -k_1 & m_1 D^2 + k_1 & 0 & 0 \\ 0 & 0 & k_2 & -k_2 \\ 0 & 0 & -k_2 & m_2 D^2 + k_2 \end{bmatrix} \begin{bmatrix} {}_{1}y_{1} \\ {}_{1}y_{2} \\ {}_{2}y_{1} \\ {}_{2}y_{2} \end{bmatrix}$$

or

$$_{L}p = {}_{L}Y\,{}_{L}y \tag{2.137}$$

the connection matrix is given by

$$\begin{bmatrix} {}_{1}y_{1} \\ {}_{1}y_{2} \\ {}_{2}y_{1} \\ {}_{2}y_{2} \end{bmatrix} = \begin{bmatrix} 1 & 0 & 0 \\ 0 & 1 & 0 \\ 0 & 1 & 0 \\ 0 & 0 & 1 \end{bmatrix} \begin{bmatrix} x_0 \\ x_1 \\ x_2 \end{bmatrix}$$

or

$$_{L}y = C\,{}_{G}y \tag{2.138}$$

Therefore

$$
\begin{aligned}
{}_G\mathrm{Y} &= \mathrm{C}^{\mathrm{T}}\,{}_L\mathrm{YC} \\
&= \begin{bmatrix} 1 & 0 & 0 & 0 \\ 0 & 1 & 1 & 0 \\ 0 & 0 & 0 & 1 \end{bmatrix} [{}_L\mathrm{Y}] \begin{bmatrix} 1 & 0 & 0 \\ 0 & 1 & 0 \\ 0 & 1 & 0 \\ 0 & 0 & 1 \end{bmatrix} \\
&= \begin{bmatrix} k_1 & -k_1 & 0 \\ -k_1 & m_1\mathrm{D}^2 + k_1 + k_2 & -k_2 \\ 0 & -k_2 & m_2\mathrm{D}^2 + k_2 \end{bmatrix}
\end{aligned}
\tag{2.139}
$$

Let us examine more generally the process of joining two subsystems of this type. The state of each subsystem was described by two coordinates, and in the complete system they each had a common coordinate. Thus for each subsystem the dynamic stiffness matrix is of the form

$$
\begin{bmatrix} {}_ip_1 \\ {}_ip_2 \end{bmatrix} = \begin{bmatrix} {}_iY_{11} & {}_iY_{12} \\ {}_iY_{21} & {}_iY_{22} \end{bmatrix} \begin{bmatrix} {}_iy_1 \\ {}_iy_2 \end{bmatrix} \tag{2.140}
$$

and the connection matrix was of the form

$$
C = \begin{bmatrix} 1 & 0 & 0 \\ 0 & 1 & 0 \\ 0 & 1 & 0 \\ 0 & 0 & 1 \end{bmatrix} \tag{2.141}
$$

giving the global dynamic stiffness matrix as

$$
{}_G\mathrm{Y} = \begin{bmatrix} {}_1Y_{11} & {}_1Y_{12} & 0 \\ {}_1Y_{21} & {}_1Y_{22} + {}_2Y_{11} & {}_2Y_{12} \\ 0 & {}_2Y_{21} & {}_2Y_{22} \end{bmatrix} \tag{2.142}
$$

As one would expect the direct dynamic stiffness on the common coordinate ${}_GY_{22}$ is the sum of the direct dynamic stiffnesses of the two subsystems at their junction ${}_2Y_{11} + {}_1Y_{22}$.

Kron formulated the above method in terms of the geometry of abstract spaces. Dealing only with simple harmonic oscillations we may write

$$
\mathrm{F} = \mathrm{YA} \tag{2.143}
$$

where:

F is the force vector, magnitudes being force amplitudes;
A is the displacement vector, magnitudes being displacement amplitudes;
and Y is the appropriate dynamic stiffness array.

This expression transforms as before. Kron postulated a space which described the displacements as a contravariant vector and the forces as a covariant vector, the metric tensor of the space being the dynamic stiffness tensor, that is $F_i = Y_{ij}A^j$. If the space represents the state space of the complete system, the state spaces of elements or subsystems are sub-spaces of it. The process of embedding a sub-space A^{α} in the system space A^{i} is

$$A^{i} = C_{\alpha}{}^{i}A^{\alpha} \tag{2.144}$$

where $C_{\alpha}{}^{i}$ is correctly a mixed tensor, see Sokolnikoff (1951). Thus in tensor notation the complete transformation is represented by

$$Y_{ij} = Y_{\alpha\beta}C_{i}{}^{\alpha}C_{j}{}^{\beta} \tag{2.145}$$

In chapter 3 we shall return to this topic in relation to more complex systems.

Miscellaneous exercises

1. A rigid bar AB, 2 m long, mass 50 kg and radius of gyration about its centre of mass G 0·5 m, is supported at its ends on springs AC and BD, each of stiffness 80 kN/m. Figure 2.26 shows the configuration. Find the system natural frequencies and normal modes for small vibrations, neglect any horizontal motion of the bar. Check that the normal modes are orthogonal.

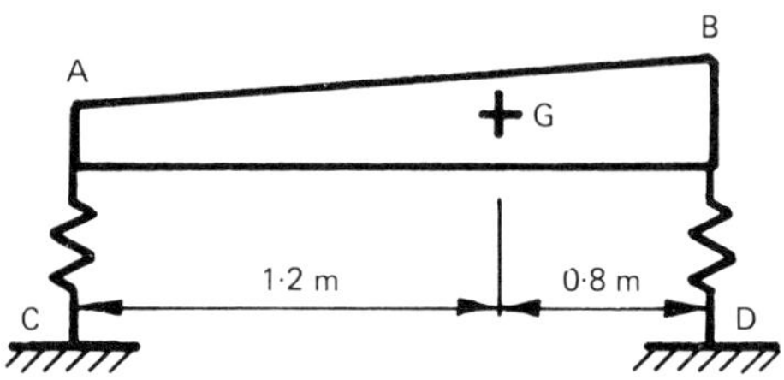

Fig. 2.26

   If now the bar is acted upon by the transverse force, 20 sin ωt N, at the centre of its length, find the exciting frequency for which it has no rotational motion.
2. If acting on each end of the bar, of exercise 1, in parallel with the springs there is a viscous damper, with a damping coefficient 15 N s/m, find the amplitude of points A and G when the frequency of the exciting force is equal to each of the natural undamped system frequencies.
3. The system shown in Fig. 2.27 is comprised of the following:

   A body of mass M, which is supported on two rollers is constrained by frictionless guides to move along a straight path. The motion of the mass M is restrained by a spring of stiffness K which is attached to a rigid frame. The rollers, which are also supported on the rigid frame, each have a mass m, external radius r and radius of gyration k. To the mass M is attached a double pendulum consisting of a simple pendulum of length L_1 and mass m_1 which is suspended from the mass M and a simple pendulum of length L_2 and mass m_2 which is suspended from the centre of mass m_1.

Find, for small amplitude oscillations, the equations of motion for free vibrations of the system. Use each of the methods presented in this chapter.

4. The system shown diagrammatically in Fig. 2.28 is used to demonstrate the relationship between centripetal force, radius and velocity of rotation of a body. It is constituted as follows:

   A pair of solid spherical weights A, each of mass M and moment of inertia about a diameter $\mathcal{J}$, are carried by the rigid arms AB of length L. The rigid arms are hinged to a vertical spindle BC in such a way that they are free to move about B in a plane containing both arms. The spindle BC, which is free to rotate in frictionless bearings, has attached to it a heavy disc of moment of inertia I. Between I and B the spindle has a torsional stiffness S. The weights of the spindle and arms may be neglected.

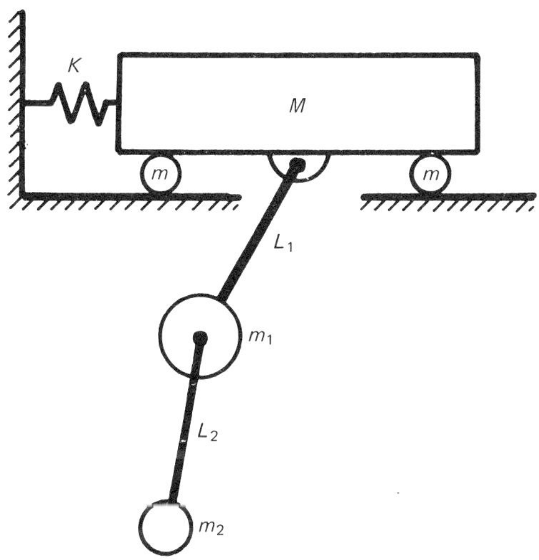

Fig. 2.27

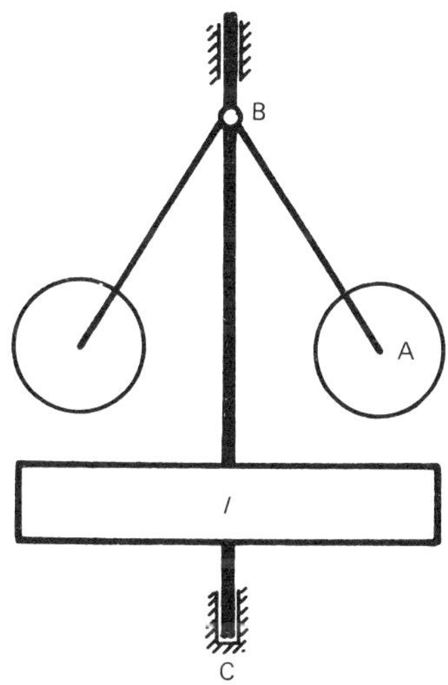

Fig. 2.28

   Assume that the system is rotating with small amplitude oscillations superimposed upon the rigid body motion. Derive the equations of motion for each of the following conditions:

   (*a*) when the weights A are prevented from rotating relative to the arms
   (*b*) when the weights A are supported on frictionless bearings in the ends of the arms.

   Give the physical significance of each term in the equations of motion.

5. Figure 2.29 shows, diagrammatically, the essential parts of a vibratory rate gyro.

   The rigid tube A is rigidly connected to an elastic shaft B of torsional stiffness S, which in turn is rigidly fixed to a rotating instrument frame C. The masses M are attached to a pin which is fixed at the axis of rotation, through springs of stiffness K. These subsystems, that is the masses M and their associated springs are excited electromagnetically at their natural frequencies. You may take the amplitude of the motion of the masses M as small and constant, otherwise neglect the effect of the friction between the masses M and the tube. Air forces on the spinning tube A exert a damping torque proportional to and in phase with

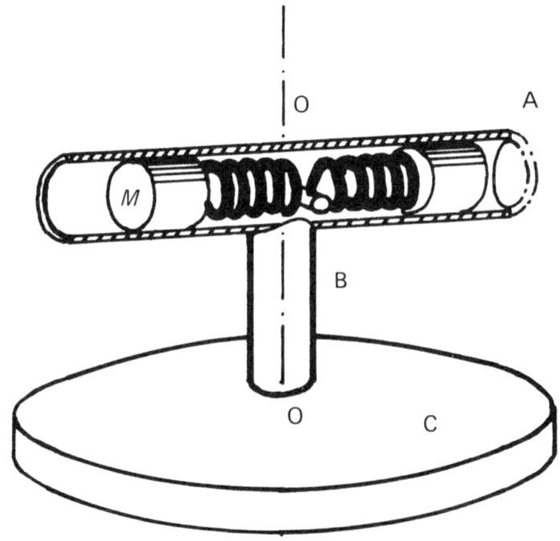

Fig. 2.29

its instantaneous angular velocity about the axis O–O. The moment of inertia of the tube about O–O is I.

You are required to derive the equation of motion for the system for the case in which the frame is given a constant angular velocity Ω about the axis O–O.

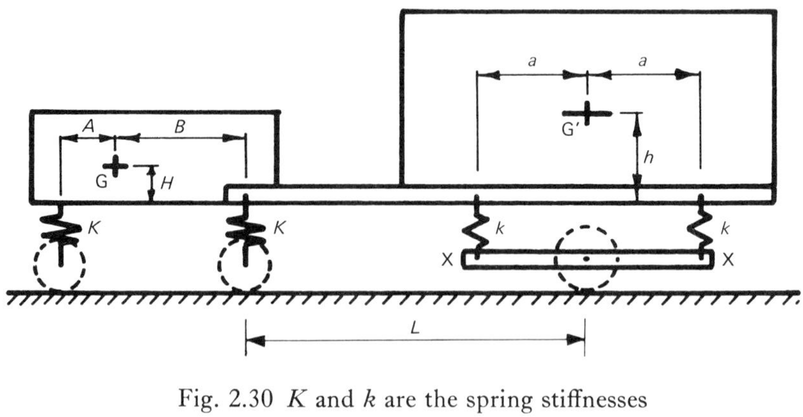

Fig. 2.30 K and k are the spring stiffnesses

Hence, or otherwise, show the amplitude of the vibratory motion of the tube about the axis O–O, provided that it is small, is a measure of the angular velocity Ω.

6. Taking the system shown in Fig. 2.30 as a model representing a vehicle towing a caravan, write down the equations of motion for small oscillations about the static equilibrium position and give the frequency equation in determinantal form.

   G and G′ are the centres of gravity of the towing vehicle and caravan respectively. The relevant moments of inertia about axes through these points are I

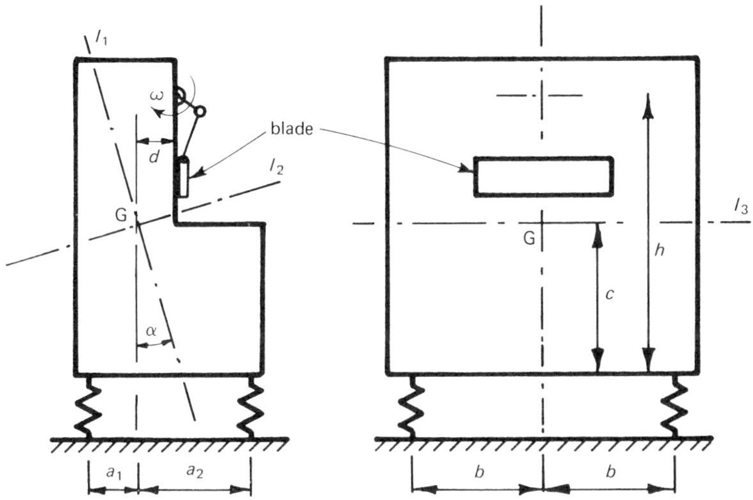

Fig. 2.31 (G is the centre of mass; I_1, I_2, and I_3 are the principal moments of inertia about the axes indicated)

and $\mathcal{J}$ respectively. Take the mass of the vehicle as M and that of the caravan as m. The vertical stiffnesses of the springs are given on the diagram. Take the horizontal stiffnesses of the springs as infinite.

   You may neglect the mass of the bar XX of the caravan. Assume that the caravan is pin-jointed to the towing vehicle at a point immediately above the rear spring support.

7. In a machine for slicing pipe tobacco the blade is driven by means of a slider-crank mechanism. The machine of total mass M, is mounted on four light springs, of stiffness K vertically and k horizontally. The slider-crank mechanism may be represented as a reciprocating mass m_r, and a rotating mass m acting on the crank radius r. The coupler bar is of length ℓ. Figure 2.31 shows the directions of the principal axes and the other relevant dimensions.

   Derive the equations of motion for the machine on its mountings. How would you check that the running speed of the machine is in the most suitable range for the supports provided?

8. In the torsional system shown in Fig. 2.32 the subsystem, including the gear,

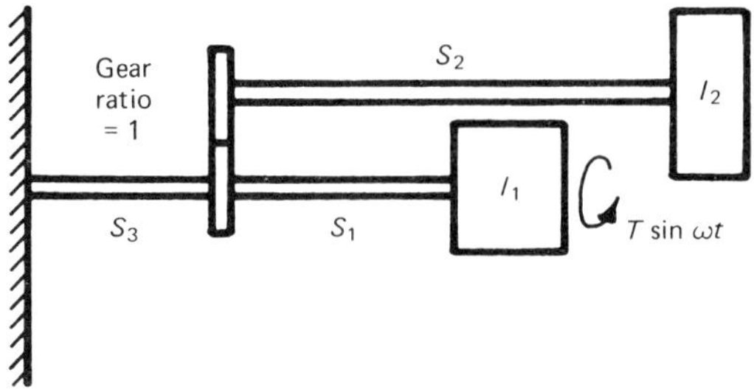

Fig. 2.32

shaft S_2, and disc I_2, was added to act as a vibration absorber. Show that the amplitude of I_1 is zero when

$$\omega^2 = \frac{S_2(S_1 + S_3)}{I_2(S_1 + S_2 + S_3)}$$

Plot the response of the system as a function of ω.

References

AITKEN, A. C.
(*a*) *Determinants and matrices*, Oliver & Boyd, 1946, p. 64.
(*b*) *Ibid.*, p. 80.

BISHOP, R. E. D. and JOHNSON, D. C.
The Mechanics of vibration, Cambridge University Press, 1960.

KRON, G.
Tensor analysis of networks, John Wiley, 1939.

KRON, G.
Diakoptics, Macdonald, London, 1963.

RAYLEIGH, LORD
Theory of sound, 2nd edn., Macmillan, 1894, Vol. 1, pp. 102–3.

SOKOLNIKOFF, I. S.
Tensor analysis, John Wiley, 1951, p. 173.

EASTHOPE, C. E.
Three dimensional dynamics, 2nd edn., Butterworth, 1964, chap. 10.

Chapter 3

Systems having distributed mass and elasticity

In many engineering structures, the modes of vibration and the distributions of mass and elasticity are such that the mass and elasticity cannot, even as an approximation, be considered concentrated in distinct components or parts of components. This situation will be illustrated for such simple cases as predicting the vibratory behaviour of bars executing longitudinal vibrations, of beams undergoing transverse vibrations, and of plates.

When its mass and elasticity are distributed throughout a system, a description of its configuration requires an infinite number of coordinates. Therefore, a system with these attributes may be regarded as having an infinite number of degrees of freedom which, in turn, means that it has an infinite number of natural frequencies.

In this chapter we shall follow through methods for setting up and solving the equations of motion for the above systems, assuming them to be composed of homogeneous, isotropic, elastic material.

3.1 Longitudinal vibration of a uniform bar

The first of the simple systems with distributed mass and elasticity we shall study is a uniform bar, vibrating so that groups of particles constituting it undergo longitudinal motion relative to one another. It is presented first because, under the assumptions made, it is one of the easiest systems to analyse mathematically and by means of it many of the concepts studied in chapter 2 may be extended to apply to continuously distributed systems.

In the analysis which follows it is assumed that during longitudinal oscillations the cross-sections of the bar remain plane and that lateral motion of the particles in these cross-sections, due to Poisson's ratio effects, is negligible. Consider the bar to be of length L, to have cross-sectional area A, mass density ρ and Young's modulus E. The distance to the equilibrium position of a typical cross-section from an arbitrary origin, along the centre line of the bar, will be described by the coordinate x, and the displacement of this typical cross-section from its equilibium position, at time t whilst the bar is vibrating, will be described by the coordinate

$u(x, t)$. Whilst the displacement is $u(x, t)$ the stress at the typical cross-section is $p(x, t)$, tensile stress being taken as positive.

In order to apply Newton's laws we consider the forces acting on the mass of an elemental length of the bar, causing acceleration. Therefore, we must determine the displacement of and stress at a cross-section a small distance, δx, from the typical cross-section in the equilibrium state. Figure 3.1 will help you visualise the displacements and forces relating to the element considered.

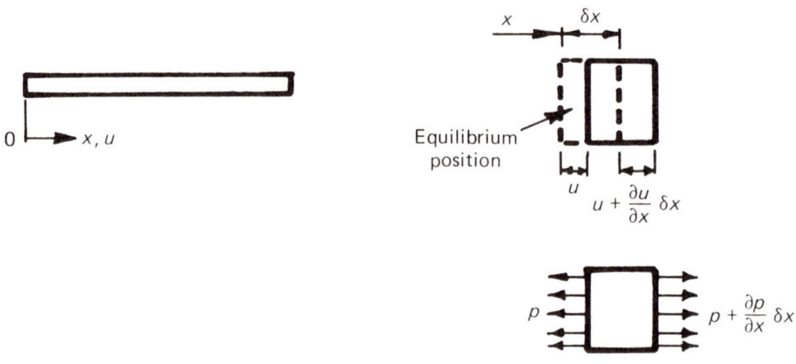

Fig. 3.1

For a uniform bar, both $u(x, t)$ and $p(x, t)$ will be continuous functions of x and of t. They may, therefore, be expanded by Taylor's theorem to give

$$u(x + \delta x, t) = u(x, t) + \frac{\partial u(x, t)}{\delta x}\,\delta x + \frac{1}{2}\,\frac{\partial^2 u(x, t)}{\partial x^2}\,(\delta x)^2 + \cdots \tag{3.1}$$

and

$$p(x + \delta x, t) = p(x, t) + \frac{\partial p(x, t)}{\partial x}\,\delta x + \frac{1}{2}\,\frac{\partial^2 p(x, t)}{(\partial x)^2}\,(\delta x)^2 + \cdots \tag{3.2}$$

The acceleration at $x + 0{\cdot}5\ \delta x$ will also be required, and is

$$\frac{\partial^2 u(x + 0{\cdot}5\,\delta x, t)}{\partial t^2} = \frac{\partial^2 u(x, t)}{\partial t^2} + \frac{1}{2}\,\frac{\partial}{\partial x}\,\frac{\partial^2 u(x, t)}{\partial t^2}\,\delta x + \frac{1}{8}\,\frac{\partial^2}{\partial x^2}\,\frac{\partial^2 u(x, t)}{\partial t^2}\,(\delta x)^2 + \cdots \tag{3.3}$$

Taking terms containing δx to the second and higher orders as negligibly small gives

$$\left.\begin{aligned} u(x + \delta x, t) &= u(x, t) + \frac{\partial u(x, t)}{\partial x}\,\delta x \\ p(x + \delta x, t) &= p(x, t) + \frac{\partial p(x, t)}{\partial x}\,\delta x \end{aligned}\right\} \tag{3.4}$$

and mass × acceleration for the element becomes

$$\rho A \, \delta x \, \frac{\partial^2 u(x + 0{\cdot}5 \, \delta x, t)}{\partial t^2} = \rho A \, \delta x \, \frac{\partial^2 u(x, t)}{\partial t^2} \tag{3.5}$$

The stress at a cross-section is related to the strain at the cross-section through Young's modulus. The strain is

$$\frac{\partial u(x, t)}{\partial x},$$

therefore the equation of state is

$$p(x, t) = E \, \frac{\partial u(x, t)}{\partial x} \tag{3.6}$$

and the rate of change of stress

$$\frac{\partial p(x, t)}{\partial x} = E \, \frac{\partial^2 u(x, t)}{\partial x^2} \tag{3.7}$$

Applying Newton's laws to the mass of the element of length δx gives

$$A \left\{ -p(x, t) + p(x, t) + E \, \frac{\partial^2 u(x, t)}{\partial x^2} \, \delta x \right\} = \rho A \, \delta x \, \frac{\partial^2 u(x, t)}{\partial t^2} \tag{3.8}$$

Thus the equation of motion for the particles in any cross-section of the bar is

$$\frac{E}{\rho} \, \frac{\partial^2 u(x, t)}{\partial x^2} - \frac{\partial^2 u(x, t)}{\partial t^2} = 0 \tag{3.9}$$

If the bar were infinitely long equation (3.9) would be satisfied by a solution of the form

$$u(x, t) = u_1(x - at) + u_2(x + at) \tag{3.10}$$

u_1 and u_2 being arbitrary functions. Obviously the initial form of the displacement is

$$u(x, 0) = u_1(x) + u_2(x) \tag{3.11}$$

At time $t = t_1$,

$$u(x, t_1) = u_1(x - at_1) + u_2(x + at_1) \tag{3.12}$$

If we let $at_1 = x_1$ then

$$u(x, t_1) = u_1(x - x_1) + u_2(x + x_1) \tag{3.13}$$

That is the components of the form (3.11) at time $t = 0$ exist unchanged at time t_1 except for translations, $u_1(x)$ being translated in the positive direction by a distance x_1, and $u_2(x)$ being translated in the negative direction by the same distance. The velocity of propagation of the displacement disturbances and the associated direct stress is a, where $a^2 = E/\rho$. The velocity of propagation of direct stress waves in a substance is known as the velocity of sound in the substance.

3.1.1 Reflection of stress waves at the ends of a bar

Bars must terminate and we shall follow now what happens when one of these disturbances u_1 or u_2 meets an end of a bar. The case of a rigidly supported end will be considered first. Since the support is rigid, the particles of the cross-section of the bar in contact with it must remain permanently at rest, that is, it is a node. Hence, when a tensile stress wave reaches the support, the only way in which the end layer of the material of the bar may be relieved of this stress is for the tensile stress to be reflected back along the bar. The displacement of the particles in the reflected stress wave will be of opposite sign to that of the incident stress wave. This state could be achieved analytically by assuming that an infinitely long bar is fixed permanently at $x = 0$ and arranging for a negatively propagating positive displacement disturbance $u_2(x + at)$ to be annulled completely at $x = 0$ by synchronising a positively propagating negative displacement disturbance $u_1(x - at) = -u_2(-x + at)$ so that

$$\begin{aligned} u(0, t) &= u_1(0 - at) + u_2(0 + at) \\ &= -u_2(at) + u_2(at) \\ &= 0 \end{aligned} \tag{3.14}$$

$u_2(x + at)$ is the incident displacement wave and $-u_2(-x + at)$ is the reflected displacement wave. The stress at $x = 0$ is obviously

$$2E \frac{\partial u_2(0 + at)}{\partial x}$$

which is twice that due to the incident wave alone.

The second type of end condition of interest at present is that in which the end of the bar is unrestrained. Since the end of the bar is unrestrained the particles in its end plane are permanently unstressed. Hence, when a tensile stress wave reaches the end plane it will cause the end plane to accelerate in the direction in which the tensile force acts. As the end plane of the bar remains stress free the velocity resulting from this acceleration may only be reduced by applying a compressive force. Thus at a free end an incident tensile stress wave is reflected as a compressive stress wave. Applying the reasoning as before, but this time to the stress we have, analytically, at end $x = 0$:

$$\frac{\partial u(0, t)}{\partial x} = \frac{\partial u_1(0 - at)}{\partial x} + \frac{\partial u_2(0 + at)}{\partial x} = 0 \tag{3.15}$$

so that

$$u_1(x - at) = u_2(-x + at)$$

In this case the displacement at the end will be twice that due to the incident wave alone.

3.1.2 Free vibration of a short bar

Let us suppose it possible to displace, statically, the cross-sections of a uniform bar of length L which is fixed rigidly at each of its ends. Let these initial displacements be according to the form

$$u(x, 0) = 2B \sin \frac{\pi x}{L} \tag{3.16}$$

from which it is released at time $t = 0$.

The previous section has indicated that this will result in two waves, one $u_1(x - at)$ progressing positively and the other $u_2(x + at)$ progressing negatively where

$$\left.\begin{aligned} u_1(x - at) &= B \sin \frac{\pi(x - at)}{L} \\ \text{and} \qquad u_2(x + at) &= B \sin \frac{\pi(x + at)}{L} \end{aligned}\right\} \tag{3.17}$$

These are shown diagrammatically for various times in Fig. 3.2, together with waves reflected from the ends and the resultant displacement of each cross-section. The initial condition $u(x, 0)$ is regained after a time $2L/a$, which is the time for a stress wave to travel twice the length of the bar. At intermediate times it maintains the sine form but the amplitude varies sinusoidally with time. Such vibrations are called stationary vibrations. A similar argument could be advanced for an initial displacement of the form

$$u(x, 0) = 2B \sin \frac{2\pi x}{L} \tag{3.18}$$

except that this condition would be achieved again at time L/a from the instant of release. Thus the frequency with which the initial shape $2B \sin (2\pi x/L)$ is regained is twice that with which the initial shape $2B \sin (\pi x/L)$ is regained. Generalising, we may say that the initial shape $2B \sin (n\pi x/L)$ would reappear with a frequency n times that of the initial shape $2B \sin (\pi x/L)$ and at any time t the distortion of the bar would be given by

$$\begin{aligned} u(x, t) &= B \sin \frac{n\pi}{L} (x - at) + B \sin \frac{n\pi}{L} (x + at) \\ &= 2B \sin \frac{n\pi x}{L} \cos \frac{n\pi at}{L} \end{aligned} \tag{3.19}$$

By analogy with the systems studied in Chapter 2, it can be said that the principal modes have the form $B_n \sin (n\pi x/L)$ and that the angular natural frequency

$$\omega_n = \frac{n\pi}{L} \sqrt{\frac{E}{\rho}}.$$

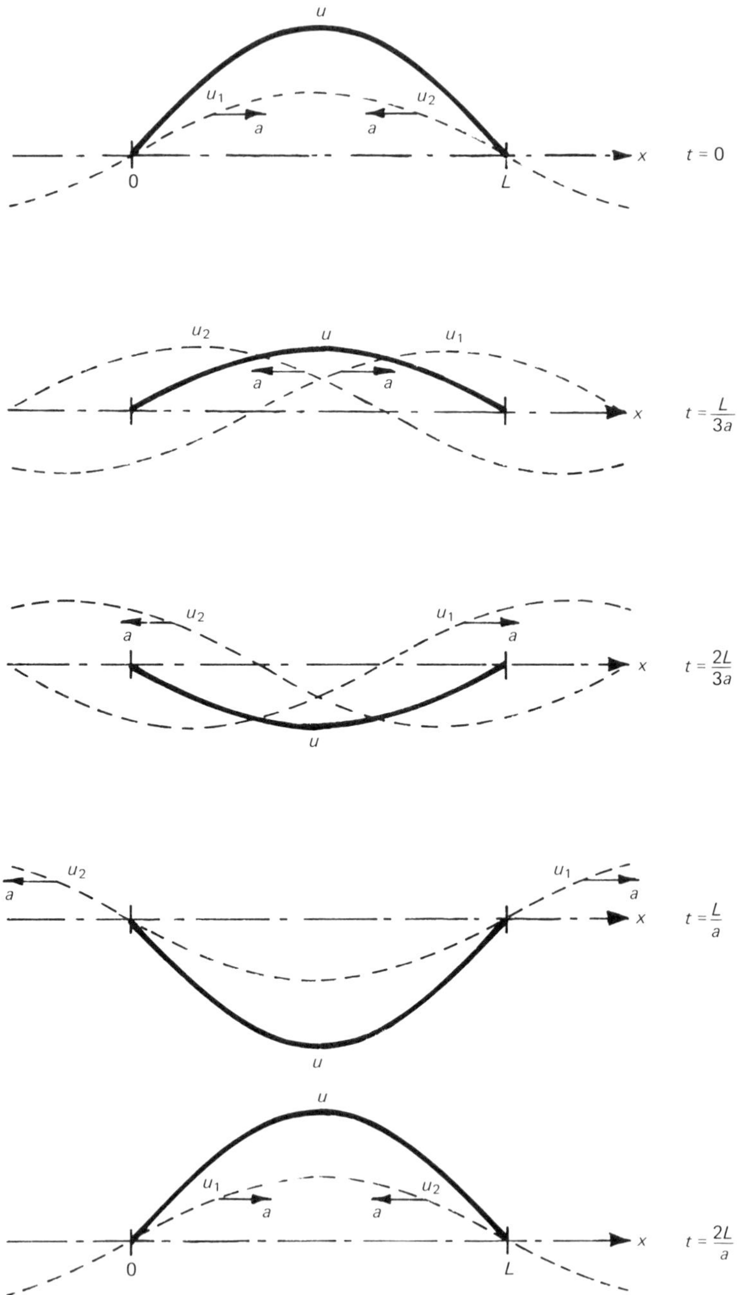

Fig. 3.2 $u_1 = B \sin \dfrac{\pi(x - at)}{L}$ $\qquad u_2 = B \sin \dfrac{\pi(x + at)}{L}$

The principal modes are obviously orthogonal because

$$\int_0^L \rho AB_n \sin \frac{n\pi x}{L} B_m \sin \frac{m\pi x}{Ldx} = 0$$

for integral values of m and n so long as $n \neq m$.

Now let us suppose it possible to displace, statically, the cross-sections of a uniform bar from their equilibrium positions according to the form

$$u(x, 0) = 2B \cos \frac{\pi x}{L} \tag{3.20}$$

where L is the length of the bar. Upon releasing it at time $t = 0$ all constraints are removed. Both of its ends will then be permanently stress free. Again this will result in two waves, one, $u_1(x - at)$, progressing positively and the other, $u_2(x + at)$ progressing negatively, where

$$\left.\begin{aligned} u_1(x - at) &= B \cos \frac{\pi(x - at)}{L} \\ \text{and} \qquad u_2(x + at) &= B \cos \frac{\pi(x + at)}{L} \end{aligned}\right\} \tag{3.21}$$

These are shown diagrammatically, in Fig. 3.3, for various times, together with waves reflected from the ends and the resultant displacement of each cross-section. The initial condition $u(x, 0)$ is regained after a time $2L/a$, that is the time for a stress wave to travel twice the length of the bar. At intermediate times it maintains the cosine form but the amplitude varies sinusoidally with time. It could also be shown that

$$u(x, t) = 2B \cos \frac{n\pi x}{L} \cos \frac{n\pi at}{L} \tag{3.22}$$

for an initial form of $u(x, 0) = 2B \cos (n\pi x/L)$. That is the angular natural frequency

$$\omega_n = \frac{n\pi}{L} \sqrt{\frac{E}{\rho}}$$

which is the same as for a bar with fixed ends, and the principal modes have the form $B_n \cos (n\pi x/L)$ which are obviously orthogonal.

The two cases considered above have shown that the principal modes depend upon the end constraints experienced by the bar. In these particular examples the frequency of vibration of the bar has been unaffected by the end constraints but this is not so in general.

As in the case of systems with lumped parameters, normal modes, that is principal modes to which have been assigned amplitudes, may be defined. The values of ω_n^2 are again referred to as eigenvalues of the system and the functions

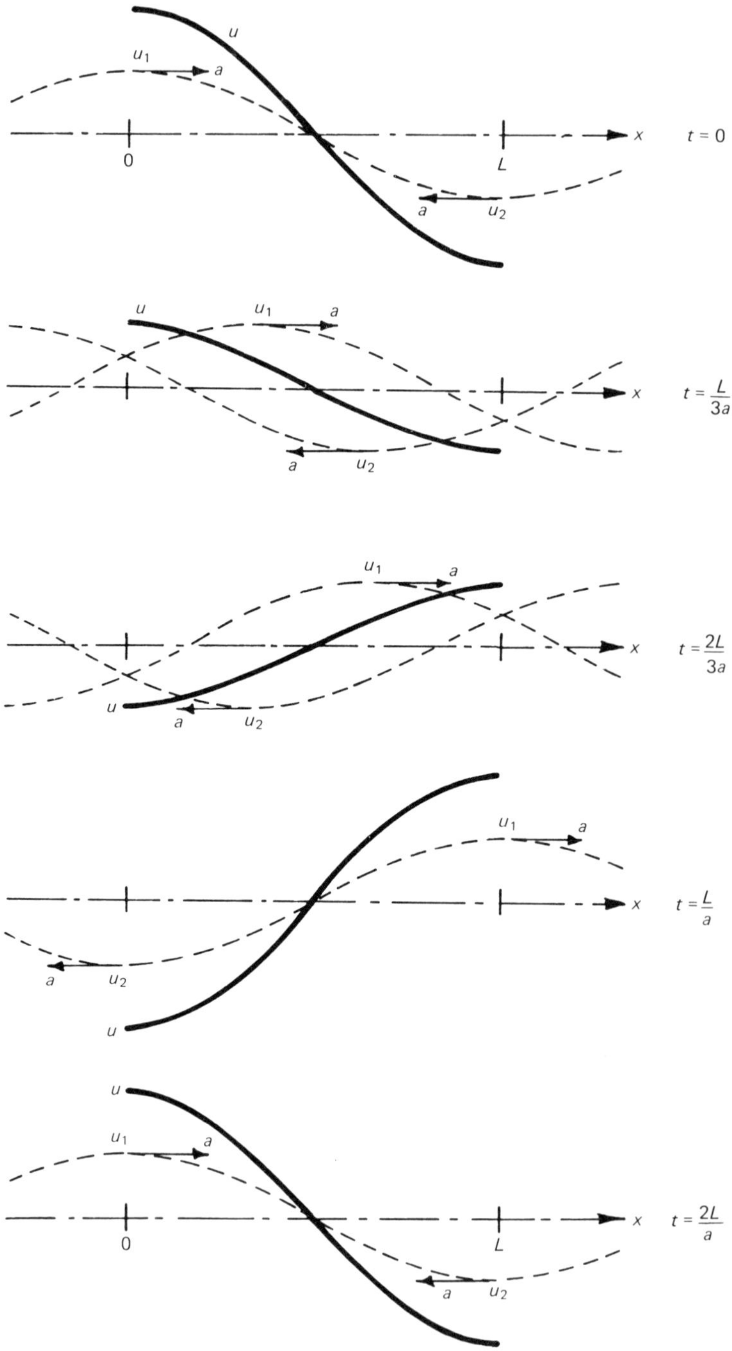

Fig. 3.3 $u_1 = B \cos \dfrac{\pi(x - at)}{L}$ $\qquad u_2 = B \cos \dfrac{\pi(x + at)}{L}$

used to describe the normal modes, for example sin $(n\pi x/L)$ and cos $(n\pi x/L)$ in the above cases, are often called characteristic functions or eigenfunctions.

The normal modes, having the form of terms of a Fourier series which satisfy the end conditions imposed upon the bar, may be used to represent any allowable initial displacement form. Thus the free vibration resulting from any initial stationary displacement form may be found. If the initial conditions take the form of imposed cross-section velocities throughout the bar, the velocities associated with the normal modes may again be used to synthesise any initial form.

Exercise

Draw diagrams of the travelling waves and their resultant for a uniform bar of length L; one end ($x = 0$) being rigidly fixed, the other ($x = L$) being permanently free from stress. Hence derive an expression for the displacement $u(x, t)$ from its equilibrium position after release from a distorted form $u(x, 0) = 2B \sin \pi x/2L$.

Show that the principal modes for this bar are orthogonal and compare its natural frequencies with those of a bar of the same length, material, and cross-section but whose ends are rigidly fixed.

3.1.3 A more convenient solution

In the previous section, it has been shown that the free vibrations of a uniform bar may be expressed as the composition of vibrations, principal modes, each correctly scaled and phased to satisfy the initial conditions. The principal modes have particular forms which satisfy the end constraints imposed by the environment upon the bar; these forms may vary harmonically with time. The free vibratory form may be expressed as

$$u(x, t) = \sum_n U_n(x)\{A_n \sin \omega_n t + B_n \cos \omega_n t\} \tag{3.23}$$

where $U_n(x)$ are independent of time and A_n and B_n, which depend upon the initial values of $u(x, t)$ and $\dot{u}(x, t)$, are independent of x. Now although this has been found from consideration of wave transmission in a uniform bar, it is in accord with our findings on undamped systems with a finite number of degrees of freedom: indeed it is general for linear undamped systems. Therefore, we may solve the equation of motion as follows

$$\frac{\partial^2 u(x, t)}{\partial x^2} - \frac{\rho}{E}\frac{\partial^2 u(x, t)}{\partial t^2} = 0 \tag{3.24}$$

Substituting for $u(x, t)$ from equation (3.23), gives

$$\frac{d^2 U_n(x)}{dx^2} + \frac{{\omega_n}^2 \rho}{E} U_n(x) = 0 \tag{3.25}$$

which is satisfied by solutions of the form

$$U_n(x) = C_n \sin \alpha_n x + D_n \cos \alpha_n x \tag{3.26}$$

where

$$\alpha_n{}^2 = \frac{\omega_n{}^2 \rho}{E}$$

and C_n and D_n are constants of integration which depend upon the end constraints on the bar.

Thus the natural frequency and principal modes of vibration for a uniform bar, rigidly fixed at $x = 0$ and free at $x = L$, are obtained as follows. From $U_n(x) = 0$ at $x = 0$ we find that $D_n = 0$, and from $E(\mathrm{d}U_n(x)/\mathrm{d}x) = 0$ at $x = L$ we find

$$C_n E \alpha_n \cos \alpha_n L = 0 \tag{3.27}$$

Equation (3.27) is the frequency equation: it is satisfied if

$$\alpha_n L = \frac{(2n - 1)}{2} \pi \tag{3.28}$$

Therefore, the angular natural frequency of the n'th mode is

$$\omega_n = \frac{(2n - 1)}{2} \frac{\pi}{L} \sqrt{\frac{E}{\rho}} \tag{3.29a}$$

and the vibratory form is

$$u(x, t) = \sum_{n=1,2}^{\infty} \sin \frac{(2n - 1)\pi x}{2L} \{A_n \sin \omega_n t + B_n \cos \omega_n t\} \tag{3.29b}$$

The values of A_n and B_n depend upon the initial conditions.

Exercises

1. A uniform bar of length L, cross-sectional area A, and mass density ρ is initially under a uniform compression due to forces of magnitude F applied to its ends. If these forces are suddenly removed, to leave it unconstrained, show that the subsequent vibration is described by

$$u(x, t) = \frac{4FL}{AE\pi^2} \sum_{n=1,3}^{\infty} \frac{(\cos n\pi x/L)(\cos n\pi at/L)}{n^2}$$

2. If a damping force proportional to strain velocity acts on the material of the bar show that the modes of vibration of a uniform bar remain uncoupled and orthogonal. Hence derive a general expression for longitudinal vibration of a uniform bar subject to this form of damping.

3.1.4 Forced longitudinal vibration of a uniform bar

Of practical interest is the case of a bar subjected to a periodic force at one or more points. For the time being we shall restrict consideration to excitation at one end point and, to illustrate the solution procedure, the bar will be regarded as having length L, being fixed rigidly at $x = 0$, and stressed by a periodic force $F \cos \omega t$ at $x = L$.

The equation of motion

$$\frac{\partial^2 u(x, t)}{\partial x^2} - \frac{\rho}{E} \frac{\partial^2 u(x, t)}{\partial t^2} = 0 \tag{3.30}$$

applies to all cross-sections of the bar and as the end cross-section at $x = L$ is acted upon by a force of period $2\pi/\omega$ we look for a steady state solution of equation (3.30) with period $2\pi/\omega$. That is, we look for a steady state solution of the form

$$u(x, t) = U(x)\{C \sin \omega t + D \cos \omega t\} \tag{3.31}$$

which, on substitution into equation (3.30), gives

$$\frac{\mathrm{d}^2 U(x)}{\mathrm{d}x^2} + \frac{\omega^2 \rho}{E} U(x) = 0 \tag{3.32}$$

Hence

$$U(x) = C_1 \sin \alpha x + D_1 \cos \alpha x \tag{3.33}$$

where

$$\alpha^2 = \omega^2 \rho / E$$

To satisfy the end conditions we have at $x = 0$,

$$U(0) = 0$$

therefore,

$$D_1 = 0$$

and at $x = L$

$$AE \frac{\partial u(x, t)}{\partial x} = F \cos \omega t$$

$$= AE\alpha \cos \alpha L\{C_3 \sin \omega t + D_3 \cos \omega t\}$$

Therefore

$$C_3 = 0 \quad \text{and} \quad D_3 = \frac{F}{AE\alpha \cos \alpha L}$$

Hence the vibratory motion of the bar may be expressed as

$$u(x, t) = \frac{F \sin \alpha x \cos \omega t}{AE\alpha \cos \alpha L} \tag{3.34}$$

Note that $u(x, t)$ will be very large when $\cos \alpha L$ becomes zero, that is when

$$\omega = \frac{(2n - 1)\pi}{2L} \sqrt{\frac{E}{\rho}}; \quad n = 1, 2, 3, \ldots \tag{3.35}$$

These are natural frequencies of a bar rigidly fixed at one end and free at the other. So, as expected, resonance occurs when the exciting force frequency coincides with a natural frequency. The dynamic magnifier $U(L)/U_{static}$, for a range of frequencies, is shown in Fig. 3.4. The results relate to a bar, fixed rigidly at one end and free at the other.

If a bar is excited by harmonic forces of the same frequency ω acting on each of its ends, it is easily shown that the relationship between the instantaneous values of the end forces and the end displacements is given by

$$\begin{bmatrix} F_1 \\ \\ F_2 \end{bmatrix} \begin{matrix} \sin \omega t \\ \text{or} \\ \cos \omega t \end{matrix} = EA\alpha \begin{bmatrix} \cot \alpha L & -\operatorname{cosec} \alpha L \\ \\ -\operatorname{cosec} \alpha L & \cot \alpha L \end{bmatrix} \begin{bmatrix} U_1 \\ \\ U_2 \end{bmatrix} \begin{matrix} \sin \omega t \\ \text{or} \\ \cos \omega t \end{matrix}$$

or
$$\mathrm{F} = \beta \mathrm{U} \tag{3.36}$$

where F_1, F_2 are the applied maximum forces on ends 1 and 2 of the bar respectively; U_1, U_2 are the resulting maximum displacements, and β is the dynamic stiffness matrix for the bar. The sign convention used in deriving equation (3.36) is that applied force and resulting displacements are positive in the positive x-direction.

When the exciting force acts at other than the end points of the bar we may proceed as follows. The shapes of the normal modes may be represented in terms of characteristic functions or eigenfunctions. For the specific case solved above, a bar fixed at one end, $x = 0$, and unconstrained at the other, $x = L$, we could take as the characteristic functions, ϕ_n, the set of orthogonal functions $\sin \alpha_n x$; any permissible vibratory form could then be expressed as

$$u(x, t) = \sum_{n=1}^{\infty} q_n \phi_n \tag{3.37}$$

q_n are generalised coordinates, independent of x. The exciting force $F(x) \sin \omega t$ may be expressed in terms of generalised forces Q_n. Obviously

$$Q_n\, \delta q_n = \int_0^L F(x)\, \delta u(x, t)\, \mathrm{d}x \sin \omega t$$

$$= \int_0^L F(x) \phi_n\, \delta q_n\, \mathrm{d}x \sin \omega t$$

giving
$$Q_n = \int_0^L F(x) \phi_n\, \mathrm{d}x \sin \omega t \tag{3.38}$$

Lagrange's equation of motion may be used to set up the equations of motion.

$$T = \frac{1}{2} \int_0^L \rho A \left(\frac{\partial u}{\partial t}\right)^2 \mathrm{d}x; \qquad V = \frac{1}{2} \int_0^L EA \left(\frac{\partial u}{\partial x}\right)^2 \mathrm{d}x$$

$$= \sum_{n=1}^{\infty} 0{\cdot}25 \rho A L \dot{q}_n^{\,2}; \qquad = \sum_{n=1}^{\infty} 0{\cdot}25 AEL \alpha_n^{\,2} q_n^{\,2}$$

Therefore
$$\ddot{q}_n + \omega_n^{\,2} q_n = \frac{2Q_n}{\rho A L} \tag{3.39}$$

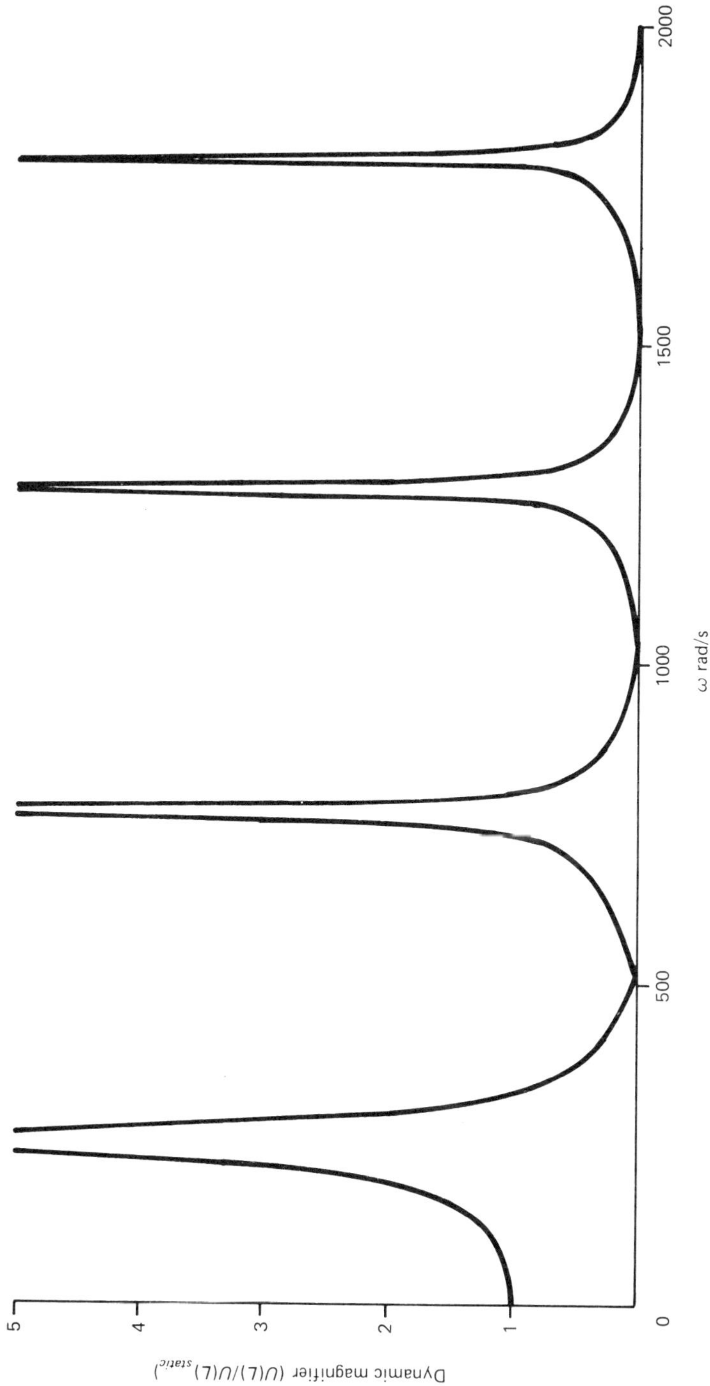

Fig. 3.4 Dynamic magnifier for a uniform bar fixed at one end and excited at the other ($L = 1$ m, $E = 207$ GN/m^2, $\rho = 7{\cdot}8$ Mg/m^3)

the steady state solution being

$$q_n = \frac{2Q_n}{\rho AL(\omega_n^2 - \omega^2)} \tag{3.40}$$

Equation (3.40) shows that the n'th mode will be excited only if there is a component Q_n in the exciting force. For example, in the present case if $F(x) = A_1 \sin \pi x/2L + A_2 \sin 3\pi x/2L$, only the first and third modes would be excited.

At this stage it is convenient to give the definition of normal coordinates which will be used in Chapter 5. If we had chosen the characteristic functions in the above analysis to be

$$\Psi_n = \left(\frac{2}{\rho AL}\right)^{1/2} \sin \alpha_n x \tag{3.41}$$

then

$$u(x, t) = \sum_{n=1}^{\infty} a_n \Psi_n \tag{3.42}$$

and the energy expressions would have been

$$2T = \sum_{n=1}^{\infty} \dot{a}_n^2; \qquad 2V = \sum_{n=1}^{\infty} \omega_n^2 a_n^2$$

Obviously a_n has the form

$$a_n = A_n \sin \omega t \tag{3.43}$$

The coefficients A_n are normal coordinates, as used in the proof of Rayleigh's principle in Chapter 5.

Exercises

1. In a hydraulic system a piston at A executes simple harmonic motion of amplitude a and angular velocity ω in one end of a rigid pipe AB. The pipe is of length L and diameter d, L being much greater than d. The end B of the pipe is closed and it is filled with a liquid of mass density ρ and bulk modulus K. Derive an expression for the fluctuation of pressure at any cross-section of the pipe.
2. A long uniform bar is connected at one of its ends to a rigid support by means of a massless spring of stiffness k, its other end is free. Show that the frequency equation for the system has the form

   $$k \cot \alpha L - EA\alpha = 0$$

   If $kL/EA = 0{\cdot}1$ plot the dynamic magnifier relating to the free end.
3. A uniform bar of length L, cross-sectional area A, mass density ρ, is fixed at one end and free at the other. It is excited in longitudinal oscillations by a force of $p \sin \omega t$ per unit length; the force is uniformly distributed over a length $0{\cdot}25L$ from the free end. Find the steady state amplitude at the free end when the exciting frequency is four times the fundamental frequency of the bar.

   If the above uniformly distributed force is disposed symmetrically about the mid-length of the bar, i.e., about $L/2$, what then is the steady state amplitude at the free end? Account for the difference.

4. Show that the equation of motion for an element of a uniform bar subjected to an exciting force $f(x) \sin \omega t$, $f(x)$ being continuous over the whole length of the bar, is

$$E \frac{\partial^2 u(x, t)}{\partial x^2} + \frac{f(x)}{A} \sin \omega t = \rho \frac{\partial^2 u(x, t)}{\partial t^2}$$

Solve this equation when $f(x)$ is a constant, that is a uniformly distributed force of magnitude f per unit length. Hence show, for a bar with fixed ends, that

$$\rho\omega^2 u = \{\cos \alpha x - (\cot \alpha L - \operatorname{cosec} \alpha L) \sin \alpha x - 1\} \frac{f}{A} \sin \omega t$$

where $\alpha^2 = \rho\omega^2/E$

3.1.5 Bars with end masses

Consider now longitudinal oscillations of the system shown in Fig. 3.5, comprised of a uniform bar to each end of which is rigidly attached a mass. At end 1

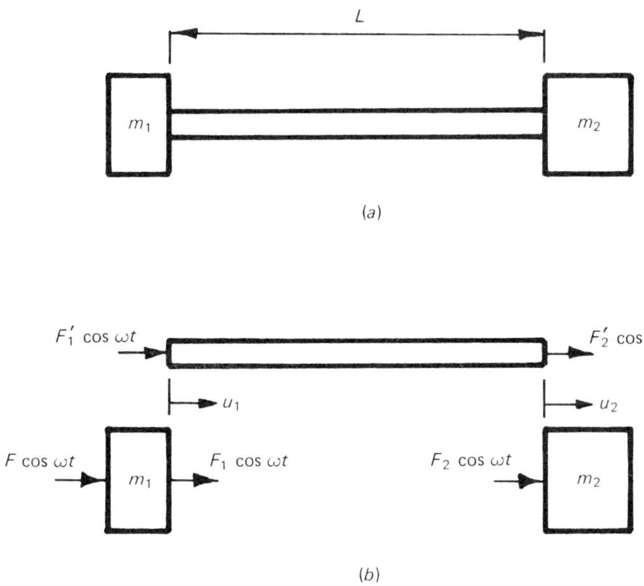

Fig. 3.5(*a*) and (*b*) Free body diagrams

the mass has value m_1 and is acted upon by a harmonically varying force of amplitude F and frequency ω. The mass at end 2 has value m_2. No other constraints act upon the system. In the steady state, Newton's laws give

1. for mass m_1, $(F + F_1) \cos \omega t = -m_1\omega^2 U_1 \cos \omega t$ (3.44)
2. for mass m_2, $F_2 \cos \omega t = -m_2\omega^2 U_2 \cos \omega t$ (3.45)

3. for the uniform bar,

$$\begin{bmatrix} F_1' \\ F_2' \end{bmatrix} \cos \omega t = EA\alpha \begin{bmatrix} \cot \alpha L & -\operatorname{cosec} \alpha L \\ -\operatorname{cosec} \alpha L & \cot \alpha L \end{bmatrix} \begin{bmatrix} U_1 \\ U_2 \end{bmatrix} \cos \omega t \tag{3.46}$$

For equilibrium at the junctions $F_1 + F_1' = 0$ and $F_2 + F_2' = 0$, therefore,

$$\begin{bmatrix} F \\ 0 \end{bmatrix} \cos \omega t = \begin{bmatrix} -m_1\omega^2 + AE\alpha \cot \alpha L & -AE\alpha \operatorname{cosec} \alpha L \\ -AE\alpha \operatorname{cosec} \alpha L & -m_2\omega^2 + EA\alpha \cot \alpha L \end{bmatrix} \begin{bmatrix} U_1 \\ U_2 \end{bmatrix} \cos \omega t \tag{3.47}$$

or

$$\mathrm{F} = \mathrm{YU}$$

giving the response $\mathrm{U} = \mathrm{Y}^{-1}\mathrm{F}$, but the inverse of a matrix is its adjoint divided by its determinant, i.e.,

$$\mathrm{Y}^{-1} = \operatorname{adj} \mathrm{Y}/|\mathrm{Y}|$$

The determinant is zero when ω is a resonant frequency. From equation (3.47) it is obvious that the determinant of Y is also the frequency equation when the system is in a state of free vibration.

Exercises

1. A uniform bar of length L, cross-sectional area A, Young's modulus E, and mass density ρ is fixed at one end and has a rigid mass M attached to the other. Show that the natural angular frequencies of the system ω_n are given by solutions to the equation

$$\rho \frac{AL}{M} = \omega_n L \sqrt{\frac{\rho}{E}} \tan\left\{\omega_n L \sqrt{\frac{\rho}{E}}\right\}$$

2. There is a possibility of the main propeller shafting of ships suffering excessive axial vibrations. In a simple representation of the system the propeller mass M_p is rigidly attached to one end of the propeller shaft which has length L, cross-sectional area A, Young's modulus E, and mass density ρ. To its other end is rigidly attached a mass M_q representing the mass of the main gear-wheel, associated pinions, thrust collar, and the thrust block casing. Between the mass M_q and the rigid support structure of the ship is a massless spring of stiffness K. Derive the frequency equation for the system: check it by considering the limiting cases, e.g., $M_p = M_q = K = 0$; $M_q = 0$, $K = \infty$; etc.
3. If in the system described in exercise 1 the total mass of the bar is one quarter that of the attached rigid mass, derive an expression for the vibration resulting from the application of an impulsive force to the rigid mass. Take the system as initially quiescent.
4. Show that a constant longitudinal force has no influence on the natural frequencies of a uniform bar in longitudinal vibration.

3.2 Torsional vibration of circular shafts

Circular shafts have been included as elements of systems studied in earlier chapters of this book, but there the mass of the shafts was neglected. However, treating torsional vibrations in a similar manner to that used for studying longitudinal vibrations of uniform bars, we may allow for the mass of the shaft being distributed along its length.

The assumptions usually made in engineering calculations on the torsion of circular shafts will be made, namely that each transverse cross-section remains plane during the torsional oscillations and that the radii of these cross-sections remain straight.

Let G be the modulus of rigidity of the shaft material
I_p be the polar second moment of area of the cross-section of the shaft
ρ be the mass density of the shaft material

then by consideration of the twisting moments on the two faces of an elemental length of the shaft it is easy to show that the equation of motion for the torsional oscillations when the shaft is free from surface tractions will be

$$GI_p \frac{\partial^2 \theta}{\partial x^2} - \rho I_p \frac{\partial^2 \theta}{\partial t^2} = 0 \tag{3.48}$$

where θ is the angle of twist at a cross-section distant x from the origin. The positive direction of θ is such that it increases in a clockwise sense as x increases positively (right-hand screw rule).

If we let $G/\rho = a^2$, equation (3.48) takes the same form as equation (3.24) for longitudinal vibrations. G/ρ is therefore the square of the velocity of propagation of torsional waves in uniform circular shafts, and the findings relating to longitudinal oscillation also apply to the present case

Exercises

1. A steel shaft AB of length 20 m and diameter 0·1 m, supported in frictionless bearings, is coupled at end A to a rotor of moment of inertia 3 kg m^2. A harmonic torque of amplitude 20 N m is applied to B at a frequency of 50 Hz, the moment of inertia of the attachment at B being negligible. Find the amplitude of vibration of end B and the position at which the shear stress is a maximum.
2. Derive and solve the frequency equation for small torsional vibrations of a system consisting of two discs of moment of inertia I_1 and I_2 connected by a uniform shaft of mass density ρ, length L, and polar second moment of area $\mathcal{J}$.

   When calculating the natural frequencies of practical systems such as diesel engine driven alternator systems it is usual to lump the inertias of the shafts with those of equivalent rotors. Show that this is a valid assumption when the lowest natural frequencies of the individual shafts is much higher than any system natural frequency of interest. Using the results found above, derive expressions for the increases in I_1 and I_2 needed in the present case.

3. Derive the dynamic stiffness matrix for a uniform circular shaft, relating end couples with end rotations.
4. Two uniform circular shafts each of length L and mass density ρ, but differing in diameter, one being of diameter d and the other D, are welded together. In service, during a fault condition, an impulsive torque was applied at the junction of the shafts; investigate the subsequent oscillations, when they are free to rotate in bearings.

3.3 Free lateral vibration of thin beams

The oscillatory behaviour of beams and structures composed of beams or beam-like elements is of great interest and importance to engineers, because in practice it is found that their lower natural frequencies often have values which give rise to the possibility of resonance occurring when machinery is running. The bending oscillations of steel railway bridges and the bending oscillations of aircraft wings are examples which have caused considerable concern in the past.

3.3.1 Thin beams with simple end conditions

In the discussion which follows we shall adopt the assumptions usual in simple bending theory, that is, the Bernoulli-Euler beam theory: namely that the beam is initially straight; that the depth of the beam is small compared with its radius of curvature at its maximum displacement; that plane sections remain plane at all phases of an oscillation, and that the deformation due to shearing of one cross-section relative to an adjacent one is negligible. In addition, we shall assume that one principal axis of a typical cross-section is perpendicular to the direction of motion in the vibration; that the beam is free from longitudinal force, and that its mass is concentrated at its neutral axis. The latter implies that we shall neglect the moments necessary to accelerate the mass of the beam between adjacent cross-sections, as the cross-sections rotate about their principal axes which lie in the neutral plane. Later we shall examine the influence of longitudinal force, shear deformation, and rotary inertia on the natural frequencies.

Let E be Young's modulus for the material of the beam
ρ be its mass density
A be the cross-sectional area of the beam
I be the second moment of area of a cross-section of the beam about the principal axis which lies in the neutral plane
L be the length of the beam, and
v be the displacement, in the y-direction, of the neutral axis at a point x due to the vibration

The sign convention adopted for this discussion is shown in Fig. 3.6, together with the forces and moments acting on an elemental length of the beam in its distorted form. The forces and moments on the right-hand end of the element have been found by assuming that the shear force and bending moment are continuous functions of x and that terms containing δx to the second and higher powers are negligibly small.

Applying Newton's laws gives

(*a*) for motion in the y-direction

$$-F + F + \frac{\partial F}{\partial x}\,\delta x = \rho A\,\delta x\,\frac{\partial^2 v}{\partial t^2}$$

therefore
$$\frac{\partial F}{\partial x} = \rho A\,\frac{\partial^2 v}{\partial t^2} \tag{3.49}$$

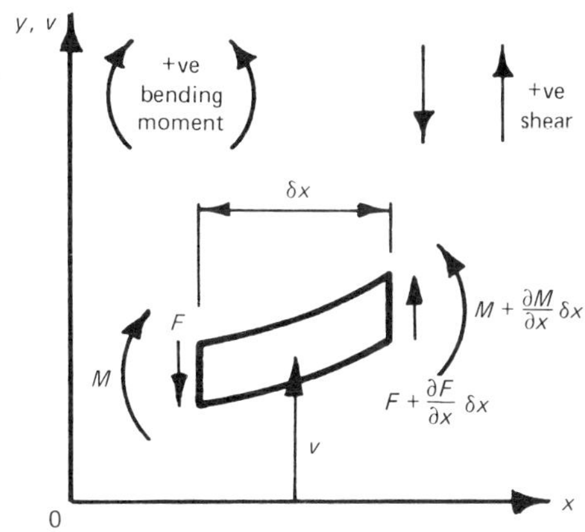

Fig. 3.6 Positive sign convention

(*b*) for rotational motion

$$-M + M + \frac{\partial M}{\delta x}\,\delta x + F\,\delta x = 0$$

therefore
$$F = -\,\frac{\partial M}{\partial x} \tag{3.50}$$

Using this value for F in equation (3.49) gives

$$\frac{\partial^2 M}{\partial x^2} + \rho A\,\frac{\partial^2 v}{\partial t^2} = 0 \tag{3.51}$$

Under the sign convention adopted here, the Bernoulli-Euler beam theory gives the following relationships between applied moment and curvature

$$M = EI\,\frac{\partial^2 v}{\partial x^2} \tag{3.52}$$

This value of M together with equation (3.5) gives

$$EI\,\frac{\partial^4 v}{\partial x^4} + \rho A\,\frac{\partial^2 v}{\partial t^2} = 0 \tag{3.53}$$

as the equation of motion for the element.

As equation (3.53) relates to wave motion it is not unreasonable to expect its solution to conform to the pattern found for longitudinal and torsional oscillations, namely

$$v(x, t) = \sum_n V_n(x)\{A_n \sin \omega_n t + B_n \cos \omega_n t\} \tag{3.54}$$

where $V_n(x)$ describe the forms of the principal modes. Hence equation (3.53) becomes

$$\frac{d^4 V_n}{dx^4} - \frac{\rho A \omega_n{}^2}{EI} V_n = 0 \tag{3.55}$$

the general solution of which has the form

$$V_n = C_{1n} \sin \lambda_n x + C_{2n} \cos \lambda_n x + C_{3n} \sinh \lambda_n x + C_{4n} \cosh \lambda_n x \tag{3.56}$$

where C_{in} are constants depending upon the end conditions imposed upon the beam and

$$\lambda_n{}^4 = \frac{\rho A \omega_n{}^2}{EI}.$$

The end of a beam may be constrained so that no lateral movement or rotation is possible, it is then referred to as an encastré or a fixed or clamped end. If lateral movement is not possible, that is it is held rigidly in position but there is no restraint on rotation, the end is said to be pinned or hinged or simply supported. When there is no restraint on it at all the end is said to be free, and when rotation is prevented but lateral movement is unrestrained the end condition is described as a sliding constraint. Each combination of these end conditions has its own set of values for the C_{in} in equation (3.56).

As an example we shall find the frequency equation and some of the mode shapes for the uniform cantilevered beam shown in Fig. 3.7. From the end conditions shown, that is fixed at $x = 0$, free at $x = L$, we obtain the following relationships

$$\begin{bmatrix} 0 & 1 & 0 & 1 \\ \lambda_n & 0 & \lambda_n & 0 \\ -EI\lambda_n{}^2 \sin \lambda_n L & -EI\lambda_n{}^2 \cos \lambda_n L & EI\lambda_n{}^2 \sinh \lambda_n L & EI\lambda_n{}^2 \cosh \lambda_n L \\ -EI\lambda_n{}^3 \cos \lambda_n L & EI\lambda_n{}^3 \sin \lambda_n L & EI\lambda_n{}^3 \cosh \lambda_n L & EI\lambda_n{}^3 \sinh \lambda_n L \end{bmatrix} \begin{bmatrix} C_{1n} \\ C_{2n} \\ C_{3n} \\ C_{4n} \end{bmatrix} = [0] \tag{3.57}$$

That is

$$\mathrm{YC} = 0$$

For this to be true either C must be equal to zero or Y must be singular.

If C were equal to zero there would be no vibration, therefore, the matrix Y must be singular, that is its determinant must be zero; giving the frequency equation

$$\det \mathrm{Y} = 0 \tag{3.58}$$

Using the elementary properties of determinants (Aitken, 1946) the frequency equation may be put into the following more convenient form

$$\cos \lambda_n L \cosh \lambda_n L + 1 = 0 \tag{3.59}$$

This equation may be solved numerically for the characteristic values, $\lambda_n L$, and hence the natural frequencies may be found. The first five characteristic values

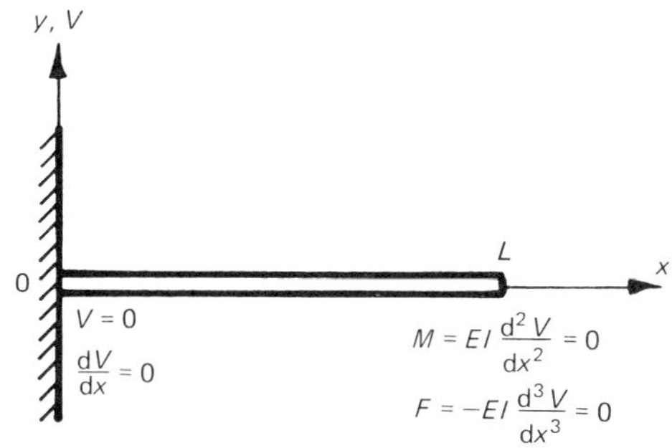

Fig. 3.7

(Rayleigh, 1894a) are $\lambda_n L = 1{\cdot}875,\ 4{\cdot}694,\ 7{\cdot}855,\ 10{\cdot}996,\ 14{\cdot}137$, for n greater than five the values are given approximately by

$$\lambda_n L = \frac{(2n - 1)}{2} \pi \tag{3.60}$$

Knowing the characteristic values or eigenvalues, we may proceed to find the principal mode shapes. These are of arbitrary amplitude therefore any of the constants C_{in} may be given an arbitrarily chosen value and the other constants found in terms of it. For example, let C_{4n} be K_n, then equation (3.57) becomes

$$\begin{bmatrix} 0 & 1 & 0 \\ \lambda_n & 0 & \lambda_n \\ -EI\lambda_n^2 \sin \lambda_n L & -EI\lambda_n^2 \cos \lambda_n L & EI\lambda_n^2 \sinh \lambda_n L \end{bmatrix} \begin{bmatrix} C_{1n} \\ C_{2n} \\ C_{3n} \end{bmatrix} = \begin{bmatrix} -K_n \\ 0 \\ -K_n EI\lambda_n^2 \cosh \lambda_n L \end{bmatrix}$$

or

$$\mathrm{T}\ {}_{\mathrm{S}}\mathrm{C} = \mathrm{S} \tag{3.61}$$

giving the values ${}_{\mathrm{S}}\mathrm{C}$ as $\mathrm{T}^{-1}\mathrm{S}$. Therefore, V_n has the form

$$V_n = K_n(\cosh \lambda_n x - \cos \lambda_n x) - \frac{K_n(\cos \lambda_n L + \cosh \lambda_n L)}{\sin \lambda_n L + \sinh \lambda_n L} (\sinh \lambda_n x - \sin \lambda_n x)$$

or

$$V_n = K_n(\cosh \lambda_n x - \cos \lambda_n x) - \frac{K_n(\sinh \lambda_n L - \sin \lambda_n L)}{(\cosh \lambda_n L + \cos \lambda_n L)} (\sinh \lambda_n x - \sin \lambda_n x) \tag{3.62}$$

where K_n is a constant for each value of n. The forms of these characteristic functions or eigenfunctions are illustrated in Fig. 3.8 for the first four modes of vibration. The distance of each node from the fixed end is also given.

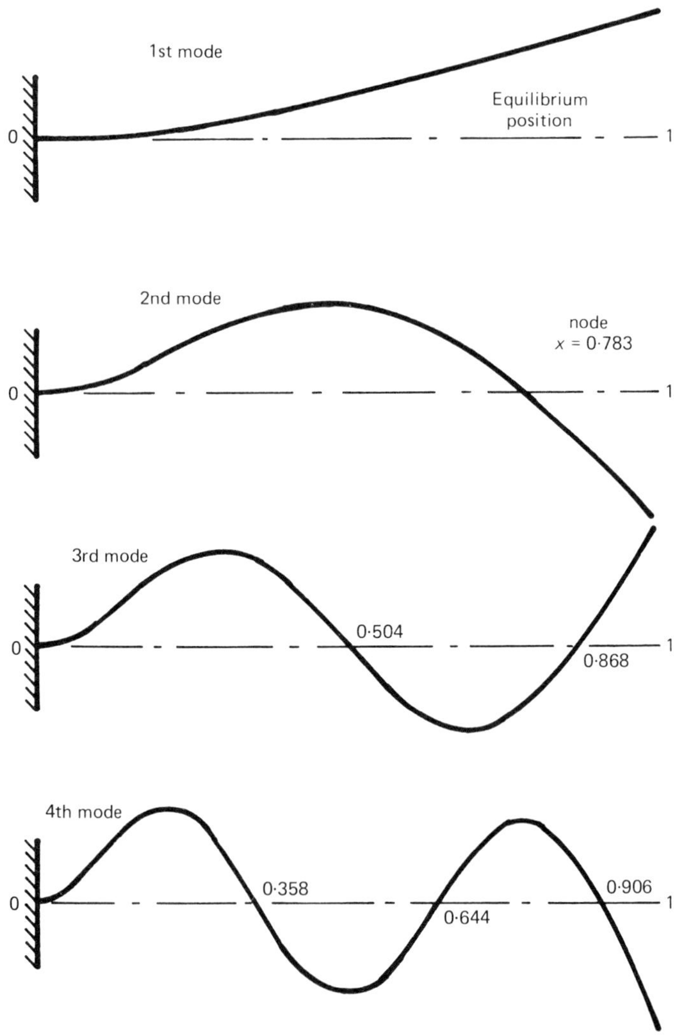

Fig. 3.8 Characteristic functions for a uniform cantilever

We have seen that the characteristic values $\lambda_n L$ are such that a transcendental function, the frequency function, is made zero as demanded by equation (3.59). The form of the frequency function whose zeros must be found depends upon the end conditions of the beam. Combinations of these functions may be used to describe the frequency equations of more complex beam systems, consequently tabulations of them were regarded desirable before the universal availability of the automatic digital computer. Tabulations of both the frequency functions and the characteristic functions may be found in the literature. For the frequency functions Bishop (1955) devised the following convenient notation:

$$\begin{aligned}
F_1 &= \sin \lambda L \sinh \lambda L; \quad & F_2 &= \cos \lambda L \cosh \lambda L \\
F_3 &= F_2 - 1; & F_4 &= F_2 + 1 \\
F_5 &= \cos \lambda L \sinh \lambda L - \sin \lambda L \cosh \lambda L; \\
F_6 &= \cos \lambda L \sinh \lambda L + \sin \lambda L \cosh \lambda L; \\
F_7 &= \sin \lambda L + \sinh \lambda L; & F_8 &= \sin \lambda L - \sinh \lambda L \\
F_9 &= \cos \lambda L + \cosh \lambda L; & F_{10} &= \cos \lambda L - \cosh \lambda L
\end{aligned} \tag{3.63}$$

Hohenemser and Prager (1933) tabulated F_1 to F_6 inclusive over $\lambda L = 0$ by 0·02 to 10; Bishop and Johnson (1956) tabulated F_1 to F_{10} inclusive over $\lambda L = 0$ by 0·05 to 11. Young and Felgar (1949) tabulated the characteristic functions for thin beams at fifty equal intervals for the first five modes of vibration: Bishop and Johnson (1956) gave similar tabulations. In both cases the tabulations related to the following end conditions: clamped-clamped, clamped-free, clamped-pinned, free-free, and free-pinned.

Exercises

1. Show that the first four eigenvalues (λL) of a beam of length L with pinned ends are 3·1416, 6·283, 9·425, and 12·566. Derive expressions for the corresponding mode shapes and show that, in addition to the nodes at the ends, nodes occur as follows: in the second mode at $L/2$, in the third mode at $L/3$ and $2L/3$, and in the fourth mode at $L/4$, $L/2$, and $3L/4$.
2. Show that the frequency equation of a beam of length L, rigidly clamped at each end, has the form

$$\cos Z \cosh Z - 1 = 0$$

   Given that the first four solutions of this equation are $Z = 4{\cdot}730$, 7·853, 10·996, and 14·137, calculate the first four natural frequencies of a uniform steel beam of span 6 m, second moment of area 0·00025 m^4 and mass per metre length 75 kg. Also calculate and plot the form of the principal modes. Hence show that nodes occur as follows: in the second mode at $0{\cdot}5L$, in the third mode at $0{\cdot}359L$ and $0{\cdot}641L$, and in the fourth mode at $0{\cdot}278L$, $0{\cdot}5L$, and $0{\cdot}722L$.
3. A uniform steel beam is 0·4 m long and has a cross-section 20 mm wide by 8 mm deep. At one end it is rigidly clamped and at the other it is pinned, show that the frequency equation has the form

$$\tan Z - \tanh Z = 0$$

   and find the first four natural frequencies of transverse vibration.

   For steel take the density as 7·8 Mg/m^3 and Young's modulus at 207 GN/m^2. The first four roots of the frequency equation are $Z = 3{\cdot}927$, 7·069, 10·210, and 13·352.
4. For each of the beams in questions 1, 2, and 3 derive expressions for the stresses in the surfaces which are parallel to the neutral plane.

5. If the free vibrations of the beam described in question 2 were initiated by displacing the neutral axis at the centre of the beam a distance 1 mm from its equilibrium position and suddenly releasing it, derive an expression for the subsequent vibratory motion. Take only the first four natural modes into account.
6. Consider an infinitely long, thin beam and show that the velocity of propagation of flexural waves varies inversely as the wavelength.

3.3.2 Thin beams with more complex end conditions

Rarely in practical situations do uniform beams occur with the relatively simple end conditions studied in the previous section; for example they may carry masses, or be embedded in yielding foundations, or be attached to other beams.

As an illustration of an infallible method for setting up the frequency equation in instances such as these, we shall now derive the frequency equation for the cantilevered beam shown in Fig. 3.9(*a*). A second, more routine method will be given later.

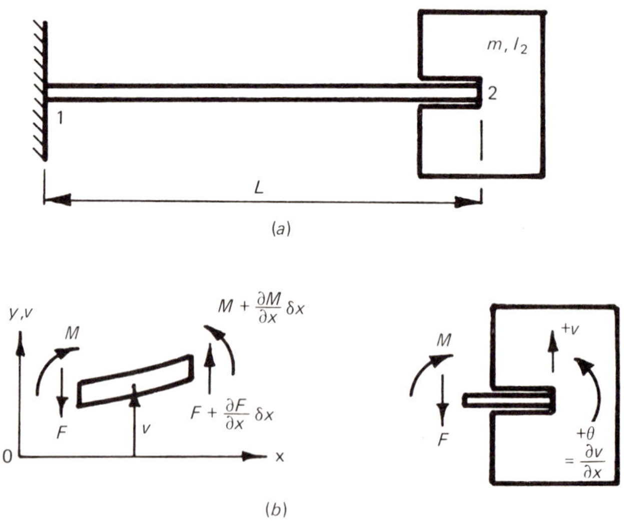

Fig. 3.9(*a*), (*b*)

The beam of Fig. 3.9(*a*) is of length L, bending rigidity EI, mass density ρ, and cross-sectional area A. It is clamped at end 1 and carries a metal block at end 2. The metal block is supported by the beam at its centre of gravity, has mass m and moment of inertia I_2, about a principal axis through 2 perpendicular to the plane of the paper. In setting up the frequency equation for this type of problem the strict adherence to a specified sign convention is absolutely essential. The sign convention used in this discussion is given in Fig. 3.9(*b*). The positive bending moment and shear are shown on the diagram of the element, and in the free body diagram of the mass an elemental length of the beam has been shown attached to it. This makes quite clear the direction of the force and moment acting upon the mass, that is, with the right-hand end of the element attached rigidly to the mass, the force and moment acting on this subsystem must be those associated with the left-hand end of the element. Until you become very familiar with this branch of vibrations, you are advised to adopt this procedure for setting up the equations of motion and/or equilibrium relating to the end conditions of beams. For the particular system under consideration the end conditions to be satisfied by the general solution, that is equation (3.56)

$$V_n = C_{1n} \sin \lambda_n x + C_{2n} \cos \lambda_n x + C_{3n} \sinh \lambda_n x + C_{4n} \cosh \lambda_n x$$

where

$$\lambda_n^4 = \frac{\rho A \omega_n^2}{EI},$$

are at $x = 0$:

$$V_n = 0; \qquad \frac{dV_n}{dx} = 0 \tag{3.64}$$

at $x = L$:

$$-M = I_2 \frac{\partial^2 \theta}{\partial t^2}; \qquad -F = m \frac{\partial^2 v}{\partial t^2}$$

that is

$$-EI \frac{d^2 V_n}{dx^2} = -\omega_n^2 I_2 \frac{dV_n}{dx}; \qquad EI \frac{d^3 V_n}{dx^3} = -\omega_n^2 m V_n \tag{3.65}$$

Thus the frequency equation is

$$\begin{vmatrix} 0 & 1 & 0 & 1 \\ \lambda_n & 0 & \lambda_n & 0 \\ (EI\lambda_n^2 \sin \lambda_n L & (EI\lambda_n^2 \cos \lambda_n L & (-EI\lambda_n^2 \sinh \lambda_n L & (-EI\lambda_n^2 \cosh \lambda_n L \\ + \omega_n^2 I_2 \lambda_n \cos \lambda_n L) & - \omega_n^2 I_2 \lambda_n \sin \lambda_n L) & + \omega_n^2 I_2 \lambda_n \cosh \lambda_n L) & + \omega_n^2 I_2 \lambda_n \sinh \lambda_n L) \\ (-EI\lambda_n^3 \cos \lambda_n L & (EI\lambda_n^3 \sin \lambda_n L & (EI\lambda_n^3 \cosh \lambda_n L & (EI\lambda_n^3 \sinh \lambda_n L \\ + \omega_n^2 m \sin \lambda_n L) & + \omega_n^2 m \cos \lambda_n L) & + \omega_n^2 m \sinh \lambda_n L) & + \omega_n^2 m \cosh \lambda_n L) \end{vmatrix} = 0 \tag{3.66}$$

Therefore,

$$(EI\lambda_n^2)^2 F_4 + EI\lambda_n \omega_n^2 m F_5 - EI\lambda_n^3 \omega_n^2 I_2 F_6 - \omega_n^4 I_2 m F_3 = 0 \tag{3.67}$$

The correctness of part of this expression may be checked as follows:

(*i*) if $I_2 = m = 0$, we have clamped-free conditions on the beam and the frequency equation becomes $F_4 = 0$, which is correct;
(*ii*) if $I_2 = 0$ and $m = \infty$, we have clamped-pinned end conditions and the frequency equation becomes $F_5 = 0$, which is correct;
(*iii*) if $I_2 = \infty$ and $m = 0$, we have clamped-sliding end conditions and the frequency equation becomes $F_6 = 0$, which is correct;

(*iv*) if $I_2 = m = \infty$, we have clamped-clamped end conditions and the frequency equation becomes $F_3 = 0$, which again is correct.

The units of each part of equation (3.67) are also consistent. Using a digital computer the first few eigenvalues, $\lambda_n L$, may easily be found by a systematic search procedure, and hence the mode shapes may be computed. In using a digital computer to find the eigenvalues, one has a choice of programming the left-hand side of equation (3.66) and using a standard determinant-evaluation routine to find its value at each step of the search, or of programming the evaluation of the left-hand side of equation (3.67) at each step. The possibility of errors in the hand reduction of the determinant, and the fact that the computer time for numerical evaluation differs very little between the methods, makes the former the more attractive. Therefore, in the exercises which follow it is the determinantal form of the frequency equation which is required unless otherwise stated.

Exercises

1. Derive the frequency equation for the free-free modes of vibration of a system comprised of a uniform beam of length L to each end of which is firmly attached a metal block of mass m. Take the bending rigidity of the beam to be EI, its cross-sectional area as A, and its mass density as ρ.
2. Write down the frequency equation for the bending oscillations of a uniform beam of length L, mass density ρ, cross-sectional area A, and flexural rigidity EI, subject to the following end conditions. To one end of the beam is attached a metal object of mass m and rotational inertia J about an axis, perpendicular to the plane of vibration, through its centre of gravity which coincides with the end of the beam. The other end of the beam is embedded in an elastic medium with properties such that it applies a moment to the beam proportional to the rotation of the neutral axis of the beam. Linear displacements of this end of the beam are negligible, the elastic medium being very stiff in these directions.
3. Show that for the symmetrical modes of vibration of a uniform beam to which is attached, at the middle of its span, a rigid block of mass m the frequency equation has the form
$$2\,\frac{\rho AL}{m} = \frac{\lambda L}{2}\left(\tan\frac{\lambda L}{2} - \tanh\frac{\lambda L}{2}\right)$$
if the beam has pinned ends. Where ρ, A, and $L/2$ are the mass density, cross-sectional area, and length from the rigid block to each pinned end, of the beam respectively.
4. If, instead of being at the ends, the pinned supports for the system described in question 3 were at the nodes of the first free-free mode of the system, show that the first natural frequency may be calculated from the lowest root of the equation.
$$-\frac{\rho AL}{m} = \frac{\lambda L}{2}\,\frac{\left(1 + \cos\frac{\lambda L}{2}\cosh\frac{\lambda L}{2}\right)}{\sin\frac{\lambda L}{2}\cosh\frac{\lambda L}{2} + \cos\frac{\lambda L}{2}\sinh\frac{\lambda L}{2}}$$

3.3.3 Thin beams with many supports or section changes

Equation (3.56) is the general solution of the equation of motion (3.51) for a uniform beam. This solution relies upon the shape of the normal mode being continuous together with its first three derivatives. Therefore, if there is a discontinuity in the amplitudes of one or more of displacement, slope, bending moment, and shear in the length of a uniform beam, equation (3.56) is not applicable to the complete length of the beam. At supports bending moment and/or shear are generally discontinuous. Also at section changes or concentrated masses one or more of these functions are discontinuous.

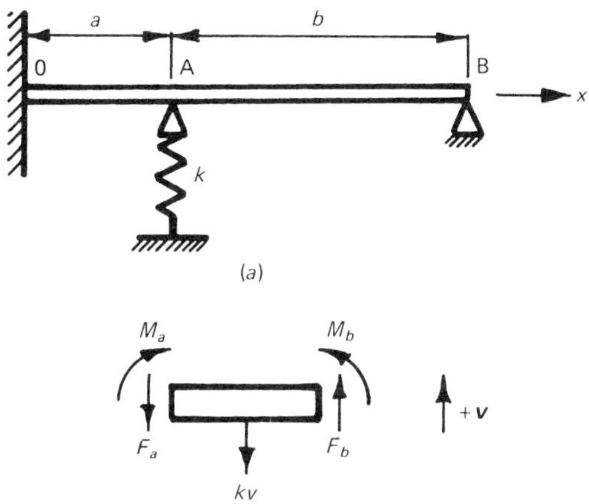

(*b*) Forces and moments acting on an element at A.

Fig. 3.10(*a*); (*b*) Forces and moments acting on an element at A.

A uniform beam over three supports must, therefore, be treated as two separate beams. Depending upon the type of support, the usual end conditions apply at the two end supports. At the intermediate support, displacement and slope must be continuous at the junction between the beams. Also the forces and moments due to bending moments and shears in the beams and to the restraints imposed by the support must be in equilibrium at the junction between the beams.

To illustrate the method of approach consider free vibration of the system shown in Fig. 3.10. The beam has uniform geometrical, mass, and elastic properties over the total length $a + b$. It is rigidly clamped at one end and pinned at the other, with an intermediate pinned support resting on a spring of stiffness k at a distance a from the clamped end.

From the general solutions of the equations of motion (3.56) for beam OA, $0 \leqslant x_a \leqslant a$,

$$ {}_aV_n = {}_aC_{1n} \sin \lambda_n x_a + {}_aC_{2n} \cos \lambda_n x_a + {}_aC_{3n} \sinh \lambda_n x_a + {}_aC_{4n} \cosh \lambda_n x_a \tag{3.68} $$

and for beam AB, $0 \leqslant x_b \leqslant b$, a similar expression, with subscript a replaced by subscript b, applies. The end conditions for the subsystems are as follows:

$$\left.\begin{array}{llr}
\text{at} & x = 0;\ x_a = 0;\ {}_aV_n = 0 & \text{(i)} \\
& \dfrac{\mathrm{d}_aV_n}{\mathrm{d}x_a} = 0 & \text{(ii)} \\
\text{at} & x = a;\ \ x_a = a;\ \ x_b = 0 & \\
\text{Continuity gives} & & \\
& {}_aV_n = {}_bV_n & \text{(iii)} \\
\text{and} & \dfrac{\mathrm{d}_aV_n}{\mathrm{d}x_a} = \dfrac{\mathrm{d}_bV_n}{\mathrm{d}x_b} & \text{(iv)} \\
\text{Equilibrium gives} & & \\
& -F_a + F_b - kv = 0 & \text{(v)} \\
\text{and} & -M_a + M_b = 0 & \text{(vi)} \\
\text{At} & x = a + b;\ \ x_b = b & \\
& {}_bV_n = 0 & \text{(vii)} \\
\text{and} & EI\,\dfrac{\mathrm{d}^2{}_bV_n}{\mathrm{d}x_b^{\,2}} = 0 & \text{(viii)}
\end{array}\right\} \quad (3.69)$$

Upon substituting for v, F, and M, equations (v) and (vi) become

$$EI\left(\frac{\mathrm{d}^3{}_aV_n}{\mathrm{d}x_a^{\,3}} - \frac{\mathrm{d}^3{}_bV_n}{\mathrm{d}x_b^{\,3}}\right) - k_bV_n = 0 \qquad \text{(v. } a\text{)}$$

and

$$-EI\left(\frac{\mathrm{d}^2{}_aV_n}{\mathrm{d}x_a^{\,2}} - \frac{\mathrm{d}^2{}_bV_n}{\mathrm{d}x_b^{\,2}}\right) = 0 \qquad \text{(vi. } a\text{)}$$

Note that in equation (v. a) the term k_bV_n could have been written k_aV_n, by equation (iii).

For each value of n there are eight equations (i), (ii), (iii), (iv), (v. a), (vi. a), (vii), and (viii) for the eight unknown constants ${}_aC_{in}$, ${}_bC_{in}$, hence we may set up the frequency equation.

Exercises

1. Set up, in determinantal form, the frequency equation for the above system.
2. A uniform beam is simply supported at its ends and at its mid-point: show that its first four natural frequencies in bending are in the ratios 1:1·57:4:5·07.

3.4 Forced lateral vibration of thin beams

Harmonic excitation of a beam may arise from concentrated forces or couples acting at any point of its length. It may also be caused by distributed forces or couples, acting over all or a portion of its length and varying harmonically with time. The oscillations resulting from concentrated forces or couples may be found by a method similar to that given on page 89 for the forced longitudinal vibration of a bar, whereas the response to distributed forcing terms may be found in terms of normal coordinates by a method similar to that given on page 90.

3.4.1 Excitation due to end forces and couples only

In this section a method for calculating the response of a uniform beam subjected to forces and/or couples concentrated at its ends only will be presented. They will be assumed to vary as a simple harmonic function of time, with a common angular

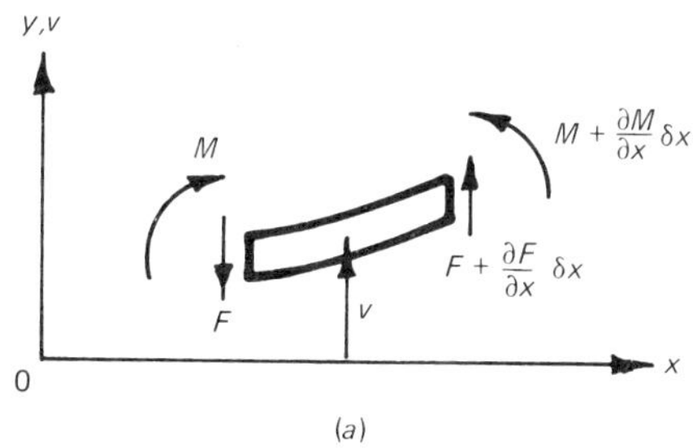

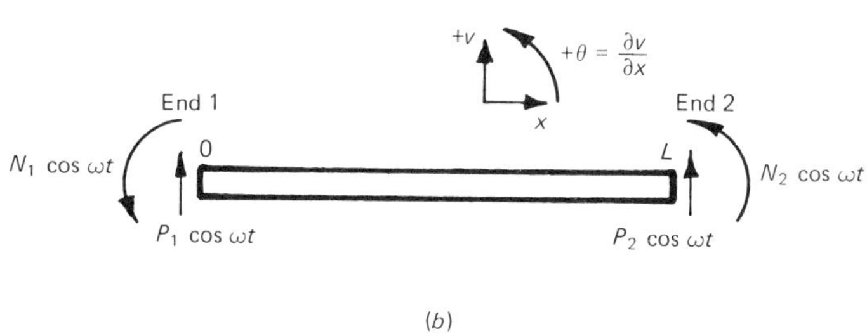

Fig 3.11 Positive sign convention:
(*a*) Bending moment and shear force sign convention
(*b*) Positive directions for end forces and couples

frequency ω. Referring to Fig. 3.11, let the maximum value of the forces be F_1 and F_2 and of the couples be N_1 and N_2. As they act only on the ends of the beam, they may be accounted for in the end conditions and, therefore, do not enter the equations of motion. That is, the equation of motion is of the same form as for free vibrations, namely

$$EI \frac{\partial^4 v}{\partial x^4} + \rho A \frac{\partial^2 v}{\partial t^2} = 0 \tag{3.70}$$

and we expect a steady state solution of the form

$$v = V(x)\{a \sin \omega t + b \cos \omega t\} \tag{3.71}$$

where ω is the angular frequency of the exciting forces and couples. Substitution for v in equation (3.70) gives

$$\frac{\mathrm{d}^4 V}{\mathrm{d}x^4} - \frac{\rho A \omega^2}{EI} V = 0 \tag{3.72}$$

giving

$$V = B_1 \sin \lambda x + B_2 \cos \lambda x + B_3 \sinh \lambda x + B_4 \cosh \lambda x \tag{3.73}$$

where
$$\lambda^4 = \frac{\rho A \omega^2}{EI}$$

Referring again to Fig. 3.11 for the positive sign convention adopted and letting $x = 0$ at end 1 and $x = L$ at end 2, we have the end conditions:

(*a*) *at end* 1

$$\left.\begin{aligned} N_1 \cos \omega t &= -EI \frac{\partial^2 v}{\partial x^2} && \text{(i)} \\ P_1 \cos \omega t &= EI \frac{\partial^3 v}{\partial x^3} && \text{(ii)} \end{aligned}\right.$$

(*b*) *at end* 2

$$\left.\begin{aligned} N_2 \cos \omega t &= EI \frac{\partial^2 v}{\partial x^2} && \text{(iii)} \\ P_2 \cos \omega t &= -EI \frac{\partial^3 v}{\partial x^3} && \text{(iv)} \end{aligned}\right\} \tag{3.74}$$

Substituting for v from equation (3.71) it is seen that $a = 0$; then letting $b = 1$ and expressing the result in matrix form gives

$$\begin{bmatrix} P_1 \\ P_2 \\ N_1 \\ N_2 \end{bmatrix} = EI \begin{bmatrix} -\lambda^3 & 0 & \lambda^3 & 0 \\ \lambda^3 \cos \lambda L & -\lambda^3 \sin \lambda L & -\lambda^3 \cosh \lambda L & -\lambda^3 \sinh \lambda L \\ 0 & \lambda^2 & 0 & -\lambda^2 \\ -\lambda^2 \sin \lambda L & -\lambda^2 \cos \lambda L & \lambda^2 \sinh \lambda L & \lambda^2 \cosh \lambda L \end{bmatrix} \begin{bmatrix} B_1 \\ B_2 \\ B_3 \\ B_4 \end{bmatrix}$$

That is,
$$\mathrm{P} = \mathrm{DB} \tag{3.75}$$

We may also express the unknown coefficients B_i in terms of the maximum end deflections and slopes, thus:

$$\begin{bmatrix} V_1 \\ V_2 \\ \Theta_1 \\ \Theta_2 \end{bmatrix} = \begin{bmatrix} 0 & 1 & 0 & 1 \\ \sin \lambda L & \cos \lambda L & \sinh \lambda L & \cosh \lambda L \\ \lambda & 0 & \lambda & 0 \\ \lambda \cos \lambda L & -\lambda \sin \lambda L & \lambda \cosh \lambda L & \lambda \sinh \lambda L \end{bmatrix} \begin{bmatrix} B_1 \\ B_2 \\ B_3 \\ B_4 \end{bmatrix}$$

or

$$\mathrm{V} = \mathrm{CB} \tag{3.76}$$

therefore,

$$\mathrm{B} = \mathrm{C}^{-1}\mathrm{V} \tag{3.77}$$

Equations (3.75) and (3.77) give

$$\mathrm{P} = \mathrm{DC}^{-1}\mathrm{V} = \beta \mathrm{V} \tag{3.78}$$

where β is a dynamic stiffness matrix for the beam.

$$\beta = \frac{EI\lambda^2}{F_3} \begin{bmatrix} -\lambda F_6 & \lambda F_7 & -F_1 & F_{10} \\ \lambda F_7 & -\lambda F_6 & -F_{10} & F_1 \\ -F_1 & -F_{10} & F_5/\lambda & F_8/\lambda \\ -F_{10} & F_1 & F_8/\lambda & F_5/\lambda \end{bmatrix} \tag{3.79}$$

If the response to applied end forces and couples is required, then

$$\mathrm{V} = \beta^{-1}\mathrm{P} = \alpha \mathrm{P} \tag{3.80}$$

α is a receptance matrix for the beam. When the corresponding instantaneous deflected form is required equations (3.71), (3.73), and (3.75) may be combined to give

$$v = [\sin \lambda x \cos \lambda x \sinh \lambda x \cosh \lambda x]\mathrm{D}^{-1}\mathrm{P} \cos \omega t \tag{3.81}$$

Rarely are all end forces and moments specified; it is more usual to specify combinations of forces, moments, displacements, and rotations. To illustrate the procedure then required, we shall derive the response for a particular system.

Consider the particular case of a uniform cantilever, excited at its free end by a force $P \sin \omega t$ and a couple $N \sin \omega t$. That is, $V_1 = \Theta_1 = 0$, $P_2 = P$, $N_2 = N$. So that β may be conveniently partitioned, it is necessary to rearrange equation (3.78) to the following form

$$\begin{bmatrix} P_1 \\ N_1 \\ \hline P_2 \\ N_2 \end{bmatrix} = \left[\begin{array}{c|c} \beta_{11} & \beta_{12} \\ \hline \beta_{21} & \beta_{22} \end{array}\right] \begin{bmatrix} 0 \\ 0 \\ \hline V_2 \\ \Theta_2 \end{bmatrix} \tag{3.82}$$

Hence

$$\begin{bmatrix} P \\ N \end{bmatrix} = \beta_{22} \begin{bmatrix} V_2 \\ \Theta_2 \end{bmatrix} = \frac{EI\lambda^2}{F_3} \begin{bmatrix} -\lambda F_6 & F_1 \\ F_1 & F_5/\lambda \end{bmatrix} \begin{bmatrix} V_2 \\ \Theta_2 \end{bmatrix}$$

giving

$$\begin{bmatrix} V_2 \\ \Theta_2 \end{bmatrix} = \frac{-F_3}{EI\lambda^2(F_1{}^2 + F_5F_6)} \begin{bmatrix} F_5/\lambda & -F_1 \\ -F_1 & -\lambda F_6 \end{bmatrix} \begin{bmatrix} P \\ N \end{bmatrix}$$

$$= \frac{1}{EI\lambda^2 F_4} \begin{bmatrix} -F_5/\lambda & F_1 \\ F_1 & \lambda F_6 \end{bmatrix} \begin{bmatrix} P \\ N \end{bmatrix} \tag{3.83}$$

The instantaneous deflection of any point on the beam is

$$v = \frac{[\sin \lambda x \cos \lambda x \sinh \lambda x \cosh \lambda x]}{2EI\lambda^3 F_4} \begin{bmatrix} F_9 & -F_8 \\ -F_7 & -F_9 \\ -F_9 & F_8 \\ F_7 & F_9 \end{bmatrix} \begin{bmatrix} P \\ N \end{bmatrix} \sin \omega t \tag{3.84}$$

Exercises

1. The portal frame shown in Fig. 3.12 consists of three uniform beams, welded together at their junctions and fixed into the foundations. The cross-sectional dimensions of the beams are small compared with their lengths and under load the angles between them at their junctions remain right angles.

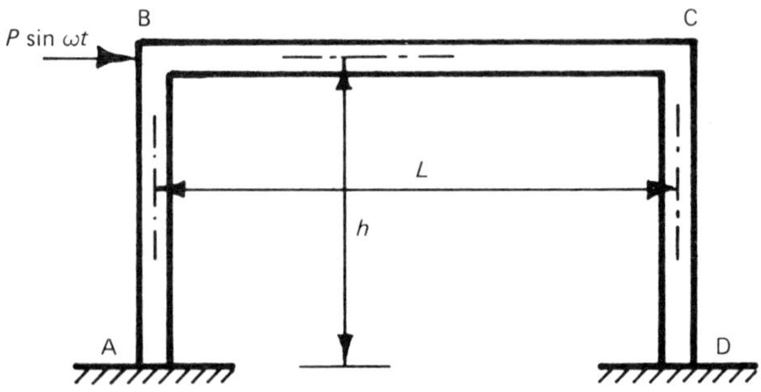

Fig. 3.12

Consider bending vibrations in the plane of the frame only and write a computer program for calculating the horizontal amplitude δ as a function of the exciting frequency ω when the frame is acted upon by a sinusoidally varying force as shown.

2. For the frame described in exercise 1, write a computer program for the case in which the exciting force acts vertically on beam BC at a distance $L/3$ from B.

3.4.2 Excitation due to distributed forces

When the exciting forces are distributed the equation of motion for an element of the beam becomes

$$EI\frac{\partial^4 v}{\partial x^4} + \rho A\frac{\partial^2 v}{\partial t^2} = f(x, t) \tag{3.85}$$

where $f(x, t)$ is the time varying force distribution acting on the beam, being positive in the positive direction of v. Consider first the case when $f(x, t)$ is discontinuous within the span.

We may change the coordinate system and it is convenient that we do so. As our new coordinate system, we shall choose the coefficients of a series composed of the normalised orthogonal characteristic functions ϕ_i which satisfy the end conditions of the beam. That is, we represent v as follows:

$$v = \sum_{i=1}^{\infty} q_i\phi_i \tag{3.86}$$

q_i, the generalised coordinates, are a function of time. Similarly, as shown in section 3.1.4, we may represent $f(x, t)$ by a set of generalised forces

$$Q_i = \int_0^L f(x, t)\phi_i \, dx \tag{3.87}$$

The equation of motion then becomes

$$\frac{d^2 q_i}{dt^2} + \omega_i^2 q_i = Q_i \tag{3.88}$$

If

$$Q_i = B_i \sin \omega t$$

the steady state solution of equation (3.88) becomes

$$q_i = \frac{B_i \sin \omega t}{(\omega_i^2 - \omega^2)}$$

$$= A_i \sin \omega t \tag{3.89}$$

where A_i are the normal coordinates.

To illustrate the procedure, the response will be found for a uniform simply supported beam excited by a uniformly distributed force acting on the mid-quarter of its span. Let the exciting force be $p \sin \omega t$ per unit length. For a simply supported beam the normalised characteristic functions are

$$\phi_i = \left(\frac{2}{\rho AL}\right)^{1/2} \sin (i\pi x/L) \tag{3.90}$$

Therefore,

$$Q_i = -\left(\frac{2}{\rho AL}\right)^{1/2} \frac{2Lp}{i\pi} \cos \frac{i5\pi}{8} \sin \omega t$$

when i is odd and

$$Q_i = 0 \tag{3.91}$$

when i is even. Consequently,

$$v = -\frac{4p}{\rho A\pi} \sum_{r=1}^{\infty} \frac{\cos\{(2r-1)5\pi/8\} \sin\{(2r-1)\pi x/L\}}{(2r-1)(\omega_{2r-1}^2 - \omega^2)} \sin \omega t \tag{3.92}$$

Note that as the exciting force is symmetrically disposed about the mid-span only the 1st, 3rd, 5th, etc., modes are excited.

When over the complete span of a beam $f(x, t)$ is a continuous function of x, and it is a sinusoidal function of time, the equation of motion becomes

$$EI \frac{d^4 V}{dx^4} - \rho A\omega^2 V = F(x) \tag{3.93}$$

The solution of this equation requires a particular integral and a complementary function. The latter is obviously

$$V_1 = C_1 \sin \lambda x + C_2 \cos \lambda x + C_3 \sinh \lambda x + C_4 \cosh \lambda x \tag{3.94}$$

where
$$\lambda^4 = \frac{\rho A\omega^2}{EI}$$

The particular integral depends upon the distribution of the exciting force. For example, if we consider a uniform cantilever fixed at $x = 0$ and loaded so that $F(x) = px/L$, the particular integral is

$$V_2 = -\frac{px}{\rho AL\omega^2} \tag{3.95}$$

The complete solution $V = V_1 + V_2$ and the values of C_i depend upon the boundary conditions. Hence

$$\begin{bmatrix} 0 & 1 & 0 & 1 \\ \lambda & 0 & \lambda & 0 \\ -\lambda^2 \sin \lambda L & -\lambda^2 \cos \lambda L & \lambda^2 \sinh \lambda L & \lambda^2 \cosh \lambda L \\ -\lambda^3 \cos \lambda L & \lambda^3 \sin \lambda L & \lambda^3 \cosh \lambda L & \lambda^3 \sinh \lambda L \end{bmatrix} \begin{bmatrix} C_1 \\ C_2 \\ C_3 \\ C_4 \end{bmatrix} = \begin{bmatrix} 0 \\ p/(\rho AL\omega^2) \\ 0 \\ 0 \end{bmatrix} \tag{3.96}$$

which is easily solved on a digital computer. This method may be generalised to include end forces and moments as exciting functions.

Exercises

1. A uniformly distributed harmonic force of intensity p per unit length and angular frequency ω acts upon the whole span of a uniform simply supported beam of length L and bending rigidity EI. Determine the steady state amplitude at mid-span.
2. A uniform simply supported beam of length L and bending rigidity EI is subjected to a concentrated harmonic force $P \sin \omega t$, applied at a distance l from the left-hand support. Derive an expression for the steady state amplitude at at any point on the beam.

   If the exciting force is applied at mid-span and 3ω equals ω_1, the frequency of the fundamental mode, show that the dynamic deflection is about 12 per cent greater than the static deflection under the same load.
3. Formulate in matrix notation the relationship between the end forces and moments and the end slopes and deflections for a uniform beam, when it is subjected to a uniformly distributed forcing function of intensity $p \sin \omega t$ over its whole length.
4. Show, for a uniform cantilever of length L subjected to a transverse force $P \sin \omega t$ at its free end, that the steady state amplitude of the maximum bending stress at a distance x from the fixed end is proportional to

$$\sum_r \frac{\dfrac{d^2\phi_r}{dx^2}\,\phi_r(L)}{\omega_r{}^2 - \omega^2}$$

   $\phi_r(x)$ and ω_r are the normal mode shape and natural angular frequency respectively of the r'th mode.

3.5 Free lateral vibration of a thin beam subjected to a constant longitudinal force

When a constant longitudinal compressive or tensile force acts on the ends of a beam, its natural frequencies in bending change. An increase in tensile force increases the frequency, conversely a compressive force decreases it. This is in keeping with the frequency increase experienced as the tension is increased in strings of musical instruments, which is common experience. An expression for the magnitude of the change for a pinned-pinned beam may be found easily.

Plane cross-sections of the beam remain plane under the longitudinal forces. Therefore, on the assumptions quoted in section 3.3.1 we may consider the forces applied to an element of the beam deflected transversely from the equilibrium state. Figure 3.13 shows the forces involved. P is the constant longitudinal force, tensile force assumed positive. Applying Newton's second law in the y-direction gives

$$\frac{\partial F}{\partial x} = \rho A \frac{\partial^2 v}{\partial t^2} \tag{3.97}$$

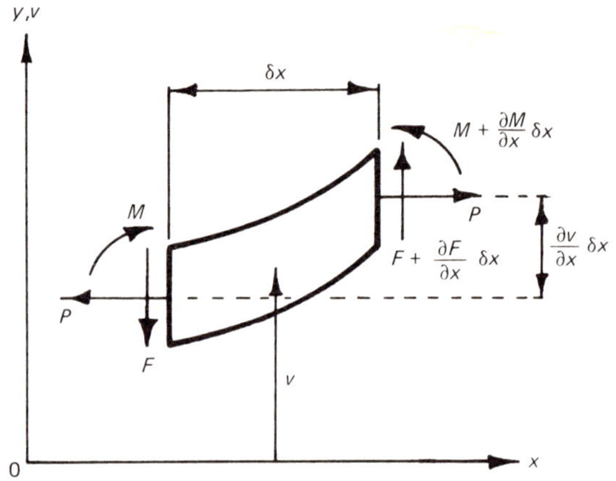

Fig. 3.13 Positive sign convention

Forces in the x-direction are balanced and equating moments about the centre of the element gives

$$-M - P\frac{\partial v}{\partial x}\,\delta x + F\,\delta x + \frac{\partial F}{\partial x}\frac{(\delta x)^2}{2} + M + \frac{\partial M}{\partial x}\,\delta x = 0$$

or, neglecting second and higher order terms,

$$F - P\frac{\partial v}{\partial x} + \frac{\partial M}{\partial x} = 0 \tag{3.98}$$

From the Bernoulli-Euler beam theory, the bending moment–curvature relationship is

$$M = EI\frac{\partial^2 v}{\partial x^2} \tag{3.99}$$

Differentiating (3.98) with respect to x and substituting from (3.97) and (3.99) gives

$$EI\frac{\partial^4 v}{\partial x^4} - P\frac{\partial^2 v}{\partial x^2} + \rho A\frac{\partial^2 v}{\partial t^2} = 0 \tag{3.100}$$

As this is a fourth order linear homogeneous differential equation, by the same arguments as on page 98, we may take the solution in the form

$$v = \sum_n V_n\{C \sin \omega_n t + D \cos \omega_n t\} \tag{3.101}$$

where V_n is a function of x only.

Substituting (3.101) in (3.100) gives

$$EI\frac{\mathrm{d}^4 V_n}{\mathrm{d}x^4} - P\frac{\mathrm{d}^2 V_n}{\mathrm{d}x^2} - \rho A\omega_n^2 V_n = 0 \tag{3.102}$$

The solution of this ordinary differential equation is

$$V_n = A_{1n} \sin \alpha_n x + A_{2n} \cos \alpha_n x + A_{3n} \sinh \beta_n x + A_{4n} \cosh \beta_n x \qquad (3.103)$$

where
$$\alpha_n^2 = -\left\{\frac{P - \sqrt{(P^2 + 4EI\rho A\omega_n^2)}}{2EI}\right\}$$

$$\beta_n^2 = \left\{\frac{P + \sqrt{(P^2 + 4EI\rho A\omega_n^2)}}{2EI}\right\}$$

The values of the coefficients A_{in} are determined by the end conditions. The simplest set of end conditions to satisfy are those associated with pinned ends, and to find the magnitude of the influence of longitudinal force on the natural flexural frequency we shall solve this case, for a beam of length L.

With pinned ends, the end deflections and couples are permanently zero, therefore it may be shown that the frequency equation is

$$\sin \alpha_n L = 0 \text{ or } \alpha_n L = n\pi \qquad (3.104)$$

Hence
$$\omega_n^2 = \frac{n^2\pi^2}{\rho A L^2}\left\{P + \frac{n^2\pi^2 EI}{L^2}\right\} \qquad (3.105)$$

The case of a strut is of particular interest. Elastic stability theory predicts that a strut with pinned ends will buckle when the load is

$$\frac{-\pi^2 EI}{L^2}$$

This is known as the Euler buckling load. If P is put equal to the Euler buckling load in equation (3.105) it is found that the frequency of the first flexural mode is reduced to zero.

Exercise

If the strut has its ends clamped, show that the frequency equation for symmetrical modes of vibration is

$$\alpha \tan \frac{\alpha L}{2} + \beta \tanh \frac{\beta L}{2} = 0$$

3.6 Free lateral vibration of deep beams

When the depth of a beam is a significant proportion of the distance between two adjacent nodes of its vibratory form, transverse-shear deformation makes a significant contribution to its lateral deflection. Consequently, in the vibration analysis of deep beams shear deformation must be allowed for, even for the lower frequencies; at very high frequencies it can also account for a considerable shift even in the case of thin beams.

Another factor which must be accounted for in vibration studies of deep beams and of higher modes of thin beams is rotary inertia. In the previous sections of this book, we have specifically ignored the moment of inertia of the beam element, and consequently the couples required for its angular acceleration.

Both of the above factors will now be considered.

3.6.1 Shear deformation

From statics it is well known that the equilibrium deflection of a loaded beam, due to the transverse shearing of an elemental length of the beam relative to neighbouring elements is small for beams of small depth/effective-length ratio. Its significance increases as that ratio increases, until for high values of it deflection due to shear predominates. Naturally, we may expect a similar trend to hold for the vibratory behaviour of beam systems. For shallow beams, usually referred to as thin beams, we expect the inclusion of shear deformation not to affect the frequencies up to the frequency range in which the beam can no longer be considered thin in relation to the distance between adjacent nodes.

Shear deformation gives rise to an increase in the flexibility of the system, therefore its inclusion in the analysis leads to the prediction of lower values for natural frequencies than those calculated without it. The magnitude of the effect may be found easily for beams with pinned ends, under the assumption that plane-cross-sections remain plane throughout the oscillations. The warping of the sections present in reality is allowed for approximately by introducing a coefficient which depends upon the form of the cross-section.

Refer to Fig. 3.6 for the sign convention and

let A = cross-sectional area
EI = bending rigidity
G = modulus of rigidity
v_f = lateral deflection due to flexure alone
v_s = lateral deflection due to shearing alone
v = total lateral deflection $(v_f + v_s)$
ρ = mass density for the material of the beam.

Newton's second law for motion in y-direction gives

$$\frac{\partial F}{\partial x} = \rho A \frac{\partial^2 v}{\partial t^2} \tag{3.106}$$

Equating moments about the centre of the element and neglecting second and higher order terms gives

$$F + \frac{\partial M}{\partial x} = 0 \tag{3.107}$$

From the Bernoulli-Euler theory of bending we have that

$$M = EI \frac{\partial^2 v_f}{\partial x^2} \tag{3.108}$$

We may also say that to a sufficient approximation

$$F = kAG \frac{\partial v_s}{\partial x} \tag{3.109}$$

where k is a constant, the value of which depends upon the cross-sectional shape of the beam. k allows approximately for the fact that the assumption that plane sections remain plane in simple bending theory is no longer true.

Substituting for M and F from equations (3.108) and (3.109) into equations (3.106) and (3.107) gives

$$kAG\left(\frac{\partial^2 v}{\partial x^2} - \frac{\partial^2 v_f}{\partial x^2}\right) = \rho A \frac{\partial^2 v}{\partial t^2} \tag{3.110}$$

and

$$kAG\left(\frac{\partial v}{\partial x} - \frac{\partial v_f}{\partial x}\right) + EI \frac{\partial^3 v_f}{\partial x^3} = 0 \tag{3.111}$$

from which v_f may be eliminated to give

$$EI \frac{\partial^4 v}{\partial x^4} - \frac{EI\rho}{kG} \frac{\partial^4 v}{\partial x^2\, \partial t^2} + \rho A \frac{\partial^2 v}{\partial t^2} = 0 \tag{3.112}$$

This is a fourth-order linear homogeneous partial differential equation, the solution of which has the form

$$v = V_n\{C \sin \omega_n t + D \cos \omega_n t\} \tag{3.113}$$

where V_n is a function of x only. Substituting for v in equation (3.112) gives

$$\frac{\mathrm{d}^4 V_n}{\mathrm{d}x^4} + 2r_n \frac{\mathrm{d}^2 V_n}{\mathrm{d}x^2} - s_n V_n = 0 \tag{3.114}$$

where

$$r_n = \frac{\rho \omega_n^{\ 2}}{2kG} \text{ and } s_n = \frac{\rho A \omega_n^{\ 2}}{EI}$$

The solution of equation (3.114) has the form

$$V_n = A_{1n} \sin \alpha_n x + A_{2n} \cos \alpha_n x + A_{3n} \sinh \beta_n x + A_{4n} \cosh \beta_n x \tag{3.115}$$

where

$$\begin{matrix}\alpha_n^{\ 2} \\ \beta_n^{\ 2}\end{matrix} = \pm r_n + (r_n^{\ 2} + s_n)^{1/2}$$

The easiest case to analyse is that in which both ends are pinned. It can be shown that the solution is then

$$v = \sum_n A_{1n} \sin \alpha_n x\{C \sin \omega_n t + D \cos \omega_n t\} \tag{3.116}$$

where the characteristic values α_n are $n\pi/L$

therefore,

$$\omega_n^{\ 2} = \frac{n^4\pi^4 EI}{\rho A L^4 \left(1 + \dfrac{n^2\pi^2 EI}{L^2 kAG}\right)} \tag{3.117}$$

therefore,
$$\omega_n \simeq \omega_{tn}\left(1 - \frac{n^2\pi^2 EI}{2L^2kAG}\right) \tag{3.118}$$

ω_{tn} being the angular natural frequency found on thin beam assumptions.

For a steel beam of rectangular cross-section, 0·3 m deep by 0·15 m wide, of length 3 m, the term

$$\frac{\pi^2 EI}{L^2kAG}$$

is approximately 0·03. Therefore, the natural frequencies predicted with an allowance for shear deformation will be 1·5 per cent lower for the first mode, 6 per cent lower for the second mode, and so on, compared with predictions on thin beam assumptions.

3.6.2 Rotary inertia

When the depth of the beam is not small compared with the distance between adjacent nodal planes, in a particular mode of vibration, the kinetic energy due to rotation of its cross-sectional planes about the neutral axis will be a significant proportion of its total kinetic energy at a particular instant of time, say on passing through the static equilibrium position. In the earlier sections, when we considered thin beams, we did, in effect, assume this to be negligible by ignoring the couple required to accelerate the beam element in rotational motion about the neutral axis. For deep beams, this couple may no longer be assumed negligible. Obviously, this has the effect of increasing the inertia of the beam, therefore, we should expect it to lower the predicted natural frequencies from those found assuming the beam to be thin. An expression for the magnitudes of these changes will be derived for a beam with pinned ends.

Refer to Fig. 3.6 for the sign convention and

let A = cross-sectional area
E = Young's modulus
I = second moment of area
ρ = mass density.

Then the moment of inertia of an element of the beam of unit length about an axis in the neutral plane at mid-length, perpendicular to the length is ρI. By Newton's laws, neglecting second and higher order quantities, we obtain

(*i*) for linear transverse motion

$$\frac{\partial F}{\partial x} = \rho A \frac{\partial^2 v}{\partial t^2} \tag{3.119}$$

(*ii*) for rotational motion

$$\frac{\partial M}{\partial x} + F = \rho I \frac{\partial^3 v}{\partial x\, \partial t^2} \tag{3.120}$$

Also from the Bernoulli-Euler beam theory

$$M = EI \frac{\partial^2 v}{\partial x^2} \tag{3.121}$$

Equations (3.119), (3.120), and (3.121) may be combined to give

$$EI \frac{\partial^4 v}{\partial x^4} - \rho I \frac{\partial^4 v}{\partial x^2 \, \partial t^2} + \rho A \frac{\partial^2 v}{\partial t^2} = 0 \tag{3.122}$$

the solution of which has the form

$$v = \sum_n V_n(C_n \sin \omega_n t + D_n \cos \omega_n t) \tag{3.123}$$

On substitution for v equation (3.122) becomes

$$\frac{d^4 V_n}{dx^4} + 2r_n \frac{d^2 V_n}{dx^2} - s_n V_n = 0 \tag{3.124}$$

which is satisfied by

$$V_n = A_{1n} \sin \alpha_n x + A_{2n} \cos \alpha_n x + A_{3n} \sinh \beta_n x + A_{4n} \cosh \beta_n x \tag{3.125}$$

where

$$\left.\begin{matrix} \alpha_n^2 \\ \beta_n^2 \end{matrix}\right\} = \pm r_n + (r_n^2 + s_n)^{1/2}$$

and

$$r_n = \frac{\rho \omega_n^2}{2E}, \qquad s_n = \frac{\rho A \omega_n^2}{EI}$$

For a beam with pinned ends it may be shown that $A_{2n} = A_{3n} = A_{4n} = 0$ and that

$$A_{1n} \sin \alpha_n L = 0 \tag{3.126}$$

where L is the length of the beam between the pinned supports. Hence

$$\alpha_n = n\pi/L$$

and

$$\omega_n^2 = \frac{n^4 \pi^4 EI}{\rho A L^4 \left(1 + \dfrac{n^2 \pi^2 I}{AL^2}\right)} \tag{3.127}$$

therefore,

$$\omega_n \simeq \omega_{tn} \left(1 - \frac{n^2 \pi^2 I}{2AL^2}\right) \tag{3.128}$$

where ω_{tn} is the corresponding natural frequency calculated on thin beam assumptions.

For the steel beam, 0·3 m deep by 0·15 m wide by 3 m long, discussed previously, the term $\pi^2 I/2AL^2$ is approximately 0·004. Therefore, to allow for rotary inertia a correction factor may be applied to the frequencies predicted on a thin beam basis. In this particular case they will be lower by 0·4 per cent in the first mode, 1·6 per cent in the second mode and so on.

Rayleigh (1894b) was the first to derive the correction for rotary inertia, and Timoshenko (1921, 1922) derived and solved the equations of motion allowing for both shear deformation and rotary inertia effects. The analysis allowing for both is generally referred to as the Timoshenko beam theory.

The influences of shear deformation and rotary inertia are often small in practical cases and they may then be combined to give correction factors which simplify the problem of estimating natural frequencies. That is, the natural frequencies for the beam being investigated are estimated on thin beam assumptions and corrected to allow for the influence of shear deformation and rotary inertia. Combining the shear and rotary inertia expressions for the beam with pinned ends gives

$$\omega_n \simeq \omega_{tn}\left(1 - \frac{\pi^2}{2}\frac{n^2}{L^2}\frac{I}{A}\left[1 + \frac{E}{kG}\right]\right) \tag{3.129}$$

Thus there are two dimensionless correction factors

$$\frac{\pi^2 n^2}{2L^2}\frac{I}{A} \quad \text{and} \quad \frac{E}{kG}$$

the former is a measure of the importance of the distance between adjacent nodes and the latter expresses the relative importance of shear deformation to that of rotary inertia. For the steel beam we considered previously E/kG is approximately 4.

Correction factors of the above form may be derived for other end conditions; Sutherland and Goodman (1951) give correction factors for the pinned-pinned and clamped-free end conditions; Huang (1964) gives them for pinned-pinned, free-free, clamped-clamped, clamped-free, clamped-pinned, and pinned-free beams.

Rigorously it is difficult to derive a value for k, the shear coefficient, but a number of workers have suggested values. For the case of a rectangular cross-section Timoshenko (1921) assumed 0·667, Sutherland and Goodman (1951) gave 0·870 when Poisson's ratio equals $\frac{1}{3}$, and Cowper (1966) gave 0·851 for the same value of Poisson's ratio. Of the above approaches that of Cowper is the most rigorous. Taking the deflection of a cross-section as its average deflection and its rotation as its mean rotation about the neutral axis, he integrated the equations of three-dimensional elasticity theory to give a new derivation of the equations of Timoshenko's beam theory.

Exercises

Show for a uniform prismatic beam of length L, cross-sectional area A, and second moment of area about the flexural axis of I, that the following frequency equations are found from the associated end conditions when the influence of both shear deformation and rotary inertia are allowed for.

(*i*) fixed-free:

$$2\,\frac{\gamma_1\gamma_2}{\alpha\beta}+\left(\frac{\gamma_1^2+\gamma_2^2}{\alpha\beta}\right)(\cos\alpha L\cosh\beta L)+\gamma_1\gamma_2\left(\frac{1}{\alpha^2}-\frac{1}{\beta^2}\right)\sin\alpha L\sinh\beta L=0$$

(*ii*) free-free:

$$\frac{2\gamma_1\gamma_2}{\alpha\beta}(\cos\alpha L\cosh\beta L-1)+\left(\frac{\gamma_2^2}{\alpha^2}-\frac{\gamma_1^2}{\beta^2}\right)\sin\alpha L\sinh\beta L=0$$

(*iii*) pinned-fixed:

$$\frac{\gamma_1}{\alpha}\cos\alpha L\sinh\beta L-\frac{\gamma_2}{\beta}\sin\alpha L\cosh\beta L=0$$

where

$$\gamma_1=\alpha-\frac{\rho\omega^2}{kG}$$

$$\gamma_2=\beta+\frac{\rho\omega^2}{kG}$$

All other symbols are as defined in the text.

3.7 A generalised approach for complex bar, shaft, and beam systems

A general method for building up the system dynamic stiffness matrix of a complex lumped parameter system was presented in section 2.8.3; this will now be extended to systems having distributed mass and elasticity.

Obviously the method given in section 2.8.3 may be applied directly to any system so long as we can describe the dynamic behaviour of the subsystems as a relationship between generalised forces and generalised displacements at their extremities. For longitudinal oscillations of a bar it was shown, section 3.1.4, that the maximum terminal forces F_i and the resulting maximum terminal displacements U_i are related by an expression of the form

$$\mathrm{F}=\boldsymbol{\beta}\mathrm{U}$$

The matrix $\boldsymbol{\beta}$ is the dynamic stiffness matrix for the bar.

Similarly in section 3.4 it was shown that the maximum of the end forces and moments F_i acting on a thin beam are related to the end displacements and rotations V_i by

$$\mathrm{F}=\boldsymbol{\beta}\mathrm{V}$$

Again $\boldsymbol{\beta}$ is the dynamic stiffness matrix for the beam.

Using methods outlined in this chapter the dynamic stiffness matrices may be derived for shafts, for thin beams under constant longitudinal force, and for thick beams. Having obtained these the procedures of section 2.8.3 may be used to set up the overall system dynamic stiffness matrix for any prescribed combination of them; concentrated masses, rigid bodies, and massless springs may of course be included.

Exercises

1. Repeat exercises 1 and 2 of section 3.4.1 using this coordinate transformation method.
2. The portal frame shown in Fig. 3.12 has the following characteristics. The vertical members are identical, having length l_1 and bending rigidity E_1I_1: they are each set in an elastic foundation which has a rotational stiffness k and an infinite stiffness to lateral and longitudinal deflections. The angles between the vertical and horizontal members remain right angles. The geometrical and physical parameters of the horizontal beam may differ from those of the vertical beams: for the horizontal beam use the subscript 2. Show that the frequency equation is given by

$$\{MK^3\Phi_1 + \Phi_4 - \Phi_5\}\,\{MK^3\Phi_1 + \Phi_4 + \Phi_5)\left(2\Phi_3 + \frac{\lambda l_1}{M^4L}\right) + 2MK^3\Phi_2{}^2\} = 0$$

where $K^4 = E_1I_1/E_2I_2$

$M^4 = \rho_1A_1/\rho_2A_2$

$L = l_1/l_2$

and Φ_i are functions of the frequency functions relating to the horizontal beams and a factor R.

$$R = kl_1/E_1I_1$$

Show that the term in the first brackets is the frequency equation relating to symmetrical modes and that in the second bracket is for unsymmetrical modes.

3.8 Deriving equations of motion by energy methods

So far in this chapter, Newtonian methods have been used to set up the equations of motion relating to the vibration of bars, shafts, and beams. They have helped to give a physical understanding of the phenomena involved but we have had to exercise care in manipulating the vector quantities displacement, acceleration, force, stress, etc. The governing differential equation and the boundary conditions may be obtained from the scalar energy expressions together with the work done by the non-conservative generalised forces, using Lagrange's equations of motion.

As an example, consider the bending oscillations of a uniform thin beam of length L, bending rigidity EI, mass density ρ, and cross-sectional area A. Let it be subjected at its ends to the generalised forces shown in Fig. 3.11.

Then

$$2T = \int_0^L \rho A \dot{v}^2 \, dx \tag{3.130}$$

$$2V = \int_0^L EI\left(\frac{\partial^2 v}{\partial x^2}\right)^2 \mathrm{d}x$$

$$= EI\,\frac{\partial^2 v}{\partial x^2}\,\frac{\partial v}{\partial x}\Big]_0^L - EI\,\frac{\partial^3 v}{\partial x^3}\,v\Big]_0^L + \int_0^L EI\,\frac{\partial^4 v}{\partial x^4}\,v\,\mathrm{d}x \qquad (3.131)$$

As generalised coordinates we take v, v_0, v_L, $(\partial v/\partial x)_0$, $(\partial v/\partial x)_L$. There are an infinite number of v's. When the motion is simple harmonic we may say that

$$\frac{\partial^2 v}{\partial x^2} = k_1\,\frac{\partial v}{\partial x}$$

$$\frac{\partial^3 v}{\partial x^3} = k_2 v$$

and

$$\frac{\partial^4 v}{\partial x^4} = k_3 v$$

Lagrange's equations of motion then give

(*a*)

$$\int_0^L \left(\rho A\,\frac{\partial^2 v}{\partial t^2} + EIk_3 v\right)\mathrm{d}x = 0$$

or

$$\rho A\,\frac{\partial^2 v}{\partial t^2} + EI\,\frac{\partial^4 v}{\partial x^4} = 0 \qquad (3.132)$$

which is identical with that found on page 97

$$\left.\begin{aligned}
&(b) \qquad EI\,\frac{\partial^3 v}{\partial x^3}\Big]_0 = P_1 \cos \omega t \\
&(c) \qquad -EI\,\frac{\partial^3 v}{\partial x^3}\Big]_L = P_2 \cos \omega t \\
&(d) \qquad -EI\,\frac{\partial^2 v}{\partial x^2}\Big]_0 = N_1 \cos \omega t \\
&(e) \qquad EI\,\frac{\partial^2 v}{\partial x^2}\Big]_L = N_2 \cos \omega t
\end{aligned}\right\} \qquad (3.133)$$

which are the boundary conditions for this system.

Exercises

Using Lagrange's equations of motion check the equations of motion and end conditions for each of the cases: longitudinal, torsional, and bending oscillations, considered earlier in this chapter.

Another energy approach occasionally used in the literature for setting up the equations of motion and boundary conditions relating to continuous elastic bodies is the application of Hamilton's principle.

Hamilton's principle states that in moving from a point $\boldsymbol{r}_1$ at time t_1 to a point $\boldsymbol{r}_2$ at time t_2 a particle takes a path such that

$$\int_{t_1}^{t_2} (\delta T + \textstyle\sum_i P_i \,\delta q_i)\, \mathrm{d}t = 0 \tag{3.134}$$

where q_i define the path taken by the particle, $q_i + \delta q_i$ define another path between $\boldsymbol{r}_1$ at time t_1 and $\boldsymbol{r}_2$ at time t_2 a small distance from q_i. δT is the variation in the kinetic energy due to the variation in δq_i and the associated velocity variation $\delta \dot{q}_i$, and P_i are the generalised forces associated with the coordinates q_i.

Now
$$\delta T = \sum_i \left(\frac{\partial T}{\partial \dot{q}_i}\,\delta \dot{q}_i + \frac{\partial T}{\partial q_i}\,\delta q_i\right) \tag{3.135}$$

It is easily shown that Hamilton's principle and Lagrange's equations of motion are equivalent, as follows. Substituting, from equation (3.135), the value of δT in equation (3.134) gives

$$\int_{t_1}^{t_2} \sum_i \left(\frac{\partial T}{\partial \dot{q}_i}\,\delta \dot{q}_i + \frac{\partial T}{\partial q_i}\,\delta q_i + P_i\,\delta q_i\right) \mathrm{d}t = 0 \tag{3.136}$$

On integrating by parts the first term becomes

$$\int_{t_1}^{t_2} \sum_i \frac{\partial T}{\partial \dot{q}_i}\,\delta \dot{q}_i\, \mathrm{d}t = \left[\sum_i \frac{\partial T}{\partial \dot{q}_i}\,\delta q_i\right]_{t_1}^{t_2} - \int_{t_1}^{t_2} \sum_i \frac{\mathrm{d}}{\mathrm{d}t}\left(\frac{\partial T}{\partial \dot{q}_i}\right) \delta q_i\, \mathrm{d}t \tag{3.137a}$$

$\delta q_i = 0$ at time t_1 and time t_2 by definition. Therefore equation (3.136) becomes

$$\int_{t_1}^{t_2} \sum_i \left\{-\frac{\mathrm{d}}{\mathrm{d}t}\left(\frac{\partial T}{\partial \dot{q}_i}\right) + \frac{\partial T}{\partial q_i} + P_i\right\} \delta q_i\, \mathrm{d}t = 0 \tag{3.137b}$$

The variations δq_i are independent, therefore, we have

$$\frac{\mathrm{d}}{\mathrm{d}t}\left(\frac{\partial T}{\partial \dot{q}_i}\right) - \frac{\partial T}{\partial q_i} = P_i \tag{3.138}$$

which are Lagrange's equations of motion.

When dealing with a conservative system such as a continuous elastic body a more usual form of Hamilton's principle is

$$\int_{t_1}^{t_2} \left\{\delta(T - V) + \sum_i P_i\,\delta q_i\right\} \mathrm{d}t = 0 \tag{3.139}$$

where V is the potential energy of the system and

$$\sum_i P_i\,\delta q_i$$

is the work done by boundary forces and couples in the variation.

Exercises

1. Using Hamilton's principle set up the equations of motion and boundary conditions for a thick beam excited at each end by a lateral harmonic force and a harmonic couple. From these equations derive the dynamic stiffness matrix relating to transverse oscillations of such a beam when the end displacements and rotations are taken as generalised coordinates.
2. Use Hamilton's principle to set up the equations of motion for longitudinal vibration of a uniform circular bar, taking into account the lateral contraction of a cross-section due to the effect of Poisson's ratio. Assume plane sections in the quiescent state remain plane during vibration.

   Check your result by Newtonian methods.

   For a steel bar 30 mm diameter by 1 m long, free at both ends, find the percentage changes in predicted natural frequencies when this effect is allowed for, compared with those found neglecting it.
3. Using Hamilton's principle, show that the equation of motion for an element of a uniform cantilever vibrating in bending when rotating about a fixed centre, so that its longitudinal axis whilst stationary is radial, has the form

$$\frac{\mathrm{d}^4 V}{\mathrm{d}x^4} = \frac{\rho A \omega^2}{EI}\left[(1+\mu^2)V - \mu^2 R\,\frac{\mathrm{d}V}{\mathrm{d}x} + 0{\cdot}5\mu^2(L^2 + 2rL - 2rx - x^2)\,\frac{\mathrm{d}^2 V}{\mathrm{d}x^2}\right]$$

where $\mu = \Omega/\omega$
Ω = the rotational velocity about the fixed centre
R = radius to element
r = radius to fixed end of cantilever
L = length of cantilever, the free end being at $r + L$ in the stationary state
x = distance from fixed end to element
I = second moment of area about a principal axis of the cross-section, the axis being parallel to the axis of rotation.

3.9 Free lateral vibration of thin plates

Like beams, plates are very frequently used elements of engineering components and structures. To understand, fully, many of the vibratory phenomena which occur with plates it is necessary to treat their mass and elasticity as distributed. When this is done, the resulting equations of motion are partial differential equations in two space dimensions and time, which lead to closed form solutions for only a limited number of boundary condition combinations.

In this section we shall consider the vibration of thin rectangular plates with some simple boundary conditions. In chapters 4 and 5 we shall study methods which may be used for more complex boundary conditions. Firstly, we must derive the equation of motion.

3.9.1 The equation of motion in rectangular coordinates

The coordinate directions and sign convention we shall work in are illustrated in Fig. 3.14. By analogy with thin-beam bending theory, let us assume the middle plane to be a neutral plane, that is it remains unstrained in the x and y directions. We shall also assume that planes normal to the neutral plane remain plane in the deformed state. Then if in the undeformed state $z = 0$ is the neutral plane, we may write

$$u = -z\frac{\partial w}{\partial x} \quad \text{and} \quad v = -z\frac{\partial w}{\partial y} \tag{3.140}$$

Hence we may write the direct and shearing strains due to flexure as

$$\begin{aligned}
\epsilon_{xx} &= \frac{\partial u}{\partial x} = -z\frac{\partial^2 w}{\partial x^2}\\
\epsilon_{yy} &= \frac{\partial v}{\partial y} = -z\frac{\partial^2 w}{\partial y^2}\\
\epsilon_{xy} &= \frac{1}{2}\left(\frac{\partial v}{\partial x} + \frac{\partial u}{\partial y}\right) = -z\frac{\partial^2 w}{\partial x\,\partial y}
\end{aligned} \tag{3.141}$$

For thin plates it is regarded as reasonable to assume that the direct stress in the z-direction is zero, that is $\tau_{zz} = 0$, the stress–strain relations then become

$$\begin{aligned}
\tau_{xx} &= \frac{E\nu}{1-\nu^2}(\epsilon_{xx} + \epsilon_{yy}) + \frac{E}{1+\nu}\epsilon_{xx}\\
\tau_{yy} &= \frac{E\nu}{1-\nu^2}(\epsilon_{xx} + \epsilon_{yy}) + \frac{E}{1+\nu}\epsilon_{yy}\\
\text{and} \qquad \tau_{xy} &= \frac{E}{1+\nu}\epsilon_{xy}
\end{aligned} \tag{3.142}$$

where E is Young's modulus and ν is Poisson's ratio for the plate material. Substituting from equations (3.141) into equations (3.142) gives the stresses due to flexure as

$$\left.\begin{aligned}
\tau_{xx} &= -\frac{E\nu}{1-\nu^2}z\nabla^2 w - \frac{Ez}{1+\nu}\frac{\partial^2 w}{\partial x^2}\\
\tau_{yy} &= -\frac{E\nu}{1-\nu^2}z\nabla^2 w - \frac{Ez}{1+\nu}\frac{\partial^2 w}{\partial y^2}\\
\text{and} \qquad \tau_{xy} &= -\frac{Ez}{1+\nu}\frac{\partial^2 w}{\partial x\,\partial y}\\
\text{where} \qquad \nabla^2 &= \frac{\partial^2}{\partial x^2} + \frac{\partial^2}{\partial y^2}
\end{aligned}\right\} \tag{3.143}$$

As we are considering flexural deformations only there will be no resultant force in the x or y directions, that is

$$\int_{-d/2}^{d/2} \tau_{xx} \, dz = 0 \tag{3.144}$$

where d is the plate thickness. Similar expressions for τ_{yy} and τ_{xy} may be written.

The resulting couples per unit length acting on the edges of an element may be found by integrating the moment of the stress per unit length, thus

$$\left.\begin{aligned} M_{xa} &= -\int_{-d/2}^{d/2} z\tau_{yy} \, dz = \frac{Ed^3}{12(1+\nu)}\left\{\frac{\nu}{1-\nu}\nabla^2 w + \frac{\partial^2 w}{\partial y^2}\right\} \\ M_{ya} &= M_{xb} = -\int_{-d/2}^{d/2} z\tau_{xy} \, dz = \frac{Ed^3}{12(1+\nu)}\frac{\partial^2 w}{\partial x \, \partial y} \\ M_{yb} &= -\int_{-d/2}^{d/2} z\tau_{xx} \, dz = \frac{Ed^3}{12(1+\nu)}\left\{\frac{\nu}{1-\nu}\nabla^2 w + \frac{\partial^2 w}{\partial x^2}\right\} \end{aligned}\right\} \tag{3.145}$$

Moment equilibrium for the element gives

(*a*) about the x-axis

$$\frac{\partial M_{xa}}{\partial y} + \frac{\partial M_{xb}}{\partial x} + Q_{za} = 0 \tag{3.146}$$

(*b*) about the y-axis

$$\frac{\partial M_{yb}}{\partial x} - \frac{\partial M_{ya}}{\partial y} - Q_{zb} = 0 \tag{3.147}$$

where Q_{za} and Q_{zb} are the resultant shear forces per unit length as shown on Fig. 3.14.

Force balance gives, by Newton's laws,

$$\frac{\partial Q_{zb}}{\partial x} + \frac{\partial Q_{za}}{\partial y} = \rho d \frac{\partial^2 w}{\partial t^2} \tag{3.148}$$

giving

$$-\left\{\frac{\partial^2 M_{yb}}{\partial x^2} + 2\frac{\partial^2 M_{xb}}{\partial x \, \partial y} + \frac{\partial^2 M_{xa}}{\partial y^2}\right\} = \rho d \frac{\partial^2 w}{\partial t^2}$$

or

$$-\frac{Ed^2}{12(1-\nu^2)}\nabla^4 w = \rho \frac{\partial^2 w}{\partial t^2} \tag{3.149}$$

where ρ is the mass density of the plate material.

Equation (3.149) has solutions of the form

$$w = W_n(C \sin \omega_n t + D \cos \omega_n t) \tag{3.150}$$

where W_n are functions of x and y only. When this expression is substituted for w equation (3.149) may be reduced to the form

$$\nabla^4 W_n - \lambda^4 W_n = 0 \tag{3.151}$$

where
$$\lambda^4 = \frac{12(1 - \nu^2)\rho\omega_n^2}{Ed^2}$$

which is a partial differential equation in the two space coordinates x and y.

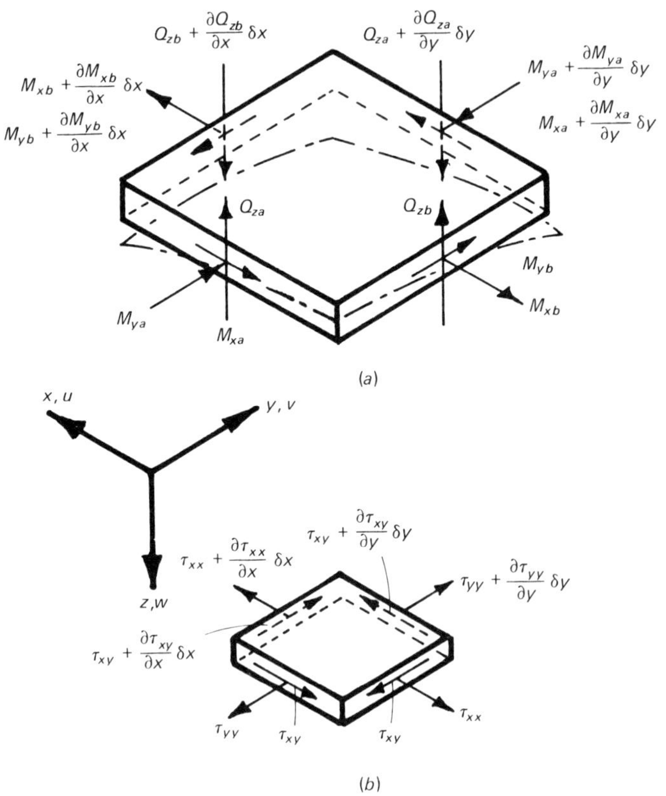

Fig. 3.14 Coordinate directions and sign convention:
(*a*) Forces and couples acting on an element
(*b*) Stresses acting on an element

3.9.2 Solutions for simple boundary conditions

If we introduce the restriction that two opposite edges of the rectangular plate are simply supported, that is their deflections are zero and they experience no restraining moment, then we may take solutions of the form

$$W_n = f(y) \sin \alpha x \tag{3.152}$$

In this case, we have assumed the edges parallel to the y-direction to be simply supported. For a plate of length a and width b, on the boundaries $x = 0$ and $x = a$ the conditions to be met are $w = 0$ and

$$M_{yb} = 0 = \frac{\nu}{1-\nu}\nabla^2 w + \frac{\partial^2 w}{\partial x^2}$$

These are each satisfied by equation (3.152) if $\sin \alpha a = 0$, that is if $\alpha a = p\pi$ where p is an integer. Hence, equation (3.152) becomes

$$W_n = f(y) \sin \frac{p\pi x}{a} \tag{3.153}$$

which may be substituted into equation (3.151) to give the ordinary differential equation

$$\frac{d^4 f(y)}{dy^4} - 2\frac{p^2\pi^2}{a^2}\frac{d^2 f(y)}{dy^2} + \left(\frac{p^4\pi^4}{a^4} - \lambda^4\right) f(y) = 0 \tag{3.154}$$

Assuming that $\lambda^2 > p^2\pi^2/a^2$ the solution has the form

$$f_p(y) = C_{1p}\sin\beta y + C_{2p}\cos\beta y + C_{3p}\sinh\gamma y + C_{4p}\cosh\gamma y \tag{3.155}$$

where

$$a^2\beta^2 = a^2\lambda^2 - p^2\pi^2$$
$$a^2\gamma^2 = a^2\lambda^2 + p^2\pi^2$$

The constants of integration C_{ip} are determined from the conditions on the boundaries $y = 0$ and $y = b$. If these edges are simply supported the frequency equation becomes

$$\sin\beta b = 0 \tag{3.156}$$

giving

$$\lambda^2 = \frac{p^2\pi^2}{a^2} + \frac{q^2\pi^2}{b^2} \tag{3.157}$$

where q is an integer. Hence

$$\omega = \left(\frac{p^2\pi^2}{a^2} + \frac{q^2\pi^2}{b^2}\right)\left(\frac{Ed^2}{12\rho(1-\nu^2)}\right)^{1/2} \tag{3.158}$$

and the vibratory form is

$$W_n = A\sin p\pi\frac{x}{a}\sin q\pi\frac{y}{b} \tag{3.159}$$

Obviously $W_n = 0$ when $\sin p\pi x/a$ or $\sin q\pi y/b$ is zero, hence in their normal modes plates have nodal lines. That is, lines on the mid-surface of the plate which are permanently at rest. For a plate in which $a/b = 1{\cdot}5$, the nodal lines in the first six modes are illustrated in Fig. 3.15.

As it merely involves fitting equation (3.155) to the boundary conditions, closed form solutions may be found for the frequency and mode shapes of rectangular

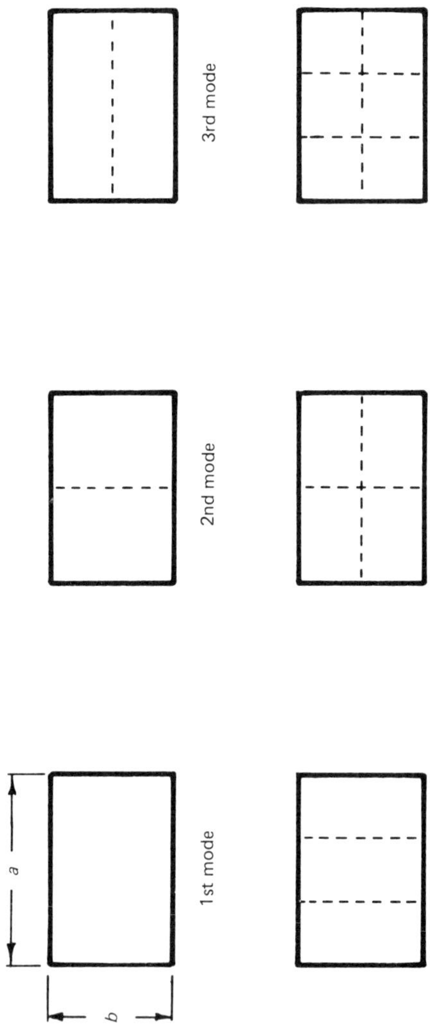

Fig. 3.15 Nodal lines (shown dotted) for the first six modes of a uniform rectangular plate, $a = 1{\cdot}5b$, simply supported on all edges

plates with two opposite edges simply supported. However, when the plate is supported in any other manner closed form solutions do not exist and other means must be employed in estimating their natural frequencies. Some of these will be discussed in chapters 4 and 5.

Exercises

1. Derive the frequency equation for a rectangular plate with two opposite edges simply supported and the other two edges supported as follows:
   (*a*) both clamped
   (*b*) both free
   (*c*) one clamped, the other simply supported
2. For symmetrical vibrations of a circular plate show that the governing differential equation is

$$\left(\frac{d^2}{dr^2} + \frac{1}{r}\frac{d}{dr} + \lambda^2\right)\left(\frac{d^2}{dr^2} + \frac{1}{r}\frac{d}{dr} - \lambda^2\right)W = 0$$

   Hence show that

$$W = C_1 J_0(\lambda r) + C_2 Y_0(\lambda r) + C_3 J_0(i\lambda r) + C_4 Y_0(i\lambda r)$$

   where $J_0(x)$, $Y_0(x)$ are zero order Bessel functions of the first and second kind respectively.
3. Show, for a continuous uniform circular plate of radius a clamped on its perimeter, that the frequency of its fundamental mode is given by

$$\omega_1 = \left(\frac{3{\cdot}20}{a}\right)^2 \left(\frac{Ed^2}{12\rho(1 - \nu^2)}\right)^{1/2}$$

4. Show that the strain energy for a rectangular plate, with sides of length a and b and of uniform thickness d, is given by

$$V = \frac{D}{2}\int_0^a \int_0^b \left\{\left(\frac{\partial^2 w}{\partial x^2}\right)^2 + \left(\frac{\partial^2 w}{\partial y^2}\right)^2 + 2\nu\left(\frac{\partial^2 w}{\partial x^2}\right)\left(\frac{\partial^2 w}{\partial y^2}\right) + 2(1 - \nu)\left(\frac{\partial^2 w}{\partial x\,\partial y}\right)^2\right\} dx\,dy$$

   where

$$D = \frac{Ed^3}{12(1 - \nu^2)}$$

   Hence, using Hamilton's principle, derive the equations of motion and the boundary conditions for a thin plate simply supported on two opposite edges and firmly clamped on the other two.

References

AITKEN, A. C.
Determinants and matrices, Oliver & Boyd, 1946, p. 36.
BISHOP, R. E. D.
'The analysis of vibrating systems which embody beams in flexure', *Proc. Inst. Mech. Engrs.*, **169**, 1955, 1031.

BISHOP, R. E. D. and JOHNSON, D. C.
Vibration analysis tables, Cambridge University Press, 1956.
COWPER, G. R.
'The shear coefficient in Timoshenko's beam theory', *J. Appl. Mech. Trans. A.S.M.E.*, **33**, June 1966, 335.
HOHENEMSER, K. and PRAGER, W.
Dynamik der stabwerke, Springer, Berlin, 1933.
HUANG, T. C.
'Eigenvalues and modifying quotients of vibration of beams', *Report No. 25, Engineering Experiment Station*, University of Wisconsin, June, 1964.
RAYLEIGH, LORD
Theory of sound, 2nd edn., Macmillan, 1894, Vol. 1, (*a*) p. 278; (*b*) p. 293.
SUTHERLAND, J. C. and GOODMAN, L. E.
'Vibrations of prismatic bars including rotatory inertia and shear corrections', *Technical Report, Structural Research Laboratory* (Ser. 12), University of Illinois, April 15, 1951.
TIMOSHENKO, S.
'On the correction for shear of the differential equation for transverse vibration of prismatic bars', *Phil. Mag.* (Ser. 6), **41**, 1921, 744.
TIMOSHENKO, S.
'On the transverse vibration of bars of uniform cross-section', *Phil. Mag.* (Ser. 6), **43**, 1922, 125.
YOUNG, D. and FELGAR, R. P.
'Tables of characteristic functions representing normal modes of vibration of a beam', *Bur. Engng. Res. Bull.* 44, University of Texas, July 1, 1949.

Chapter 4

Piecewise representation of continuous systems

Many, if not most, physical systems which engineers design cannot be adequately represented by the relatively simple systems analysed in chapter 3. For example, the plates may be of irregular shape, weakened by holes, and reinforced by beams or flanges; the beams may be of varying cross-section with a curved neutral axis.

These problems were faced by aircraft designers and vibration analysts when it became necessary to predict the flutter speeds of aircraft in the 1930's. Since then, although the main advances in solution methods have resulted from the problems of the aero-space industry, designers in other branches of engineering have found it necessary to know natural frequencies of complex systems.

Here we shall deal with solution methods involving the piecewise representation of the continuous systems. The three most frequently used methods are known as (*a*) the lumped-mass, (*b*) the finite element, and (*c*) the finite difference, methods. Of these, the finite element method is probably the most popular and useful currently. We shall lead up to it through the static deformation analysis and the lumped mass vibration analysis of structures.

4.1 Lumped-mass representation

4.1.1 Static analysis of beam type structures

Vibration analysis by this method evolved directly from one of the matrix analysis methods used for calculating the distortion of structures, mainly aircraft structures, resulting from the application of steady loads. The contribution of Argyris (1954, 1955) is outstanding in this field. In this approach the force versus displacement relationship for an element of the structure, say a beam, is given by

$$\mathrm{f}_i = \mathrm{s}_i \mathrm{u}_i \tag{4.1}$$

where s is the stiffness matrix of the element, u is a vector of displacements and f is the vector of forces which must be applied in the directions of the displacements to cause them. The subscript *i* indicates that these entities relate to the *i*'th element in the structure, the coordinate directions being local to the element, that is they would move with the beam if it were placed in a different orientation in space.

For a uniform straight beam subjected to forces and moments only at its ends, equation (4.1) in full becomes

$$\begin{bmatrix} N_1 \\ Q_1 \\ M_1 \\ N_2 \\ Q_2 \\ M_2 \end{bmatrix} = \begin{bmatrix} \frac{EA}{L} & 0 & 0 & -\frac{EA}{L} & 0 & 0 \\ 0 & \frac{12EI}{L^3} & \frac{6EI}{L^2} & 0 & -\frac{12EI}{L^3} & \frac{6EI}{L^2} \\ 0 & \frac{6EI}{L^2} & \frac{4EI}{L} & 0 & -\frac{6EI}{L^2} & \frac{2EI}{L} \\ -\frac{EA}{L} & 0 & 0 & \frac{EA}{L} & 0 & 0 \\ 0 & -\frac{12EI}{L^3} & -\frac{6EI}{L^2} & 0 & \frac{12EI}{L^3} & -\frac{6EI}{L^2} \\ 0 & \frac{6EI}{L^2} & \frac{2EI}{L} & 0 & -\frac{6EI}{L^2} & \frac{4EI}{L} \end{bmatrix} \begin{bmatrix} u_1 \\ v_1 \\ \theta_1 \\ u_2 \\ v_2 \\ \theta_2 \end{bmatrix} \tag{4.2}$$

where E is Young's modulus
A is the beam cross-sectional area
I is its second moment of area about the neutral axis
L is its length

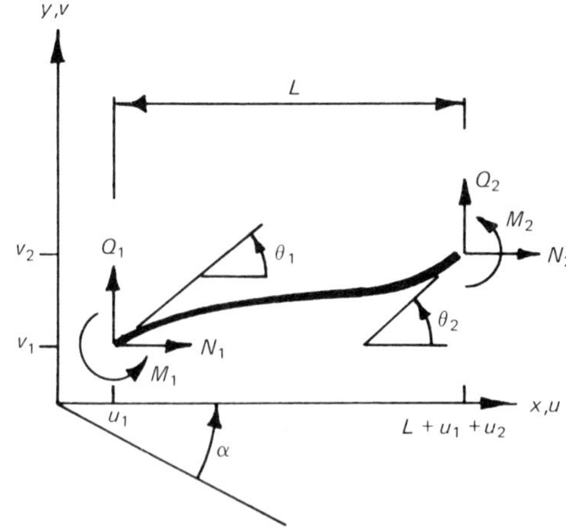

Fig. 4.1 Positive sign convention

and the directions of the forces and displacements in the local x, y coordinate system are given in Fig. 4.1. α is the angular position of the neutral axis of the beam, in another coordinate system.

Exercise

Show that the square matrix of equation (4.2) may be found from the corresponding dynamic stiffness matrix by letting the angular frequency ω approach zero.

Consider a structure composed of a number of these beam elements connected together at their end points or nodes; we must ensure that the end deflections of adjoining beams, at their common node, are compatible. This may be achieved most simply by setting up a global coordinate system for the complete structure and expressing the local coordinates of each beam in terms of it. For example, consider the portal frame shown in Fig. 4.2; comprised of three beams, each with a different stiffness, loaded at node B with a force P and at node C with a couple M. We shall take the global coordinate system as X, Y; the displacements in it being U, V, Θ. Then, in terms of the global displacements, the local displacements are given by

$$\begin{bmatrix} u_1{}^1 \\ v_1{}^1 \\ \theta_1{}^1 \\ u_2{}^1 \\ v_2{}^1 \\ \theta_2{}^1 \\ u_1{}^2 \\ v_1{}^2 \\ \theta_1{}^2 \\ u_2{}^2 \\ v_2{}^2 \\ \theta_2{}^2 \\ u_1{}^3 \\ v_1{}^3 \\ \theta_1{}^3 \\ u_2{}^3 \\ v_2{}^3 \\ \theta_2{}^3 \end{bmatrix} = \begin{bmatrix} 0 & 1 & 0 & 0 & 0 & 0 & 0 & 0 & 0 & 0 & 0 & 0 \\ -1 & 0 & 0 & 0 & 0 & 0 & 0 & 0 & 0 & 0 & 0 & 0 \\ 0 & 0 & 1 & 0 & 0 & 0 & 0 & 0 & 0 & 0 & 0 & 0 \\ 0 & 0 & 0 & 0 & 1 & 0 & 0 & 0 & 0 & 0 & 0 & 0 \\ 0 & 0 & 0 & -1 & 0 & 0 & 0 & 0 & 0 & 0 & 0 & 0 \\ 0 & 0 & 0 & 0 & 0 & 1 & 0 & 0 & 0 & 0 & 0 & 0 \\ 0 & 0 & 0 & 0 & 0 & 0 & 1 & 0 & 0 & 0 & 0 & 0 \\ 0 & 0 & 0 & 0 & 0 & 0 & 0 & 1 & 0 & 0 & 0 & 0 \\ 0 & 0 & 0 & 0 & 0 & 0 & 0 & 0 & 1 & 0 & 0 & 0 \\ 0 & 0 & 0 & 1 & 0 & 0 & 0 & 0 & 0 & 0 & 0 & 0 \\ 0 & 0 & 0 & 0 & 1 & 0 & 0 & 0 & 0 & 0 & 0 & 0 \\ 0 & 0 & 0 & 0 & 0 & 1 & 0 & 0 & 0 & 0 & 0 & 0 \\ 0 & 0 & 0 & 0 & 0 & 0 & 0 & 0 & 0 & 0 & 1 & 0 \\ 0 & 0 & 0 & 0 & 0 & 0 & 0 & 0 & 0 & -1 & 0 & 0 \\ 0 & 0 & 0 & 0 & 0 & 0 & 0 & 0 & 0 & 0 & 0 & 1 \\ 0 & 0 & 0 & 0 & 0 & 0 & 1 & 0 & 0 & 0 & 0 & 0 \\ 0 & 0 & 0 & 0 & 0 & 0 & -1 & 0 & 0 & 0 & 0 & 0 \\ 0 & 0 & 0 & 0 & 0 & 0 & 0 & 0 & 1 & 0 & 0 & 0 \end{bmatrix} \begin{bmatrix} U_A \\ V_A \\ \Theta_A \\ U_B \\ V_B \\ \Theta_B \\ U_C \\ V_C \\ \Theta_C \\ U_D \\ V_D \\ \Theta_D \end{bmatrix} \qquad (4.3)$$

or

u = CU which is the compatibility relationship.

For the local displacements, the superscripts give the beam numbers and the subscripts the node number in the local system.

Note that for beams 1 and 3 α has been taken as 90 degrees and for beam 2 it has been taken as zero.

In addition to compatibility, equilibrium must exist at each node of the complete structure. Again this is most simply achieved in the global coordinate system.

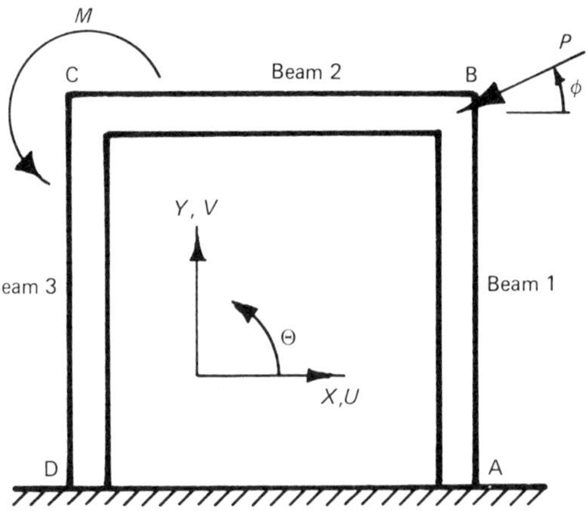

Fig. 4.2 A portal frame

The magnitude of the potential energy of a system does not depend upon the particular coordinate system it is expressed in, therefore,

$$F^TU = f^Tu$$

$$= f^TCU \tag{4.4}$$

That is,

$$F = C^Tf \tag{4.5}$$

which is the equilibrium relationship.

Exercise

For the above structure, confirm by equilibrium considerations at each node, that F, the vector of applied forces in the global coordinate system is related to f the vector of applied forces in the local coordinate system by

$$F = C^Tf$$

where C^T is the transpose of the connection matrix C in equation (4.3).

Equations (4.3) and (4.5) may be used to transform the overall force versus displacement relationship for the complete structure from that in the local coordinate system to its global coordinate system form, as follows. In local coordinates it is

$$\begin{bmatrix} f_1 \\ f_2 \\ f_3 \end{bmatrix} = \begin{bmatrix} s_1 & 0 & 0 \\ 0 & s_2 & 0 \\ 0 & 0 & s_3 \end{bmatrix} \begin{bmatrix} u_1 \\ u_2 \\ u_3 \end{bmatrix} \tag{4.6}$$

where f_i, s_i, and u_i are defined in equation (4.1). Let equation (4.6) be represented by

$$f = su \tag{4.7}$$

Transforming f and u to global coordinates we get

$$F = C^T f = C^T s C U \tag{4.8}$$

which may be written in the form

$$F = SU \tag{4.9}$$

where S the stiffness matrix in the global coordinate system is given by

$$S = C^T s C \tag{4.10}$$

Note: *this is a coordinate transformation of the type discussed in section 2.8.3.*

In general, not all of the components of the force vector F will be known initially but in that case the corresponding components of the displacement vector U will be known. Hence we may partition equation (4.9), into the form

$$\begin{bmatrix} F_\alpha \\ \hline F_\beta \end{bmatrix} = \left[\begin{array}{c|c} S_{\alpha\alpha} & S_{\alpha\beta} \\ \hline S_{\beta\alpha} & S_{\beta\beta} \end{array}\right] \begin{bmatrix} U_\alpha \\ \hline U_\beta \end{bmatrix} \tag{4.11}$$

where F_α and U_β are known initially. Equation (4.11) may be solved to give

$$\left.\begin{aligned} U_\alpha &= S_{\alpha\alpha}^{-1}[F_\alpha - S_{\alpha\beta} U_\beta] \\ \text{then} \quad F_\beta &= S_{\beta\alpha} U_\alpha + S_{\beta\beta} U_\beta \\ &= \{S_{\beta\beta} - S_{\beta\alpha} S_{\alpha\alpha}^{-1} S_{\alpha\beta}\} U_\beta + S_{\beta\alpha} S_{\alpha\alpha}^{-1} F_\alpha \end{aligned}\right\} \tag{4.12}$$

For the system shown in Fig. 4.2

$$F_\alpha = \begin{bmatrix} {}_BF_X \\ {}_BF_Y \\ {}_BM \\ {}_CF_X \\ {}_CF_Y \\ {}_CM \end{bmatrix} = \begin{bmatrix} -P\cos\phi \\ -P\sin\phi \\ 0 \\ 0 \\ 0 \\ M \end{bmatrix} \quad F_\beta = \begin{bmatrix} {}_AF_X \\ {}_AF_Y \\ {}_AM \\ {}_DF_X \\ {}_DF_Y \\ {}_DM \end{bmatrix} \quad U_\alpha = \begin{bmatrix} U_B \\ V_B \\ \Theta_B \\ U_C \\ V_C \\ \Theta_C \end{bmatrix} \quad U_\beta = \begin{bmatrix} U_A \\ V_A \\ \Theta_A \\ U_D \\ V_D \\ \Theta_D \end{bmatrix} = \begin{bmatrix} 0 \\ 0 \\ 0 \\ 0 \\ 0 \\ 0 \end{bmatrix} \tag{4.13}$$

The procedure expressed in equation (4.12) is very useful in reducing out unwanted coordinates in vibration problems; it has been used in section 3.4.1, and will be used again in the next section, for this purpose.

From the foregoing we can, in principle, now solve any structure of connected beams, given the element stiffness matrices, the compatibility relations at their connections, and the applied loads.

4.1.2 Vibration of a simple structure

We are now in a position to set up the equations of motion for a simple structure, such as a simple portal frame. Referring to Fig. 4.2, let the beams be massless and let masses m_B and m_C be attached firmly at the joints B and C respectively. Using the results of the foregoing analysis to find the restoring forces, the equations of motion, in matrix form, become

$$\mathrm{m}\ddot{\mathrm{U}} = -\mathrm{SU} \tag{4.14}$$

where m is the mass matrix for the complete system, U is the displacement vector in the global coordinate system and thus the set of generalised coordinates for the system, denotes the differentiation with respect to time. As this represents a set of linear second-order homogeneous differential equations we may take the solution in the form

$$\mathrm{U} = \mathrm{A}\sin(\omega \mathrm{t} + \phi) \tag{4.15}$$

where A is a vector of maximum displacements. Thus

$$[\mathrm{S} - \omega^2\mathrm{m}]\mathrm{A} = 0 \tag{4.16}$$

In many cases, some of the generalised coordinates are constrained to be permanently zero and some of the masses or moments of inertia are zero. We may, therefore, reduce the size of the matrices by taking cognisance of these system constraints, as follows:

Let $\quad \mathrm{U} = \begin{bmatrix} \mathrm{U}_\alpha \\ \mathrm{U}_\beta \\ \mathrm{U}_\gamma \end{bmatrix}$

where U_α are the generalised coordinates associated with mass points
U_β are the unconstrained generalised coordinates not associated with masses
and U_γ are the generalised coordinates which are constrained to be zero.

Then the rows and columns of S and m associated with U_γ may be eliminated leaving

$$\begin{bmatrix} \mathrm{S}_{\alpha\alpha} - \omega^2\mathrm{m}_{\alpha\alpha} & \mathrm{S}_{\alpha\beta} \\ \mathrm{S}_{\beta\alpha} & \mathrm{S}_{\beta\beta} \end{bmatrix} \begin{bmatrix} \mathrm{A}_\alpha \\ \mathrm{A}_\beta \end{bmatrix} = 0 \tag{4.17}$$

A_β is easily solved for, to give

$$\mathrm{A}_\beta = -\mathrm{S}_{\beta\beta}{}^{-1}\mathrm{S}_{\beta\alpha}\mathrm{A}_\alpha$$

hence equation (4.17) reduces to

$$[\mathrm{S}_{\alpha\alpha} - \mathrm{S}_{\alpha\beta}\mathrm{S}_{\beta\beta}{}^{-1}\mathrm{S}_{\beta\alpha} - \omega^2\mathrm{m}_{\alpha\alpha}]\mathrm{A}_\alpha = 0 \tag{4.18}$$

Thus we have the frequency equation

$$|\mathrm{K} - \omega^2\mathrm{m}_{\alpha\alpha}| = 0 \tag{4.19}$$

where K is the reduced stiffness matrix for the system and

$$\mathrm{K} = \mathrm{S}_{\alpha\alpha} - \mathrm{S}_{\alpha\beta}\mathrm{S}_{\beta\beta}{}^{-1}\mathrm{S}_{\beta\alpha} \tag{4.20}$$

Exercises

1. The system shown in Fig. 4.3 consists of a uniform steel plane framework which is supported on immovable pin joints at the base of each vertical beam. A rigid body of mass M and relevant moment of inertia I is carried at the free

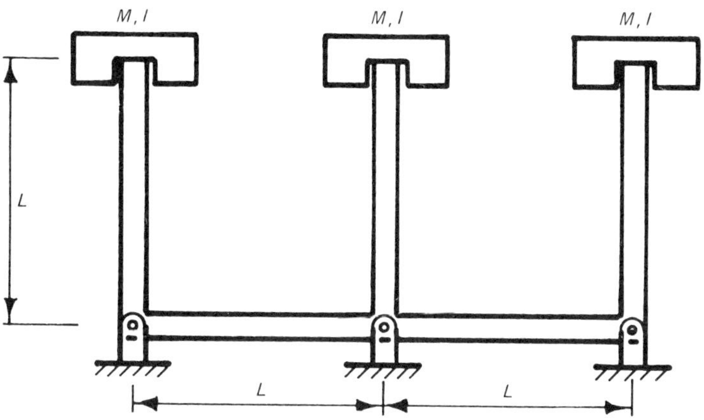

Fig. 4.3

end of each vertical beam. On the assumption that the mass of the framework may be neglected, derive, for bending oscillations in its own plane, the equations of motion for the system.

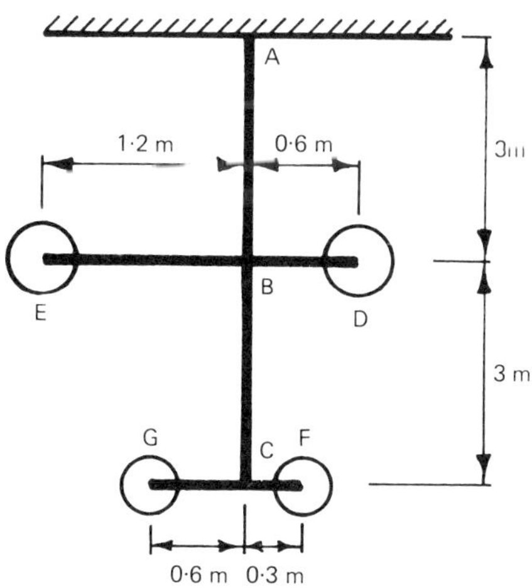

Fig. 4.4

2. The system shown in Fig. 4.4 is comprised as follows:

   (*i*) a horizontal elastic shaft ABC of length 6 m has flexural rigidity $2{\cdot}5 \times 10^7$ N m^2 and torsional rigidity $6{\cdot}2 \times 10^6$ N m^2. You may neglect the mass of this shaft.

(*ii*) masses of 500 kg at D and E which are attached to the horizontal elastic shaft EBD, EBD being rigidly attached to ABC at B and having flexural rigidity 5×10^7 N m^2.

(*iii*) masses of 250 kg at F and G which are attached to the horizontal elastic shaft FCG, FCG being rigidly attached to ABC at C and having flexural rigidity 5×10^7 N m^2.

For small amplitude vibration, the masses moving in the vertical plane only, set up the dynamic stiffness matrix for the system.

4.1.3 The lumped-mass method

Here we use the methods developed above, but instead of assuming that the beams are massless, we assume their masses to be lumped at discrete points, each connected by a massless elastic beam element. Figure 4.5 shows two ways in which a uniform cantilevered beam could be represented. That is we could assume half the

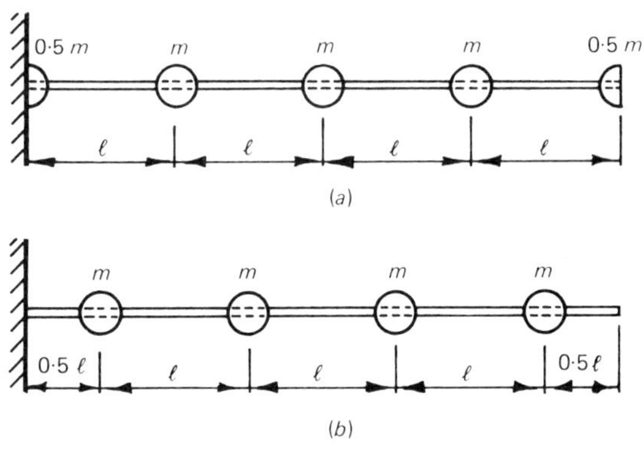

Fig. 4.5 Lumped mass representation of beams

mass of an element to be concentrated at each of its ends, as shown in Fig. 4.5(*a*) or concentrated at its centre as shown in Fig. 4.5(*b*): both are approximations. For a uniform beam, it is possible to examine the accuracy of each by finite difference calculus, this is illustrated in chapter 6.

Non-uniform beams, or structures which may be approximated by elemental uniform beams, are similarly represented. That is, the non-uniform beam is split into a number of lengths, elements, and the mass of each element calculated. Depending upon the positioning of the masses, the stiffness matrices for the intervening lengths of the beam are calculated. The system stiffness and mass matrices may then be found by the methods given in section 4.1.2.

Exercise

Derive the mass and stiffness matrices, for bending, of a simply supported uniform beam of length 3 m, mass per unit length 15 kg/m, and bending rigidity 4000 N m^2. Represent the beam by four elements and calculate the first two natural frequencies assuming the masses to be (*i*) concentrated at the mid point of the element and (*ii*) placed half at each end point. Check the natural frequencies given against those found by the methods of chapter 3.

4.2 Finite element method

In effect, we have, in the foregoing end mass method, assumed that the potential energy stored in an element is that due to generalised forces at the nodes. That is, the beam element takes up the form of a massless beam subjected only to end moments and forces, whereas the kinetic energy is based upon an arbitrary sharing of the mass between the ends, regardless of the variation of vibratory amplitude along the element. It would appear, therefore, that the mathematical model, with its distributed elasticity, should give a closer representation of the potential energy than of the kinetic energy. The kinetic energy would be better represented if the mass were considered distributed but the problem is how to accomplish this without reverting to the 'exact' method, or to approximating the eigenfunctions for use in energy solutions by such procedures as the Rayleigh-Ritz method of chapter 5. The solution is simple and involves using the so-called 'finite element' approach. Clough (1960) introduced the concept whereby continuous systems were represented by a connected set of finite elements, the synthesis being possible by Argyris's method. In Clough's original paper the method was applied to the analysis of static plane stress problems, but since then it has been applied to a wide range of static and dynamic problems.

Continuing our discussion of beam-type structures, the finite element approach is similar to the lumped-mass approach. The structure is considered to be composed of a number of elements of finite size. Stiffness and mass matrices are developed for each of these elements, being now relationships between a set of generalised forces due to potential energy and the generalised displacements at each node, and a set of generalised forces due to kinetic energy and the generalised velocities at each node, respectively. Expressing this symbolically for a simple beam element gives

Potential energy $$V = \int_0^\ell \frac{EI}{2}\left(\frac{\partial^2 y}{\partial x^2}\right)^2 \mathrm{d}x$$

Kinetic energy $$T = \int_0^\ell \frac{\rho A}{2}\left(\frac{\partial y}{\partial t}\right)^2 \mathrm{d}x \qquad (4.21)$$

Let $$y = f(\mathrm{v}, x)\sin \omega t \qquad (4.22)$$

where v is the vector of generalised displacements at the nodes of the element, then, applying Lagrange's equations of motion, for small vibrations we obtain

$$[\mathrm{S} - \omega^2\mathrm{m}]\mathrm{v} = \mathrm{Q} \qquad (4.23)$$

for the element.

If we select the deformed form $f(\mathrm{v}, x)$ to be other than the true vibratory form, the forces represented on the L.H.S. of equation (4.23) will not coincide exactly with the magnitudes which exist in the physical structure. The exact vibratory form is usually complicated and generally we utilise simple displacement functions which are approximately true, hence the above discrepancy is not uncommon. However, as will be illustrated, solutions with adequate accuracy for practical purposes are given by these simple displacement functions.

4.2.1 Vibration of beams

Consider a specific example: to find the lowest flexural natural frequency of a uniform beam. Let the maximum transverse displacement of a beam element be represented by a displacement function, a polynomial of the form

$$V = a_0 + a_1 x + a_2 x^2 + a_3 x^3 = \mathrm{Pa} \tag{4.24}$$

where $\mathrm{P} = [1, x, x^2, x^3]$

We have chosen a polynomial with four arbitrary constants, a_i, because a beam element has four generalised coordinates. The a_i may be expressed in terms of the generalised displacements V_i, thus

$$\begin{bmatrix} V_1 \\ \Theta_1 \\ V_2 \\ \Theta_2 \end{bmatrix} = \begin{bmatrix} 1 & 0 & 0 & 0 \\ 0 & 1 & 0 & 0 \\ 1 & \ell & \ell^2 & \ell^3 \\ 0 & 1 & 2\ell & 3\ell^2 \end{bmatrix} \begin{bmatrix} a_0 \\ a_1 \\ a_2 \\ a_3 \end{bmatrix} \tag{4.25}$$

where ℓ is the length of the element, or

$$\mathrm{V} = \mathrm{C}_1 \mathrm{a}$$

Therefore $\mathrm{a} = \mathrm{C}_1^{-1}\mathrm{V}$

$$= \begin{bmatrix} 1 & 0 & 0 & 0 \\ 0 & 1 & 0 & 0 \\ -\dfrac{3}{\ell^2} & -\dfrac{2}{\ell} & \dfrac{3}{\ell^2} & -\dfrac{1}{\ell} \\ \dfrac{2}{\ell^3} & \dfrac{1}{\ell^2} & -\dfrac{2}{\ell^3} & \dfrac{1}{\ell^2} \end{bmatrix} \begin{bmatrix} V_1 \\ \Theta_1 \\ V_2 \\ \Theta_2 \end{bmatrix} \tag{4.26}$$

That is

$$v = \mathrm{PC}_1^{-1}\mathrm{V} \sin \omega t \tag{4.27}$$

and

$$\frac{\mathrm{d}^2 v}{\mathrm{d}x^2} = \frac{\mathrm{d}^2\mathrm{P}}{\mathrm{d}x^2} \mathrm{C}_1^{-1}\mathrm{V} \sin \omega t$$

$$= [0\ 0\ 2\ 6x]\mathrm{C}_1^{-1}\mathrm{V} \sin \omega t \tag{4.28}$$

in the present case.

Hence the kinetic energy of the element is

$$T = 0{\cdot}5 \int_0^\ell \rho A \omega^2 \cos^2 \omega t\, \mathrm{V}^{\mathrm{T}}[\mathrm{C}_1^{-1}]^{\mathrm{T}}\mathrm{P}^{\mathrm{T}}\mathrm{P}\mathrm{C}_1^{-1}\mathrm{V}\, \mathrm{d}x$$

For a uniform beam, P alone are functions of x, therefore

$$T = 0{\cdot}5\rho A\omega^2 \cos^2 \omega t \; \mathrm{V}^{\mathrm{T}}[\mathrm{C}_1{}^{-1}]^{\mathrm{T}} \int_0^{\ell} \mathrm{P}^{\mathrm{T}}\mathrm{P} \, \mathrm{d}x \; \mathrm{C}_1{}^{-1}\mathrm{V} \tag{4.29}$$

When the integration has been performed and the matrices multiplied together we find

$$\begin{aligned} T &= 0{\cdot}5 \, \frac{\rho A \ell \omega^2 \cos^2 \omega t}{420} \, \mathrm{V}^{\mathrm{T}} \begin{bmatrix} 156 & 22\ell & 54 & -13\ell \\ 22\ell & 4\ell^2 & 13\ell & -3\ell^2 \\ 54 & 13\ell & 156 & -22\ell \\ -13\ell & -3\ell^2 & -22\ell & 4\ell^2 \end{bmatrix} \mathrm{V} \\ &= 0{\cdot}5\omega^2 \cos \omega t \; \mathrm{V}^{\mathrm{T}}\mathrm{m}\mathrm{V} \cos \omega t \end{aligned} \tag{4.30}$$

where m is the mass matrix for the element. The potential energy of the element is

$$\begin{aligned} V &= 0{\cdot}5 \int_0^{\ell} EI \sin^2 \omega t \; \mathrm{V}^{\mathrm{T}}[\mathrm{C}_1{}^{-1}]^{\mathrm{T}} \left[\frac{\mathrm{d}^2\mathrm{P}}{\mathrm{d}x^2}\right]^{\mathrm{T}} \frac{\mathrm{d}^2\mathrm{P}}{\mathrm{d}x^2} \; \mathrm{C}_1{}^{-1}\mathrm{V} \, \mathrm{d}x \\ &= 0{\cdot}5 \, \frac{EI}{\ell^3} \sin \omega t \; \mathrm{V}^{\mathrm{T}} \begin{bmatrix} 12 & 6\ell & -12 & 6\ell \\ 6\ell & 4\ell^2 & -6\ell & 2\ell^2 \\ -12 & -6\ell & 12 & -6\ell \\ 6\ell & 2\ell^2 & -6\ell & 4\ell^2 \end{bmatrix} \mathrm{V} \sin \omega t \\ &= 0{\cdot}5 \sin \omega t \; \mathrm{V}^{\mathrm{T}}\mathrm{S}\mathrm{V} \sin \omega t \end{aligned} \tag{4.31}$$

It will be noted that the coefficients of the stiffness matrix S in (4.31) are identical to the corresponding bending terms of the static matrix (4.2). This was to be expected since the assumed displacement function (4.24) corresponds with the static displacement of a beam loaded by a linearly varying bending moment.

Let our beam be simply supported at each end and be represented by one element. Then for free vibration

$$[\mathrm{S} - \omega^2\mathrm{m}]\mathrm{V} = 0 \tag{4.32}$$

and as $V_1 = V_2 = 0$ this reduces to

$$\left\{\frac{EI}{\ell}\begin{bmatrix} 4 & 2 \\ 2 & 4 \end{bmatrix} - \frac{\rho A\omega^2\ell^3}{420}\begin{bmatrix} 4 & -3 \\ -3 & 4 \end{bmatrix}\right\}\begin{bmatrix} \Theta_1 \\ \Theta_2 \end{bmatrix} = 0 \tag{4.33}$$

Letting
$$\frac{\rho A\ell^4\omega^2}{420EI} = \lambda$$

the frequency equation becomes

$$\begin{vmatrix} 4 - 4\lambda & 2 + 3\lambda \\ 2 + 3\lambda & 4 - 4\lambda \end{vmatrix} = 0 \tag{4.34}$$

therefore, $$\lambda = \frac{2}{7} \text{ or } 6$$

therefore, $$\frac{\rho A \ell^4 \omega^2}{EI} = 120 \text{ or } 2520$$

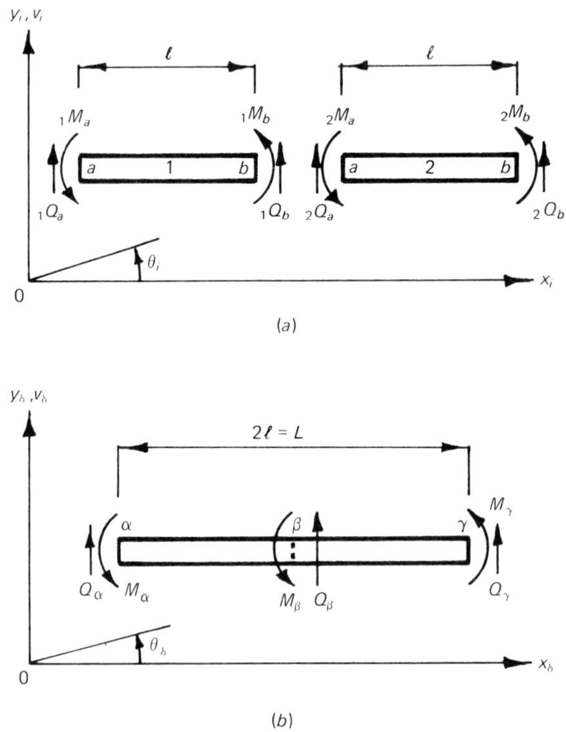

Fig. 4.6 Coordinate systems for beam finite elements:
(*a*) Local coordinate system
(*b*) Global coordinate system

The corresponding exact values are 97·41, 1559, 7890. That is, the one element approximation gives the lowest eigenvalue with an error of about 23 per cent, 10 per cent on frequency.

We may improve the above approximation by representing the beam as two equal length elements, as follows

Figure 4.6 shows the elements in both the local and the global coordinate systems, and the notation we shall use.

In the local coordinate system, the equations of motion give the dynamic force versus deformation relations

$$\begin{bmatrix} {}_1Q_a \\ {}_1M_a \\ {}_1Q_b \\ {}_1M_b \\ {}_2Q_a \\ {}_2M_a \\ {}_2Q_b \\ {}_2M_b \end{bmatrix} = \left\{ -\omega^2 \begin{bmatrix} {}_1m & 0 \\ 0 & {}_2m \end{bmatrix} + \begin{bmatrix} {}_1S & 0 \\ 0 & {}_2S \end{bmatrix} \right\} \begin{bmatrix} {}_1v_a \\ {}_1\theta_a \\ {}_1v_b \\ {}_1\theta_b \\ {}_2v_a \\ {}_2\theta_a \\ {}_2v_b \\ {}_2\theta_b \end{bmatrix}$$

or

$$\mathrm{Q}_a = \mathrm{R}\mathrm{v}_a \tag{4.35}$$

We may transform to the global coordinate system by invoking the compatability and equilibrium conditions, given by

$$\mathrm{v}_a = \begin{bmatrix} {}_1v_a \\ {}_1\theta_a \\ {}_1v_b \\ {}_1\theta_b \\ {}_2v_a \\ {}_2\theta_a \\ {}_2v_b \\ {}_2\theta_b \end{bmatrix} = \begin{bmatrix} 1 & 0 & 0 & 0 & 0 & 0 \\ 0 & 1 & 0 & 0 & 0 & 0 \\ 0 & 0 & 1 & 0 & 0 & 0 \\ 0 & 0 & 0 & 1 & 0 & 0 \\ 0 & 0 & 1 & 0 & 0 & 0 \\ 0 & 0 & 0 & 1 & 0 & 0 \\ 0 & 0 & 0 & 0 & 1 & 0 \\ 0 & 0 & 0 & 0 & 0 & 1 \end{bmatrix} \begin{bmatrix} v_\alpha \\ \theta_\alpha \\ v_\beta \\ \theta_\beta \\ v_\gamma \\ \theta_\gamma \end{bmatrix} = \mathrm{C}\mathrm{v}_\delta \tag{4.36}$$

and

$$\mathrm{Q}_\delta = \begin{bmatrix} Q_\alpha \\ M_\alpha \\ Q_\beta \\ M_\beta \\ Q_\gamma \\ M_\gamma \end{bmatrix} = \begin{bmatrix} 1 & 0 & 0 & 0 & 0 & 0 & 0 & 0 \\ 0 & 1 & 0 & 0 & 0 & 0 & 0 & 0 \\ 0 & 0 & 1 & 0 & 1 & 0 & 0 & 0 \\ 0 & 0 & 0 & 1 & 0 & 1 & 0 & 0 \\ 0 & 0 & 0 & 0 & 0 & 0 & 1 & 0 \\ 0 & 0 & 0 & 0 & 0 & 0 & 0 & 1 \end{bmatrix} \begin{bmatrix} {}_1Q_a \\ {}_1M_a \\ {}_1Q_b \\ {}_1M_b \\ {}_2Q_a \\ {}_2M_a \\ {}_2Q_b \\ {}_2M_b \end{bmatrix} = \mathrm{C}^{\mathrm{T}}\mathrm{Q}_a \tag{4.37}$$

That is

$$\mathrm{Q}_\delta = \mathrm{C}^{\mathrm{T}}\mathrm{R}\mathrm{C}\mathrm{v}_\delta \tag{4.38}$$

This may be reduced by putting $v_\alpha = v_\gamma = 0$ to give the frequency equation

$$\begin{vmatrix} \ell^2(4 - 4\lambda) & -\ell(6 + 13\lambda) & \ell^2(2 + 3\lambda) & 0 \\ -\ell(6 + 13\lambda) & 24 - 312\lambda & 0 & \ell(6 + 13\lambda) \\ \ell^2(2 + 3\lambda) & 0 & \ell^2(8 - 8\lambda) & \ell^2(2 + 3\lambda) \\ 0 & \ell(6 + 13\lambda) & \ell^2(2 + 3\lambda) & \ell^2(4 - 4\lambda) \end{vmatrix} = 0 \qquad (4.39)$$

where
$$\lambda = \frac{\rho A \ell^4 \omega^2}{420EI}$$

Equation (4.39) may be solved easily using a digital computer, to give

$$\frac{16\rho A\ell^4\omega^2}{EI} = \frac{\rho A L^4 \omega^2}{EI} = 98{\cdot}18,\ 1920,\ 12\,130,\ \text{or}\ 40\,320$$

compared with the first four exact values of

$$97{\cdot}41,\ 1559,\ 7890,\ \text{and}\ 24\,940$$

The error in the lowest natural frequency is now about 0·4 per cent high, the second natural frequency being estimated too high by about 10 per cent.

Exercises

1. Assume the cubic displacement function given in equation (4.24), and find the eigenvalues for a uniform cantilevered beam of length L, representing the beam firstly by a single element and secondly by two equal length elements. Compare your answers with the exact solutions.
2. Find, by the finite element method, estimates of the first two eigenvalues for torsional vibrations of a uniform circular rod, clamped rigidly at one end and free at the other. Represent it by 1, 2, 3, and 4 elements and compare the eigenvalues so obtained with the exact values.

4.2.2 Shape functions

In the foregoing analysis we assumed the vibratory form of the element to be a cubic function of x, the distance from one of its ends. As an alternative, we could have assumed its vibratory form as a linear combination of the individual forms resulting from a unit displacement applied to each of its generalised coordinates. For the uniform beam element, those shown in Fig. 4.7 comprise such a set. They are called shape functions. That is we could describe the vibratory form of our beam element by

$$v = \left[1 - 3\left(\frac{x}{\ell}\right)^2 + 2\left(\frac{x}{\ell}\right)^3,\ \left(\frac{x}{\ell}\right)\left(1 - \frac{x}{\ell}\right)^2,\ 3\left(\frac{x}{\ell}\right)^2 - 2\left(\frac{x}{\ell}\right)^3,\ \left(\frac{x}{\ell}\right)^2\left(\frac{x}{\ell} - 1\right)\right]\begin{bmatrix} v_1 \\ \theta_1 \\ v_2 \\ \theta_2 \end{bmatrix} \qquad (4.40)$$

which is exactly the same form as equation (4.24) combined with equation (4.26).

Complete understanding of the influence of the selected displacement function or shape function on the results of most systems has not yet been achieved. Consequently, their selection for a particular system is still an art and many workers prefer to make the selection at the shape function stage, where they may also use their physical insight in guiding the selection.

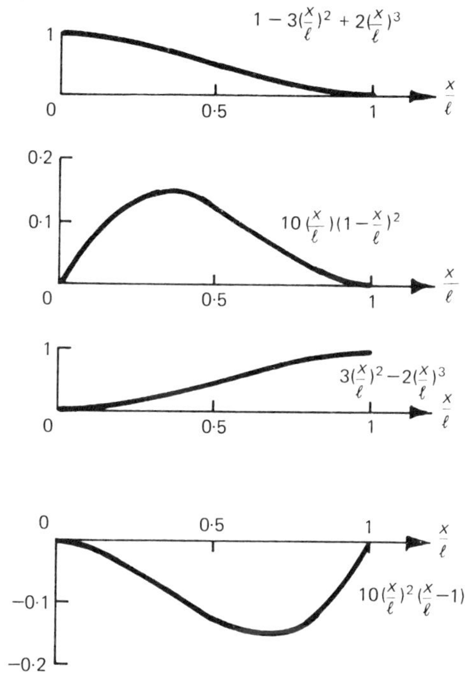

Fig. 4.7 Shape functions

Although the choice of the displacement or the shape functions is to some extent arbitrary, for the satisfactory prediction of natural frequencies and mode shapes both must allow

(*i*) each generalised displacement of the element to be varied independently,
(*ii*) each virtual rigid body displacement of the element to be imposed without changing the strain energy stored in the element,
(*iii*) the functions which define the generalised displacements to be continuous throughout the element.

Additionally, it would appear desirable to have the functions, which define the generalised displacements, continuous across boundaries between adjacent elements. If this continuity exists the functions are said to be conforming functions: if it does not, they are said to be non-conforming functions. For use in vibration problems, conforming functions have the desirable attribute of giving frequencies which converge monotonically from above to the exact values for a set of the lower natural frequencies, as the number of elements representing a particular system is increased. This is expected on the basis of the Rayleigh-Ritz principle, see section 5.1.3. That is, for an element the kinematic boundary conditions are satisfied, whether the boundary is an external system boundary or an artificial boundary created in the finite element process. Hence the frequencies are all upper bound. As the number of elements is increased, the number of degrees of freedom of the model system is increased; consequently, we would expect the lower natural frequencies of the model system to approach the lower natural frequencies of the actual system.

Non-conforming functions give eigenvalues which may be above or below the true values for the system represented, and convergence is not, in general, monotonic.

For a system represented by a given number of finite elements, the number of natural frequencies which may be found within a given accuracy increases as the accuracy with which the deflection function satisfies the governing differential equation increases. Cohen and McCallion (1969) illustrated this for beam system.

Exercise

Show that the displacement function given in equation (4.24) and the shape functions given in equation (4.40) are conforming functions and satisfy the conditions (*i*) to (*iii*) above.

Would you expect the eigenvalues found using these functions to be above or below the exact values? Check your expectation against the results found in the earlier exercise.

4.2.3 Vibration of plates

We shall consider the vibration of thin plates to illustrate, more clearly, the distinction between conforming and non-conforming functions.

It may be shown that the potential energy stored in a thin plate is

$$V = \int_A \frac{D}{2} \left\{ \left(\frac{\partial^2 w}{\partial x^2}\right)^2 + \left(\frac{\partial^2 w}{\partial y^2}\right)^2 + 2\nu \left(\frac{\partial^2 w}{\partial x^2}\right)\left(\frac{\partial^2 w}{\partial y^2}\right) + 2(1-\nu)\left(\frac{\partial^2 w}{\partial x\,\partial y}\right)^2 \right\} \mathrm{d}A \tag{4.41}$$

where $$D = \frac{Ed^3}{12\,(1-\nu^2)}, \quad \nu = \text{Poisson's ratio}$$

A is the area of the element and d is the plate thickness. The kinetic energy is obviously

$$T = \int_A \frac{\rho d}{2} \left(\frac{\partial w}{\partial t}\right)^2 \mathrm{d}A \tag{4.42}$$

$$\text{Let } w = W(x,y) \sin \omega t \tag{4.43}$$

where $W(x, y)$ is the vibratory form. We now wish to select a polynomial to represent the vibratory form, but firstly we must decide the element shape and the number of degrees of freedom it is to have. Let the element be rectangular and let it have nodes at each of its corners, that is points at which it is forced to be compatible with adjoining elements.

Let the vibratory form of the element be described in terms of twelve generalised displacements, three at each node, being the displacement, its first derivative with respect to x and its first derivative with respect to y. Therefore, we require a polynomial with twelve arbitrary constants. Dawe (1965) chose the following form,

$$W = a_0 + a_1 x + a_2 y + a_3 x^2 + a_4 xy + a_5 y^2 + a_6 x^3 + a_7 x^2 y + a_8 xy^2 + a_9 y^3 + a_{10} x^3 y + a_{11} xy^3 \quad (4.44)$$

from which we obtain

$$\frac{\partial W}{\partial x} = a_1 + 2a_3 x + a_4 y + 3a_6 x^2 + 2a_7 xy + a_8 y^2 + 3a_{10} x^2 y + a_{11} y^3 \quad (4.45)$$

and

$$\frac{\partial W}{\partial y} = a_2 + a_4 x + 2a_5 y + a_7 x^2 + 2a_8 xy + 3a_9 y^2 + a_{10} x^3 + 3a_{11} xy^2 \quad (4.46)$$

For W, $\partial W/\partial x$, or $\partial W/\partial y$ to be compatible along the whole of a common boundary between two common nodes of two elements, its variation along the boundary must be influenced only by the values of the generalised displacements at the common nodes. On any boundary parallel to the y-direction the value of W together with $\partial W/\partial y$ are determined by only four constants, that is the two nodal values of W and the two nodal values of $\partial W/\partial y$. A similar argument applies to W and $\partial W/\partial x$ for boundaries parallel to the x-direction. Therefore, in displacement and gradient along each boundary the elements conform continuously between two common nodes. However, the gradients normal to the boundaries are determined by four different constants, only two of which may be determined from the remaining two nodal generalised displacements. Thus the normal gradients depend also upon generalised displacements at non-common nodes and do not, therefore, in general, conform. Hence the function given in equation (4.44) is non-conforming.

Exercise

Show that this non-conforming displacement function, equation (4.44) satisfies the conditions laid down in section 4.2.2.

If this displacement function is substituted into the expressions for potential and kinetic energy and the necessary integration performed, we may proceed as in the case of beam elements. The element stiffness and mass matrices may again be combined to give the dynamic stiffness matrix for the element. The generalised displacements will, in this case, be the four deflections and eight first derivatives of it at the nodes. The generalised forces do not exist physically in this case; they will be abstract quantities which are sometimes referred to as fictitious generalised forces. Although abstract, they must, together with any external forces, be in equilibrium in the assemblage of elements which represent the physical plate. Thus the dynamic stiffness matrix, the generalised forces, and the generalised displacements for a complete system may be assembled in local coordinates and transformed to global coordinates exactly in the same manner as already described for beams. However, this requires an enormous amount of computation and the usefulness and power of the finite element method in solving vibration problems could not have been realised but for the development of high speed digital computers with large fast access stores. Using such a machine it was found that, for a rectangular plate 2 m by 1 m by 0·01 m thick, simply supported on all edges and represented by 16 equal-sized finite elements, the first four natural frequencies were in error by −3·6, −6·5, −5·3, and −4·4 per cent respectively. Hence, this non-conforming displacement function gives results with sufficient accuracy for most engineering purposes.

Let us now turn to consider a conforming shape function for a rectangular plate element of dimensions a by b, with three degrees of freedom per node,

$$\begin{aligned} W = {} & f_1(r)f_1(s)W_1 + f_3(r)f_1(s)\theta_{x1} + f_1(r)f_3(s)\theta_{y1} \\ & + f_2(r)f_1(s)W_2 + f_4(r)f_1(s)\theta_{x2} + f_2(r)f_3(s)\theta_{y2} \\ & + f_2(r)f_2(s)W_3 + f_4(r)f_2(s)\theta_{x3} + f_2(r)f_4(s)\theta_{y3} \\ & + f_1(r)f_2(s)W_4 + f_3(r)f_2(s)\theta_{x4} + f_1(r)f_4(s)\theta_{y4} \end{aligned} \tag{4.47}$$

where $r = x/a$, $s = y/b$. W_i is the deflection, θ_{xi} and θ_{yi} are the rotations at the i'th node. The nodes 1, 2, 3, and 4 have the coordinates (0, 0), (a, 0), (a, b) and (0, b) respectively. The functions $f_i(r)$ and $f_i(s)$ could take the following forms:

$$\left.\begin{aligned} f_1(r) &= 1 - 3r^2 + 2r^3 \\ f_2(r) &= 3r^2 - 2r^3 \\ f_3(r) &= r - 2r^2 + r^3 \\ f_4(r) &= -r^2 + r^3 \end{aligned}\right\} \tag{4.48}$$

the function $f_i(s)$ being formed in the same manner as $f_i(r)$ with r replaced by s.

Exercise

Sketch the twelve component shape functions of equation (4.46) and show that they lead to a conforming shape function. Check that they meet the conditions laid down in section 4.2.2.

If elements based upon this shape function are used in a computer program for the solution of the natural frequencies of the 2 m by 1 m by 0·01 m thick simply supported plate considered earlier, we find that the percentage errors in the first four natural frequencies are 3·2, 8·7, 12, and 1·4 when 16 finite elements are employed.

Thus, although an apparently better representation has been used, displacement and both first derivatives being continuous across the element boundaries, two of the frequencies are more in error than those found using the non-conforming displacement function.

4.3 Finite difference method

It is common knowledge that the slope of the tangent, or first derivative of the curve, at a point A on a continuous curve may be represented approximately by the slope of a chord to the curve between B and C in the vicinity of A; Fig. 4.8 illustrates it. We call the slope of the chord a finite difference approximation to the

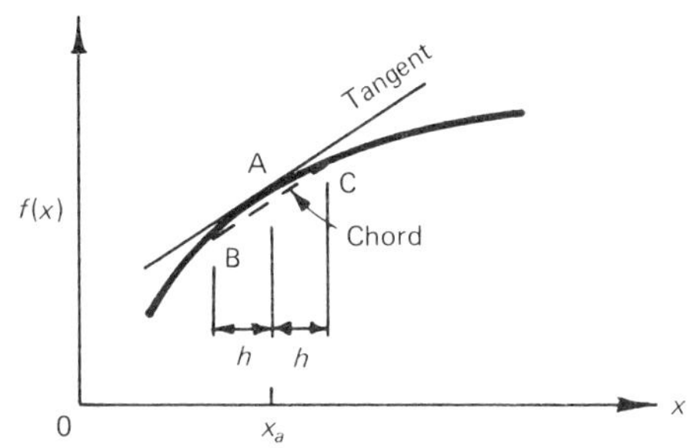

Fig. 4.8

first derivative of the function. It may be expressed symbolically by means of Taylor's expansion, as follows

$$f(x_a + h) = f(x_a) + hf'(x_a) + \frac{h^2}{2} f''(x_a) + \frac{h^3}{6} f'''(x_a) + \frac{h^4}{24} f''''(x_a) + \cdots$$

$$f(x_a - h) = f(x_a) - hf'(x_a) + \frac{h^2}{2} f''(x_a) - \frac{h^3}{6} f'''(x_a) + \frac{h^4}{24} f''''(x_a) - \cdots$$

Hence
$$f'(x_a) = \frac{f(x_a + h) - f(x_a - h)}{2h} - \frac{h^2}{6} f'''(x_a) - \cdots \tag{4.49}$$

where the primes denote differentiation with respect to x. Therefore, if h together with the third and higher derivatives of $f(x)$ are small we have as a good approximation

$$f'(x_a) = \frac{f(x_a + h) - f(x_a - h)}{2h} \tag{4.50}$$

The error terms being in order $h^2 f'''(x)$. Expression (4.50) is called a central finite difference formula for the first derivative. Sometimes in practice the use of a central difference formula is inconvenient and off-centred or one-sided difference formulae are used: for a given number of terms in the expression these are less accurate. For example, we would say that approximately

$$f'(x_a) = \frac{f(x_a + h) - f(x_a)}{h} \tag{4.51}$$

but the error terms in this case are of order $hf''(x)$.

Exercise

Show that a good finite difference approximation to the second derivative of $f(x)$ at x is

$$\{f(x + h) - 2f(x) + f(x - h)\}/h^2$$

4.3.1 Longitudinal vibration of a uniform bar

In finite difference form the equation of motion for longitudinal vibration of a uniform bar, equation (3.25), becomes

$$\frac{d^2U}{dx^2} + \alpha^2 U \simeq U_{i+1} - (2 - \alpha^2 h^2)U_i + U_{i-1} = 0 \tag{4.52}$$

To obtain this expression we imagine the bar to be marked off in n equal sections each of length h, the point i being at a typical boundary between two sections. The coordinates U_i describe the vibratory form. At the ends $x = 0$ and L let $i = 0$ and n respectively: at each end we must satisfy the boundary conditions.

Consider, as a simple example the case of a bar with fixed ends, $U_0 = U_n = 0$, then

$$\begin{bmatrix} 2-\lambda & -1 & 0 & 0 & \dots & . & . & . \\ -1 & 2-\lambda & -1 & 0 & \dots & . & . & . \\ 0 & -1 & 2-\lambda & -1 & \dots & . & . & . \\ \vdots & & & & & & & \\ . & . & . & . & \dots & -1 & 2-\lambda & -1 \\ . & . & . & . & \dots & 0 & -1 & 2-\lambda \end{bmatrix} \begin{bmatrix} U_1 \\ U_2 \\ U_3 \\ \vdots \\ U_{n-2} \\ U_{n-1} \end{bmatrix} = 0 \tag{4.53}$$

where $\lambda = \alpha^2 h^2$

or $$[A - \lambda I]U = 0 \tag{4.54}$$

A being a tridiagonal matrix with 2's on the leading diagonal and -1's on the adjacent diagonals. The eigenvalues λ may be found easily in this case, see chapter 6, but for more complicated systems, tapering bars for example, they may be found, using standard library routines, on a digital computer as described in chapter 5.

If one end has been permanently stress free, say at $i = 0$, then it could have been accounted for as follows. Using a central difference formula for higher accuracy we find

$$\left(\frac{\mathrm{d}U}{\mathrm{d}x}\right)_0 \simeq \frac{U_1 - U_{-1}}{2h} = 0 \tag{4.55}$$

but U_{-1} is beyond the end of the bar. We imagine the function $f(x)$ to be continuous beyond the end of the bar and create what is called a fictitious node at point $-h$; U_{-1} is then a fictitious displacement. The matrix equation governing the motion of this bar, free at $x = 0$, fixed at $x = L$ is then

$$\begin{bmatrix} -1 & 0 & 1 & 0 & 0 & \cdots & . & . & . \\ -1 & 2-\lambda & -1 & 0 & 0 & \cdots & . & . & . \\ 0 & -1 & 2-\lambda & -1 & 0 & \cdots & . & . & . \\ \vdots & & & & & & & & \\ . & . & . & . & & \cdots & -1 & 2-\lambda & -1 \\ . & . & . & . & & \cdots & 0 & -1 & 2-\lambda \end{bmatrix} \begin{bmatrix} U_{-1} \\ U_0 \\ U_1 \\ \vdots \\ U_{n-2} \\ U_{n-1} \end{bmatrix} = 0 \tag{4.56}$$

This may be reduced to the form (4.54), by eliminating U_{-1}, as follows. Let equation (4.56) be represented as

$$\begin{bmatrix} -1 & \mathrm{A}_{\alpha\beta} \\ \mathrm{A}_{\beta\alpha} & \mathrm{A}_{\beta\beta} \end{bmatrix} \begin{bmatrix} U_{-1} \\ \mathrm{U}_\beta \end{bmatrix} = 0 \tag{4.57}$$

Then $$-U_{-1} + \mathrm{A}_{\alpha\beta}\mathrm{U}_\beta = 0$$

giving $$U_{-1} = \mathrm{A}_{\alpha\beta}\mathrm{U}_\beta \tag{4.58}$$

hence $$\{\mathrm{A}_{\beta\alpha}\mathrm{A}_{\alpha\beta} + \mathrm{A}_{\beta\beta}\}\mathrm{U}_\beta = 0 \tag{4.59}$$

$$\begin{bmatrix} 2-\lambda & -2 & 0 & 0 & 0 & . & . & . & . \\ -1 & 2-\lambda & -1 & 0 & 0 & . & . & . & . \\ 0 & -1 & 2-\lambda & -1 & 0 & . & . & . & . \\ \vdots & & & & & & & & \\ . & . & . & . & . & 0 & -1 & 2-\lambda & -1 \\ . & . & . & . & . & 0 & 0 & -1 & 2-\lambda \end{bmatrix} \begin{bmatrix} U_0 \\ U_1 \\ U_2 \\ \vdots \\ U_{n-2} \\ U_{n-1} \end{bmatrix} = 0 \tag{4.60}$$

which has the form

$$[\mathrm{A} - \lambda\mathrm{I}]\mathrm{U} = 0$$

Note: *The points i are often referred to as mesh points and h is often called the mesh length.*

Exercise

1. Derive, in determinantal form, the finite difference analogue of the frequency equation for longitudinal vibration of a uniform bar with fixed ends.

2. Show that the fourth derivative of a continuous function may be represented by

$$\{U_{i-2} - 4U_{i-1} + 6U_i - 4U_{i+1} + U_{i+2}\}/h^4$$

Hence derive a finite difference form of equation (3.55), the equations of motion for lateral vibration of thin beams.

Note: *Collatz (1960) gives a comprehensive list of finite difference equivalents to differential operators.*

4.3.2 Vibration of beams

We have already considered the introduction of fictitious nodes to satisfy end conditions, for lateral vibration of beams up to two fictitious nodes at each end may be required. That is the function $f(x)$ is considered to exist for a distance $2h$ beyond the physical end of the beam, at $x = 0$ and $x = L$.

Consider a beam clamped at $x = i = 0$, free at $x = L$ where $i = n$.

$$\left.\begin{array}{lll} \text{Then at} & x = 0, & W_0 = 0 \\ & & W_1 - W_{-1} = 0 \\ \text{at} & x = L, & W_{n+1} - 2W_n + W_{n-1} = 0 \\ & & W_{n+2} - 2W_{n+1} + 2W_{n-1} - W_{n-2} = 0 \end{array}\right\} \quad (4.61)$$

These together with the equations of motion

$$W_{i+2} - 4W_{i+1} + 6W_i - 4W_{i-1} + W_{i-2} - h^4\lambda^4 W_i = 0 \quad (4.62)$$

may be used to give

$$\begin{bmatrix} 0 & -1 & 0 & 1 & 0 & 0 & . & . & . & . \\ 0 & 0 & 1 & 0 & 0 & 0 & . & . & . & . \\ 1 & -4 & 6-\alpha & -4 & 1 & 0 & . & . & . & . \\ 0 & 1 & -4 & 6-\alpha & -4 & 1 & . & . & . & . \\ \vdots & & & & & & & & & \\ . & . & . & . & 1 & -4 & 6-\alpha & -4 & 1 & 0 \\ . & . & . & . & 0 & 1 & -4 & 6-\alpha & -4 & 1 \\ . & . & . & . & 0 & 0 & 1 & -2 & 1 & 0 \\ . & . & . & . & 0 & -1 & 2 & 0 & -2 & 1 \end{bmatrix} \begin{bmatrix} W_{-2} \\ W_{-1} \\ W_0 \\ W_1 \\ \vdots \\ W_{n-1} \\ W_n \\ W_{n+1} \\ W_{n+2} \end{bmatrix} = 0$$

(4.63)

W_0 is zero, so the corresponding row and column may be eliminated. The remaining matrix is rearranged to the form

$$\begin{bmatrix} \mathrm{S}_{rr} & \mathrm{S}_{rf} \\ \mathrm{S}_{fr} & \mathrm{S}_{ff} \end{bmatrix} \begin{bmatrix} \mathrm{U}_r \\ \mathrm{U}_f \end{bmatrix} = 0 \quad (4.64)$$

where U_f are the fictitious displacements and U_r are those associated with points on the beam. U_f may be eliminated to give

$$\{S_{rr} - S_{rf}S_{ff}^{-1}S_{fr}\}U_r = 0 \tag{4.65}$$

which again has the form

$$[A - \alpha I]U_r = 0$$

An examination of factors affecting the accuracy of the eigenvalues will be made in chapter 6.

Exercises

Set up the determinantal form of a finite difference analogue of the frequency equation for lateral vibration of thin beams

(*i*) with clamped ends
(*ii*) with pinned ends
(*iii*) with free ends.

4.3.3 Vibration of plates

The finite difference form of equation (3.151), for transverse vibration of thin plates, may be treated in the same manner as that for beams. In setting up the finite difference equations we imagine a rectangular mesh to be drawn on the surface of the plate as illustrated in Fig. 4.9. We may approximate the biharmonic operator $\nabla^4 W$ at a typical point i, j by

$$\begin{aligned}(\nabla^4 W)_{i,j} = {} & \{W_{i-2,j} - 4W_{i-1,j} + 6W_{i,j} - 4W_{i+1,j} + W_{i+2,j}\}/h^4 \\ & + 2\{W_{i-1,j+1} + W_{i-1,j-1} + W_{i+1,j+1} + W_{i+1,j-1} \\ & \quad - 2(W_{i-1,j} + W_{i,j-1} + W_{i+1,j} + W_{i,j+1} - 2W_{i,j})\}/h^2g^2 \\ & + \{W_{i,j-2} - 4W_{i,j-1} + 6W_{i,j} - 4W_{i+1,j} + W_{i+2,j}\}/g^4\end{aligned} \tag{4.66}$$

Hence a finite difference form of equation (3.151) may be derived.

At the boundaries the deflection, slope, moment, and shear force constraints must be satisfied; finite difference forms of equations (3.145), (3.146), and (3.147) are appropriate. These lead to two fictitious nodes per boundary point, excluding corners. In some cases, at the corners between two free edges an extra fictitious node is required. The additional equation required to evaluate it is $M_{xb} = M_{ya} = 0$ at the corner of the plate giving, for example,

$$(W_{m+1,n+1} - W_{m+1,n-1}) - (W_{m-1,n+1} - W_{m-1,n-1}) = 0 \tag{4.67}$$

In relation to rectangular plates, McLean (1966) showed that eigenvalues and eigenvectors may be evaluated easily for many combinations of boundary conditions. However, he had difficulty in evaluating accurately the first torsional mode for a cantilevered plate. This appears to be due to the fact that at nodes one mesh length from the clamped boundary the finite difference form of the equation of motion demands only that the second derivative along the fixed boundary be zero, leaving the possibility of rigid body rotation about the axis of symmetry perpendicular to the fixed boundary. McLean overcame this difficulty by setting the amplitude of each point on the fixed boundary to the same value. As an example of the

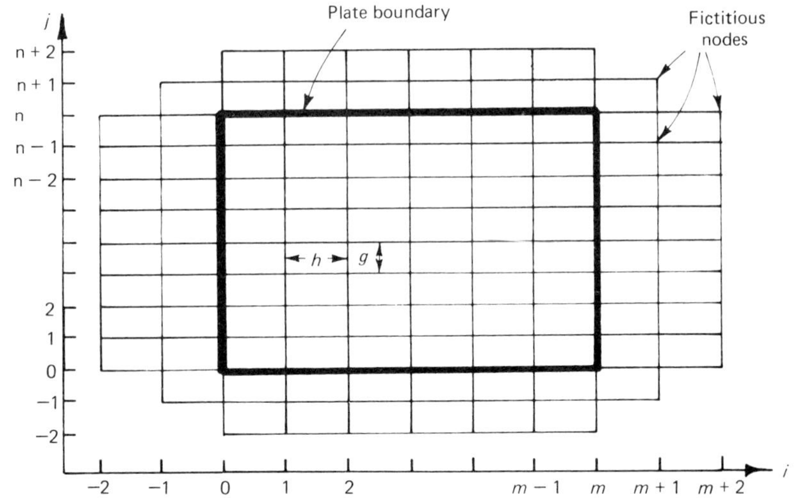

Fig. 4.9 Rectangular mesh for finite difference representation of a plate

accuracy of the method, for a simply supported square plate and a mesh which divided it into approximately 250 equal areas McLean obtained the first seven natural frequencies to within 2·5 per cent of the exact values.

Exercise

Set up in matrix form finite difference equations to represent the free vibratory behaviour of a thin rectangular plate of sides a by b

(*i*) clamped on two adjoining edges, free on the other two,
(*ii*) two opposite edges simply supported, the other two free.

References

ARGYRIS, J. H.
'Energy theorems and structural analysis', *Aircraft Engng.*, **26**, 1954, 386–94.

ARGYRIS, J. H.
'Energy theorems and structural analysis', *Aircraft Engng.*, **27**, 1955, 42–58, 80–94, 125–34, 145–58.

CLOUGH, R. W.
'Finite element method in plane-stress analysis', *Proc. 2nd Conf. on Electronic Computation, Am. Soc. Civil Engrs.*, Sept. 1960, pp. 8–9.

COHEN, E. and MCCALLION, H.
'Improved deformation functions for the finite element analysis of beam systems', *Inter. J. Numerical Meth. Engng.*, **1**, 1969, 163–7.

COLLATZ, L.
Numerical treatment of differential equations, 3rd edn., Springer-Verlag, Berlin, 1960.

DAWE, D. J.
'A finite element approach to plate vibration problems', *J. Mech. Eng. Sci.*, **7**, No. 1, 1965, 28–32.

MCLEAN, R. F.
The free vibration of thin flat plates (Ph.D. Thesis), University of Strathclyde, 1966.

Chapter 5

Natural frequencies — on finding numerical values

It is not sufficient for the mechanical engineer to know that he can set up the frequency equation correctly for a system. He wishes to predict the numerical values of the frequencies and the corresponding mode shapes so that he may modify the system, if necessary, at the design stage to avoid a dangerous operating condition, or so that he may correctly tune the system to run at resonance in resonance fatigue testing machines, vibratory conveyors, vibratory pile-drivers, etc. The natural frequencies of simple systems which have only one or two degrees of freedom and lead to not more than a quadratic equation may be solved easily; these have been dealt with in chapters 1 and 2. Beam systems with simple end conditions may be solved graphically or by simple numerical techniques to improve trial and error methods, as considered in chapter 3. A lumped-parameter representation of a diesel engine driven alternator set could easily have ten degrees of freedom which would lead to a ninth-order algebraic equation in ω^2. Two problems arise here: firstly, we would need to be very careful in deriving the coefficients of this equation from the determinantal form of the frequency equation and then by graphical or trial and error methods we would have to find the natural frequencies, that is the roots of the equation. Similarly, practical beam systems rarely constitute one beam with simple end conditions: a three-phase transformer core has a minimum of seven beams.

In this chapter we shall consider some of the problems involved in determining the numerical values of natural frequencies; the methods given all produce answers which are approximate. In some the proximity of the answer to the exact solution is very good or can be improved if desired.

Because they are powerful and useful, and because they are used later in the chapter to improve the results given by other methods, we shall consider energy methods first. They apply to conservative systems, and are based upon Rayleigh's principle.

5.1 Energy methods for conservative systems

5.1.1 Rayleigh's principle

Lord Rayleigh (1894) showed that, beginning with a rough approximation to the fundamental mode shape, we may calculate a first approximation to the lowest natural frequency by equating, the kinetic energy of the system T_{max} in passing through the mean position to its potential energy V_{max} stored at an extreme position, that is when its velocity is zero.

In chapters 2 and 3, it has been shown that the free vibratory form of a system may be expressed in terms of its normal modes. It has also been shown that we may represent the contribution of the i'th normal mode to the free vibratory form by its normal coordinate (A_i say). The constrained form F which approximates to the first mode must be that of a permissible initial form from which it could be released to vibrate freely, otherwise inherent system constraints would be violated. In terms of normal coordinates $F = \sum_i A_i$ and hence in the constrained form

$$\left.\begin{aligned} 2T_{max} &= \omega^2 \sum_i A_i^2 \\ 2V_{max} &= \sum_i \lambda_i A_i^2 \end{aligned}\right\} \tag{5.1}$$

where λ_i is the i'th eigenvalue, that is ω_i^2.

According to Rayleigh's principle

$$\omega^2 = \sum_i \lambda_i A_i^2 \Big/ \sum_i A_i^2 \tag{5.2}$$

is a first approximation to ω_1^2. What is more ω^2 is an upper bound value of ω_1^2 as shown below

$$\frac{\omega^2}{\lambda_1} = \frac{1 + \dfrac{\lambda_2}{\lambda_1}\left(\dfrac{A_2}{A_1}\right)^2 + \cdots + \dfrac{\lambda_r}{\lambda_1}\left(\dfrac{A_r}{A_1}\right)^2 + \cdots + \dfrac{\lambda_n}{\lambda_1}\left(\dfrac{A_n}{A_1}\right)^2}{1 + \left(\dfrac{A_2}{A_1}\right)^2 + \cdots + \left(\dfrac{A_r}{A_1}\right)^2 + \cdots + \left(\dfrac{A_n}{A_1}\right)^2} \tag{5.3}$$

Since $\lambda_r > \lambda_1$, $1 < r \leqslant n$, the numerator cannot be less than the denominator. Hence $\omega \geqslant \omega_1$.

This means that, if we constrain a system to vibrate in a form which approximates to its first mode, a first approximation to its lowest natural frequency may be found, and that this value is never less than the true value so long as the constrained form is a permissible initial form of the system. When the form chosen coincides with the first mode no constraints are required and the true value of $\omega = \omega_1$ is given.

As an example, reconsider the system shown in Figs. 2.13 and 2.14, with

$$m_1 = 2\ \text{Mg}; \qquad m_2 = 200\ \text{kg}$$
$$k_1 = 5\ \text{MN/m}; \qquad k_2 = 500\ \text{kN/m}$$

To find the first approximation to ω_1 let $X_1 = 1$ and $X_2 = 2$.

Then
$$T_{max} = V_{max}$$

gives
$$\omega^2 = \frac{5500}{2{\cdot}8} = 1964\ \text{rad}^2/\text{s}^2$$

therefore,
$$\omega_1 \simeq 44{\cdot}3\ \text{rad/s}$$

Solving the frequency equation gave 42·7 rad/s, so the approximate value is about 4 per cent in error: a very good result when you consider that we did not have to solve a differential equation to obtain it.

Rayleigh stated that the above estimate may be improved by using it in the equations of motion to give a new approximation to the mode shape. That is, for this example,

$$-500X_1 + (500 - 0{\cdot}2\omega^2)X_2 = 0$$

Let $X_1 = 1$ then $X_2 = 4{\cdot}67$ giving $\omega^2 = 1844$; $\omega_1 \simeq 42{\cdot}9$ rad/s which is less than 1 per cent in error.

Rayleigh's principle is also applicable to continuous systems. For example, to determine an approximate value for the first natural bending frequency of a beam let the constrained form be $Y = F(x)$. When the beam length is L, its local bending rigidity EI, local cross sectional area A, and its local mass density ρ, we have

$$2T_{max} = \omega^2 \int_0^L \rho A[F(x)]^2\, \mathrm{d}x$$

and
$$2V_{max} = \int_0^L EI \left[\frac{\mathrm{d}^2F(x)}{\mathrm{d}x^2}\right]^2 \mathrm{d}x$$

As a particular case, take $F(x) = a \sin \pi x/L$ for a uniform beam with pinned ends, and we find

$$\omega^2 = \frac{\pi^4 EI}{\rho A L^4}$$

which is the exact value. The selected form was that of the first mode, hence the form did not introduce constraints and consequently gave the exact value.

It has already been pointed out that it must be possible to deflect the actual system to the selected form by static forces and/or moments. This ensures that the boundary conditions associated with kinematic constraints, such as zero linear or angular motion, are satisfied. It is not important in Rayleigh's method to satisfy the so-called natural boundary conditions, that is the force and moment boundary conditions. When they are not satisfied, or when the selected form does not satisfy the differential equation of motion for each element or subsystem, it merely means that the selected form is comprised of components of more than one normal mode.

Exercises

1. Estimate, using Rayleigh's principle, the fundamental bending frequency for a thin uniform cantilever of length L. Take, for the constrained form, the deflected form under a static uniformly distributed load, that is

$$F(x) = a\{6\alpha^2 - 4\alpha^3 + \alpha^4\}$$

where $$\alpha = x/L$$

Show that the value found is about 0·5 per cent in error compared with that found from solution of the transcendental frequency equation.

2. When choosing a constrained form for use with Rayleigh's principle in finding the fundamental frequency of a beam, it would be convenient if simple polynomials or trigonometrical functions were permissible. For a cantilever, may the following constrained forms be used?

(*i*) $F(x) = a\left(\dfrac{x}{L}\right)^n, \; n > 1$

(*ii*) $F(x) = -3\left(\dfrac{x}{L}\right)^2 + \left(\dfrac{x}{L}\right)^3$

(*iii*) $F(x) = a[1 - \cos(\pi x/2L)]$

Give the reason for your answer.

3. Show, for a uniform cantilever, that the constrained form $F(x) = a(x/L)^n$ leads, by Rayleigh's principle, to an approximate value for the fundamental frequency given by

$$\omega^2 = \frac{n^2(n-1)^2(2n+1)}{(2n-3)} \frac{EI}{\rho AL^4}$$

4. Show, for a uniform cantilever, that the constrained form

$$F(x) = a[1 - \cos(\pi x/2L)]$$

leads to an approximate value for the fundamental frequency given by

$$\omega^2 = \frac{\pi^5}{16(3\pi - 8)} \frac{EI}{\rho AL^4}$$

5. Work out the error in frequency arising from the functions of question 3 for $n = 2$, 3, and 4, and from the function of question 4. Explain why they differ.

[**Hint:** *find the amplitude of each normal mode in each vibratory form.*]

6. Estimate the lowest natural frequency of a thin uniform rectangular plate, of sides a and b, simply supported on two opposite edges and clamped on its other two edges. Take as the vibratory form a product of the characteristic beam functions corresponding to the boundary conditions, that is take

$$W = A(\sin \pi x/a)\{\cosh \beta - \cos \beta - 0{\cdot}9825(\sinh \beta - \sin \beta)\}$$

where
$$\beta = 4{\cdot}73x/b$$

[For results obtained using this approach see Warburton (1954).]

5.1.2 Systems requiring care

Rayleigh's principle empowers us to calculate the fundamental frequency of many complicated systems, but for some, extreme care must be exercised to include all relevant energies. To illustrate this point we shall find an expression for the fundamental frequency of a pinned-ended strut, and the influence of rotor velocity on the fundamental frequency of a turbine blade will be discussed.

Consider a straight uniform thin strut with pinned ends a distance L apart, bending rigidity EI, cross-sectional area A, and mass density ρ, subjected to a constant thrust P. Assume the constrained vibratory form to be

$$F(x) = a \sin(\pi x/L)$$

then
$$T_{max} = \frac{\rho A \omega^2}{2} \int_0^L [a \sin(\pi x/L)]^2 \, dx \tag{5.4}$$

The potential energy V_{max} may be regarded as composed of two parts, the part V_1 due to bending only and the part V_2 due to the work done by the constant end

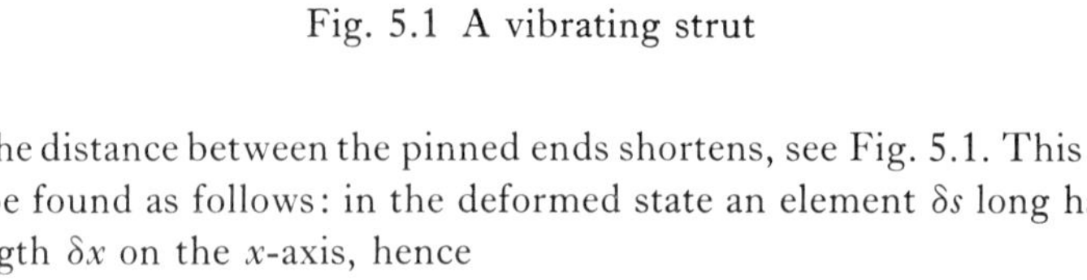

Fig. 5.1 A vibrating strut

thrust as the distance between the pinned ends shortens, see Fig. 5.1. This shortening may be found as follows: in the deformed state an element δs long has a projected length δx on the x-axis, hence

$$\delta s = \{(\delta x)^2 + (\delta y)^2\}^{1/2} = \delta x \left\{1 + \left(\frac{\delta y}{\delta x}\right)^2\right\}^{1/2}$$

$$\simeq dx \left\{1 + \frac{1}{2}\left(\frac{dy}{dx}\right)^2\right\} \tag{5.5}$$

That is the change in projected length of an element

$$\delta s - \delta x = \frac{1}{2}\left(\frac{dy}{dx}\right)^2 dx \tag{5.6}$$

Hence the potential energy $V_2 = -\frac{1}{2}P \int_0^L \left(\frac{dy}{dx}\right)^2 dx$ (5.7)

Therefore the total potential energy

$$V_{max} = \frac{1}{2}\int_0^L \left\{EI\left(\frac{d^2F(x)}{dx^2}\right)^2 - P\left(\frac{dF(x)}{dx}\right)^2\right\} dx \tag{5.8}$$

giving

$$\omega^2 = \frac{\pi^4 EI}{\rho A L^4}\left\{1 - \frac{L^2 P}{\pi^2 EI}\right\} \tag{5.9}$$

As the vibratory form chosen was that of the first mode this is the exact solution, see section 3.5.

When a bladed wheel, of say an axial flow steam turbine, is rotating and any particular blade is deflected from its equilibrium position the restoring force caused by the elasticity of the blade is supplemented by the centrifugal force. The lowest natural frequency of the blade is, therefore, increased. In design calculations this increase is allowed for approximately by an expression of the form

$$\omega_{1r}^2 = \omega_{1s}^2 + B\Omega^2$$

where ω_{1r} and ω_{1s} are the first natural angular frequencies of the blade when the wheel is rotating and stationary respectively, Ω is the angular velocity of the wheel and B depends upon the dimensions of the blade and the diameter of the wheel.

Exercise

If the blade referred to above is a uniform cantilever of height h, relevant flexural rigidity EI, cross-sectional area A, and mass density ρ, show that B has the form

$$\frac{\int_0^h (r + x)\int_0^x \left(\frac{dF(x)}{dx}\right)^2 dx\, dx}{\int_0^h (F(x))^2\, dx}$$

where r is the wheel radius.

5.1.3 The Rayleigh-Ritz method

This method extends the applicability of Rayleigh's principle to higher modes than the fundamental. In essence it is based upon the following facts.

The free vibratory form of a system with n degrees of freedom may be expressed as

$$F = \sum_i A_i$$

A_i are the normal coordinates. Let us represent this system approximately by one having two degrees of freedom corresponding approximately to its lowest two normal coordinates. The vibratory forms, F_1 and F_2 respectively, associated with these two degrees of freedom must be orthogonal. Consequently we may represent them as follows:

$$\left.\begin{aligned} F_1 &= A_1 + \text{a sum of normal modes higher than the second,} \\ &\qquad \text{excluding those in } F_2 \\ F_2 &= A_2 + \text{a sum of normal modes higher than the second,} \\ &\qquad \text{excluding those in } F_1 \end{aligned}\right\} \tag{5.10}$$

Let the maximum kinetic energy associated with a unit amplitude of F_1 be $\omega^2 T_1$, that associated with a unit amplitude of F_2 be $\omega^2 T_2$, and the corresponding potential energies be V_1 and V_2 respectively.

Initially it would be exceptional if we could choose forms which coincide with F_1 and F_2. Let us assume that we select the deflected form at any instant to be $f_1 + f_2$ where

$$f_1 = (F_1 + a_1 F_2)\alpha_1 \quad \text{and} \quad f_2 = (a_2 F_1 + F_2)\alpha_2 \tag{5.11}$$

a_1 and a_2 are constants and α_1 and α_2 are functions of time which we shall take as generalised coordinates to set up the equations of motion. The potential and kinetic energies at any instant of time will then be

$$\left.\begin{aligned} T &= (\dot{\alpha}_1 + a_2\dot{\alpha}_2)^2 T_1 + (a_1\dot{\alpha}_1 + \dot{\alpha}_2)^2 T_2 \\ V &= (\alpha_1 + a_2\alpha_2)^2 V_1 + (a_1\alpha_1 + \alpha_2)^2 V_2 \end{aligned}\right\} \tag{5.12}$$

The frequency equation

$$\begin{vmatrix} V_1 - \omega^2 T_1 & a_1(V_2 - \omega^2 T_2) \\ a_2(V_1 - \omega^2 T_1) & V_2 - \omega^2 T_2 \end{vmatrix} = 0 \tag{5.13}$$

may be found using Lagrange's equations of motion. Obviously,

$$\omega_1{}^2 = \frac{V_1}{T_1} \quad \text{and} \quad \omega_2{}^2 = \frac{V_2}{T_2}$$

are the roots of equation (5.13). The proof that Rayleigh's principle gives an upper bound to the lowest natural frequency also applies here for $\omega_1{}^2$. That is ω_1 is an upper bound for the lowest natural frequency of the n-degree of freedom system. Similarly, as V_2 and T_2 do not include energies associated with the fundamental mode, ω_2 must be an upper bound of the second natural frequency of the n-degrees-of-freedom system.

The above reasoning may be extended to an m-degree-of-freedom model of a real system with any number of degrees of freedom greater than m. If

$$\lambda_1 < \lambda_2 < \lambda_3 < \cdots < \lambda_r < \cdots < \lambda_m < \cdots < \lambda_{m+k}$$

are eigenvalues of the real system and

$$\omega_1^2 < \omega_2^2 < \omega_3^2 < \omega_4^2 < \cdots < \omega_r^2 < \cdots < \omega_m^2$$

are the eigenvalues of the model system, then

$$\omega_1^2 > \lambda_1, \ldots, \omega_r^2 > \lambda_r, \ldots, \omega_m^2 > \lambda_m$$

so long as the assumed vibratory form contains components which may represent approximately the first m eigenvectors of the real system.

As $\omega_i^2 > \lambda_i$ the smaller ω_i^2 is, the more accurate are the estimates of the natural frequencies. Hence, when the assumed vibratory form f of a system is a linear combination of functions f_i which satisfy the kinematic constraints on the system, that is

$$f = \sum_{i=1}^{n} b_i f_i$$

where b_i are constants of unknown magnitude, the minima of ω_i^2 may be found from

$$V_{max} = T_{max}$$

Let

$$T_{max} = \omega^2 \Psi$$

The minima are given by

$$\frac{\partial \omega^2}{\partial b_i} = 0 = \frac{\partial V_{max}}{\partial b_i} - \omega^2 \frac{\partial \Psi}{\partial b_i},$$

which has the form

$$(\mathrm{K} - \omega^2 \mathrm{M})\mathrm{b} = 0$$

from which the eigenvalues ω_i^2 may be found.

Exercises

1. Find approximate values for the first two natural frequencies of a uniform cantilever of length L by assuming the vibratory form

$$V = a_1(x/L)^2 + a_2(x/L)^3$$

Check your results against the exact values.

2. Repeat question 1 using

$$V = a_1(x/L)^2 + a_2(x/L)^3 + a_3(x/L)^4$$

3. Devise a set of functions, suitable for use with the Rayleigh-Ritz method, to estimate the first two natural frequencies of a uniform beam, fixed at one end and simply supported at the other. Check the accuracy of your estimates by comparing with exact values.
4. Young (1950) used the Rayleigh-Ritz method to calculate the six lowest modes of vibration of a uniform thin square plate of side a, clamped on all four edges. He found the lowest two frequencies to be

$$\omega_n = \beta_n[D/(\rho d a^4)]^{1/2}$$

where

$$\beta_1 = 35{\cdot}99 \quad \text{and} \quad \beta_2 = 73{\cdot}41$$

Using characteristic beam functions, check the above values and find the corresponding mode shapes.

5.2 Frequency equations of polynomial form

Frequency equations for undamped lumped parameter systems and for continuous systems represented in a finite element or a finite difference form may be reduced to algebraic equations in ω^2. By hand, this reduction is tedious and liable to error. Hence, in the past, methods which avoid this reduction have been sought. The Holzer and Myklestad methods are examples, but they apply to a limited range of systems only.

The frequency equations arise naturally in determinantal form, so for more general systems than may be coped with by the Holzer and Myklestad methods we could resort to evaluating the determinant at regular intervals over the frequency range of interest and locating its zeros by interpolation. Using an automatic digital computer this would be easy to perform, but if more than a very few frequencies in a very small frequency range are required it would be inefficient use of computer time. For most large computers standard library programs are available to reduce a general symmetric matrix to tridiagonal form, and then to find the eigenvalues based upon properties of the tridiagonal form. When all the frequencies in a known frequency range are required these are less time consuming than the use of a general procedure for evaluating a determinant. Another procedure for finding the eigenvalues of a general matrix which numerical analysts currently recommend involves reduction of the original matrix to diagonal form by the so-called Q-R algorithm. It has been found to be fast, whilst maintaining accuracy and stability. An outline of this method will be given: standard library programs are available on most large computers.

A method which is useful if only one or two natural frequencies are to be located, and which will be employed in section 5.3, is matrix iteration. It utilizes a property of a matrix which causes convergence to the highest eigenvalue.

5.2.1 The Holzer method

The Holzer method relates to systems composed of a chain of subsystems, two adjacent subsystems being connected by a single coordinate. Torsional systems such as diesel engine driven alternator sets are examples.

The method is best described by means of an example. Figure 5.2 is a suitable system; it represents a set of discs of moment of inertia I_r, connected by light

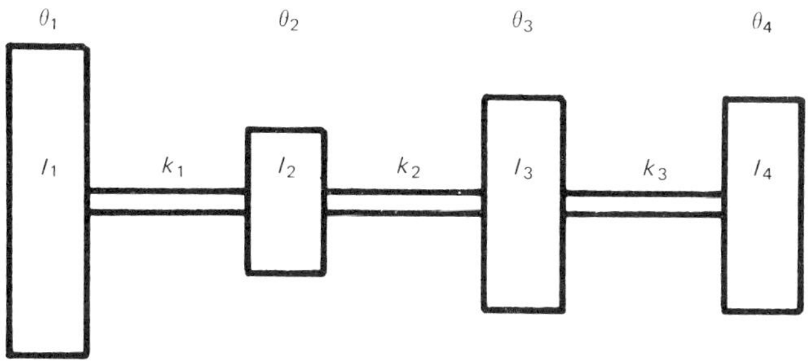

Fig. 5.2 The Holzer method; a suitable system ($I_1 = 40$ kg m^2; $I_2 = 10$ kg m^2; $I_3 = I_4 = 20$ kg m^2; $k_1 = k_2 = k_3 = 4$ kN m/rad)

elastic shafts of stiffness k_r and free to vibrate in torsion. We commence by assuming a value of the natural frequency ω and arbitrarily fixing the amplitude of I_1 to be unity by letting θ_1 be 1·0. Obviously, as there is no other torque acting on I_1, the torque T_1 in shaft 1 must be $I_1\omega^2$ and the twist in it must be $-I_1\omega^2/k_1$. Hence the angle θ_2 of I_2 must be $1 - I_1\omega^2/k_1$. The torque T_2 in shaft 2 is now easily found to be $T_1 + I_2\omega^2\theta_2$ and the twist in shaft 2 is T_2/k_2. This procedure is repeated for each inertia until the end one I_4 is reached. Except when ω coincides with a natural frequency, the torque in shaft 3 will not equal that required to balance the inertia torque required by I_4. For the last inertia we have, in general,

$$T_{n-1} + I_n\omega^2\theta_n = \sum I\omega^2\theta = \tau_n \tag{5.14}$$

where τ_n is the external torque which must be applied to the system at angular frequency ω to sustain the motion calculated. At a natural frequency τ_n is zero. One way to find natural frequencies is to evaluate τ_n at a number of points in the frequency range of interest and find its zeros by interpolation.

By hand, the above procedure is conveniently carried out using a tabular layout, as shown for the case in which

$$I_1 = 40 \text{ kg m}^2, \quad I_2 = 10 \text{ kg m}^2, \quad I_3 = I_4 = 20 \text{ kg m}^2,$$

$$k_1 = k_2 = k_3 = 4000 \text{ N m/rad}$$

As an illustration we shall evaluate the torque τ_n at $\omega^2 = 80$ rad^2/s^2. Table 5.1

indicates that this is not a natural frequency because it requires an exciting torque amplitude of 378 N m to maintain the form calculated in the θ column.

TABLE 5.1 A Holzer tabulation (estimated natural frequency corresponds to $\omega^2 = 80$ rad²/s²)

I kg m²	$\dfrac{I\omega^2}{10^3}$	θ rad	$I\omega^2\theta$ kN m	$\sum I\omega^2\theta$ kN m	K kN m/rad	$\Delta\theta$ rad
40	3·2	1·000	3·200	3·200		
					4	0·800
10	0·8	0·200	0·160	3·360		
					4	0·840
20	1·6	−0·640	−1·024	2·336		
					4	0·584
20	1·6	−1·224	−1·958	0·378		

Exercise

Repeat the above calculations for $\omega^2 = 70$ and 90 rad²/s². Is there a natural frequency in this range, and if so which one is it?

5.2.2 Improvements on the Holzer method

Instead of searching in this trial and error manner, Crandall and Strange (1957) suggested that Rayleigh's principle could be used to improve a rough approximation. Having carried out one tabulation, namely Table 5.1, we have a possible constrained form, but in general for this type of system the frequency estimates found by Rayleigh's quotient are not upper bound values of the first vibratory mode. This is an unrestrained system; it has, therefore, a zero frequency rigid body mode and Rayleigh's principle gives an upper bound to the zero frequency.

Mahalingham (1958) devised a similar improvement by considering changes in system frequency resulting from a change in the moment of inertia of one of the rotors to compensate for the residual torque τ_n.

Cohen and McCallion (1967) put forward an improvement on the Holzer method based upon the fact that, as the principal modes are orthogonal, the energies associated with them, in any vibratory form, may be found and eliminated before calculating the Rayleigh quotient. This method will be applied to find the natural frequencies of the system shown in Fig. 5.2. Let X_{ir} be the normalised amplitude of the r'th inertia I_r in the i'th principal mode of a system with $n + 1$ inertias and n shafts of torsional stiffness k_r. Principal modes are orthogonal, therefore

$$\sum_{r=1}^{n+1} I_r X_{ir} X_{jr} = 0 \quad \text{if } i \neq j$$

We shall use the normalising condition

$$\sum_{r=1}^{n+1} I_r X_{ir} X_{ir} = 1$$

Any constrained vibratory form X_r which is permissible may be represented as

$$X_r = \sum_{s=1}^{n+1} a_s X_{sr}$$

hence a_s may be found easily from

$$a_s = \sum_{r=1}^{n+1} I_r X_{sr} X_r$$

The potential energy V_{max} associated with X_r is

$$2V_{max} = \sum_{r=1}^{n} k_r(X_{r+1} - X_r)^2 = \sum_{s=1}^{n+1} a_s^2 \lambda_s \tag{5.15a}$$

and the kinetic energy T_{max} is

$$2T_{max} = \omega^2 \sum_{r=1}^{n+1} I_r X_r^2 = \omega^2 \sum_{s=1}^{n+1} a_s^2 \tag{5.15b}$$

where λ_s is the s'th eigenvalue ω_s^2.

So if we wish to find an upper bound to the t'th natural frequency Rayleigh's quotient would become

$$\omega_t^2 \simeq \frac{\sum_{r=1}^{n} k_r(X_{r\;1} - X_r)^2 - \sum_{s=1}^{t-1} a_s^2 \lambda_s}{\sum_{r=1}^{n+1} I_r X_r^2 - \sum_{s=1}^{t-1} a_s^2} \tag{5.16}$$

In the present example

$$\lambda_1 = 0 \quad \text{and } X_{1r} = 1/9{\cdot}5$$

therefore $a_1^2 = 0{\cdot}247$, thus

$$\omega_2^2 \simeq \frac{6740 - 0}{78{\cdot}58 - 0{\cdot}247} = 86 \text{ rad}^2/\text{s}^2$$

To find an approximate value of ω_3^2, guess $\omega^2 = 200$ rad^2/s^2, the calculations will give a θ vector $(1, -1, -2{\cdot}5, -1{\cdot}5)$. From this we find $a_1^2 = 27{\cdot}7$, $a_2^2 = 158$, $\sum k_r(\Delta\theta_r)^2 = 29\,000$, $\sum I_r\theta_r^2 = 220$, hence the Rayleigh quotient becomes

$$\omega_3^2 \simeq \frac{29\,000 - 0 - 158 \times 86}{220 - 27{\cdot}7 - 158} = 450 \text{ rad}^2/\text{s}^2$$

Exercises

1. Check that at $\omega^2 = 86$ rad^2/s^2 the residual torque τ_n is zero, and find the normal mode shape.
2. Using the value $\omega^2 = 450$ rad^2/s^2 find a closer approximation to $\omega_3{}^2$, check this value and find the shape of the normal mode. Hence find the highest natural frequency ($\omega_4{}^2$) of the system.

5.2.3 Transfer matrices

The Holzer method and a similar method for flexural vibration problems originally published by Myklestad (1944) may be expressed conveniently and clearly

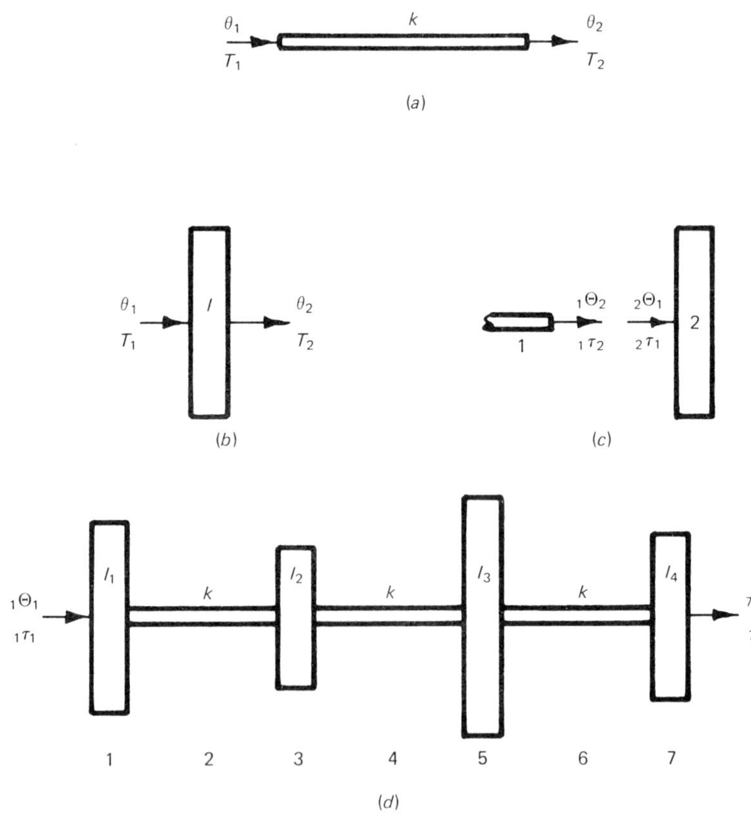

Fig. 5.3(*a*), (*b*), (*c*), and (*d*) Transfer matrices for simple and composite systems

by introducing the concept of a transfer matrix, Thomson (1950) was probably the first to do so. Essentially, a transfer matrix expresses the relationship between the generalised forces and generalised deflections at one end of a system or subsystem and those at the other. For example, this relationship for the massless torsional spring element of stiffness k, in Fig. 5.3(a) is

$$\begin{bmatrix} T_2 \\ \theta_2 \end{bmatrix} = \begin{bmatrix} -1 & 0 \\ -\dfrac{1}{k} & 1 \end{bmatrix} \begin{bmatrix} T_1 \\ \theta_1 \end{bmatrix} = \mathrm{S} \begin{bmatrix} T_1 \\ \theta_1 \end{bmatrix} \tag{5.17}$$

S is the transfer matrix.

For the disc of moment of inertia I shown in Fig. 5.3(b) it is

$$\begin{bmatrix} T_2 \\ \theta_2 \end{bmatrix} = \begin{bmatrix} -1 & I\mathrm{D}^2 \\ 0 & 1 \end{bmatrix} \begin{bmatrix} T_1 \\ \theta_1 \end{bmatrix} \tag{5.18}$$

or

$$\begin{bmatrix} \tau_2 \\ \Theta_2 \end{bmatrix} = \begin{bmatrix} -1 & -I\omega^2 \\ 0 & 1 \end{bmatrix} \begin{bmatrix} \tau_1 \\ \Theta_1 \end{bmatrix} = \mathscr{I} \begin{bmatrix} \tau_1 \\ \Theta_1 \end{bmatrix} \tag{5.19}$$

for simple harmonic oscillations, where $\mathscr{I}$ is the transfer matrix.

At a junction between two subsystems equilibrium and compatability must be maintained, hence at the junction between the spring and the disc shown in Fig. 5.3(c)

$$\begin{bmatrix} {}_2\tau_1 \\ {}_2\Theta_1 \end{bmatrix} = \begin{bmatrix} -1 & 0 \\ 0 & 1 \end{bmatrix} \begin{bmatrix} {}_1\tau_2 \\ {}_1\Theta_2 \end{bmatrix} = \mathrm{C}_2{}^1 \begin{bmatrix} {}_1\tau_2 \\ {}_1\Theta_2 \end{bmatrix} \tag{5.20}$$

The overall transfer matrix for simple harmonic oscillations of the system shown in Fig. 5.3(d) is made up as follows

$$\begin{bmatrix} {}_7\tau_2 \\ {}_7\Theta_2 \end{bmatrix} = \mathscr{I}_7 \mathrm{C}_7{}^6 \mathrm{S}_6 \mathrm{C}_6{}^5 \mathscr{I}_5 \mathrm{C}_5{}^4 \mathrm{S}_4 \mathrm{C}_4{}^3 \mathscr{I}_3 \mathrm{C}_3{}^2 \mathrm{S}_2 \mathrm{C}_2{}^1 \mathscr{I}_1 \begin{bmatrix} {}_1\tau_1 \\ {}_1\Theta_1 \end{bmatrix} \tag{5.21}$$

${}_7\tau_2$ and ${}_1\tau_1$ are applied torques, ${}_j\Theta_i$ gives the resulting vibratory form. Obviously, when ${}_1\tau_1 = 0$ this is a matrix representation of the process used in the Holzer method.

Exercises

1. Show for a single reduction spur gear that

$$\begin{bmatrix} T_2 \\ \theta_2 \end{bmatrix} = \begin{bmatrix} n & -(I_1 n + I_2/n)\mathrm{D}^2 \\ 0 & -1/n \end{bmatrix} \begin{bmatrix} T_1 \\ \theta_1 \end{bmatrix}$$

where $n = r_2/r_1$, r_i being the radii of the pitch circles; I_1 and I_2 are the moments of inertia of gear wheels 1 and 2 respectively; θ_1 and θ_2 are the angles turned through by the gear wheels, both positive in the same direction, and T_1 and T_2 are the related applied torques.

2. Find the lowest natural frequency of torsional vibration and the corresponding amplitude ratios for the systems shown in Figs. 5.4(a) and (b).

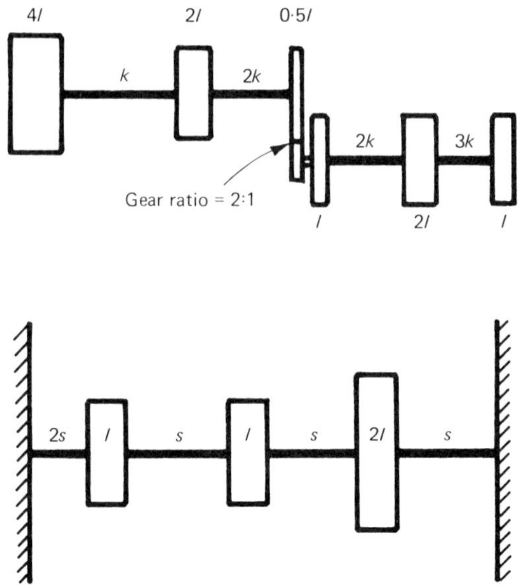

Fig. 5.4(*a*) ($I = 20$ kg m^2; $k = 200$ kN m/rad)
Fig. 5.4(*b*) ($I = 20$ kg m^2; $s = 200$ kN m/rad)

3. Calculate the amplitudes of torsional vibration of the system in Fig. 5.5 if an exciting torque of amplitude 1200 N m and frequency 15·9 Hz were applied to the end disc of inertia $3I$.

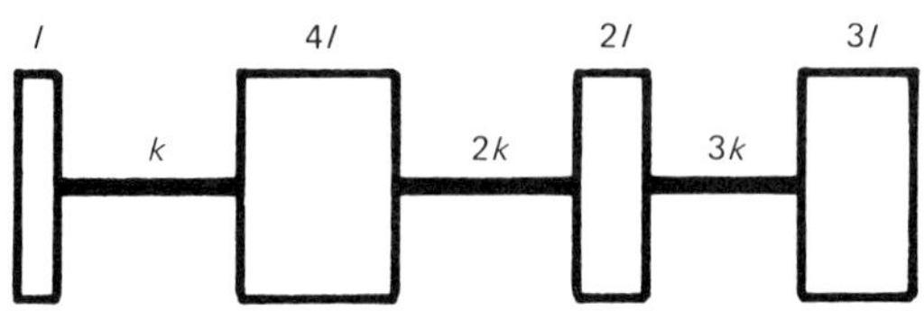

Fig. 5.5 ($I = 20$ kg m^2; $k = 200$ kN m/rad)

4. Show for a massless beam of length L and bending rigidity EI that

$$\begin{bmatrix} F_2 \\ M_2 \\ y_2 \\ \theta_2 \end{bmatrix} = \begin{bmatrix} -1 & 0 & 0 & 0 \\ L & -1 & 0 & 0 \\ L^3/6EI & -L^2/2EI & 1 & L \\ L^2/2EI & -L/EI & 0 & 1 \end{bmatrix} \begin{bmatrix} F_1 \\ M_1 \\ y_1 \\ \theta_1 \end{bmatrix}$$

in the sign convention shown on Fig. 5.6.

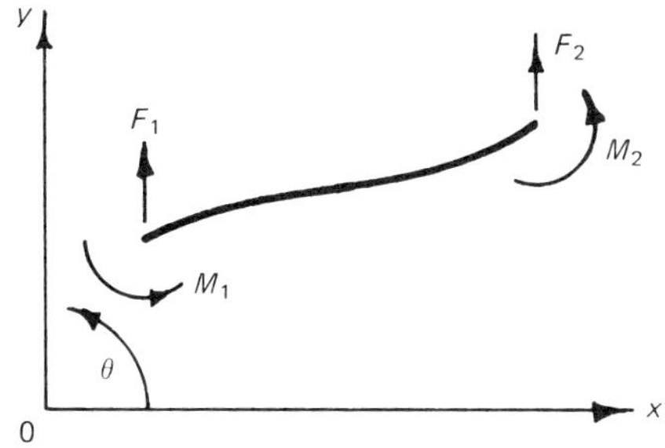

Fig. 5.6 Sign convention for the transfer matrix of a beam in bending

5.2.4 The Myklestad method for flexural vibration

The transfer matrix for a beam was given in the exercise above. This may be used to determine the natural frequencies of the system shown in Fig. 5.7, which represents a beam of total length $L_1 + L_2$ carrying two point masses m_1 and m_2, giving

$$\begin{bmatrix} {}_2F_4 \\ {}_2M_4 \\ {}_2Y_4 \\ {}_2\Theta_4 \end{bmatrix} = \mathrm{S}_4 \mathrm{C}_4{}^3 \mathscr{I}_3 \mathrm{C}_2{}^3 \mathrm{S}_2 \mathrm{C}_2{}^1 \mathscr{I}_1 \begin{bmatrix} {}_1F_1 \\ {}_1M_1 \\ {}_1Y_1 \\ {}_1\Theta_1 \end{bmatrix} \quad (5.22)$$

where S_i are the transfer matrices for the beams
$\mathscr{I}_i$ are the transfer matrices for the masses
and C_i^j are the connection matrices between the i'th and j'th subsystems

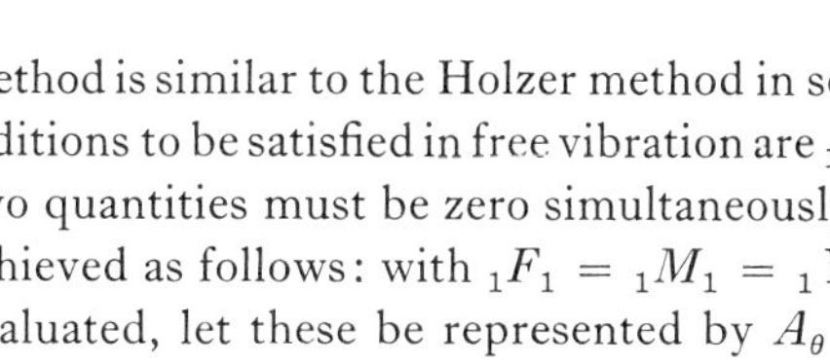

Fig. 5.7 A lumped mass flexural system

The Myklestad method is similar to the Holzer method in so much as one guesses ω^2, but here the conditions to be satisfied in free vibration are ${}_1F_1 = {}_1M_1 = {}_2Y_4 = {}_2\Theta_4 = 0$. That is two quantities must be zero simultaneously at the end 2 of subsystem 4. This is achieved as follows: with ${}_1F_1 = {}_1M_1 = {}_1Y_1 = 0$ and ${}_1\Theta_1 = 1$, ${}_2Y_4$ and ${}_2\Theta_4$ are evaluated, let these be represented by A_θ and B_θ respectively; with ${}_1F_1 = {}_1M_1 = {}_1\Theta_1 = 0$ and ${}_1Y_1 = 1$, A_y and B_y are similarly evaluated. These conditions are superimposed in an attempt to meet the condition ${}_2Y_4 = {}_2\Theta_4 = 0$. That is, at a natural frequency

$$a_1A_y + a_2A_\theta = 0$$

and

$$a_1B_y + a_2B_\theta = 0 \quad (5.23)$$

Hence

$$\begin{vmatrix} A_y & A_\theta \\ B_y & B_\theta \end{vmatrix} = 0$$

when the guessed value of ω coincides with a natural frequency. Natural frequencies may be found by varying the value of ω systematically through the frequency range of interest and plotting the value of the above determinant against ω^2.

Exercises

1. Using the above procedure, determine the natural bending frequencies of the beam system in Fig. 5.8.

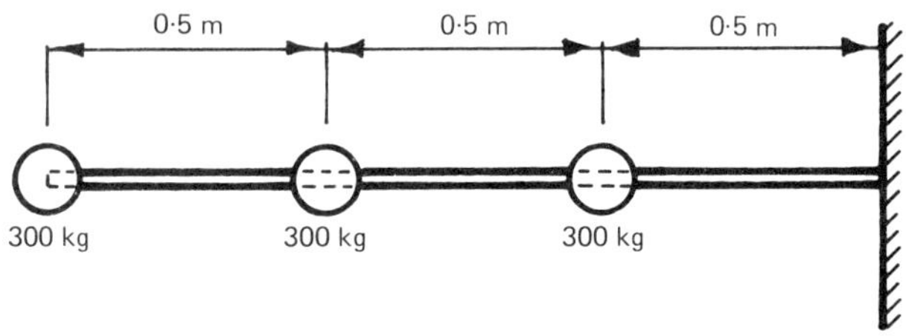

Fig. 5.8 *EI* for each section of beam is 12 MN m²

2. Repeat exercise 1, taking the initial states at the fixed end and transferring in the direction of the free end.

5.2.5 Matrix iteration

A useful technique for finding the largest eigenvalue of a matrix is matrix iteration.

To find the highest eigenvalue of the matrix S with real eigenvalues in the equation

$$\lambda^{(1)}\mathrm{x}_{(2)} = \mathrm{S}\mathrm{x}_{(1)} \tag{5.24}$$

we guess the vector $\mathrm{x}_{(1)}$ on the right-hand side and, by matrix multiplication, calculate the value of $\lambda^{(1)}\mathrm{x}_{(2)}$ on the left-hand side. An element of the guessed vector $\mathrm{x}_{(1)}$ is made equal to unity and the value of $\lambda^{(1)}$ found by making the same element in $\mathrm{x}_{(2)}$ equal to unity; $\lambda^{(1)}$ is the first approximation to the highest eigenvalue of S. A second approximation $\lambda^{(2)}$ may be found in a similar manner, using $\mathrm{x}_{(2)}$ on the right-hand side to give

$$\lambda^{(2)}\mathrm{x}_{(3)} = \mathrm{S}\mathrm{x}_{(2)} \tag{5.25}$$

This procedure may be repeated any number of times giving, in general,

$$\lambda^{(k)}\mathrm{x}_{(k+1)} = \mathrm{S}\mathrm{x}_{(k)} \tag{5.26}$$

where $\lambda^{(k)}$ is the k'th estimate of the highest eigenvalue of S and $x_{(k)}$ is the k'th estimate of the corresponding mode shape. This is known as direct iteration.

As an example we shall iterate towards the highest eigenvalue of the system shown in Fig. 5.2. It is convenient to eliminate the rigid body degree of freedom, momentum considerations give

$$I_1\theta_1 + I_2\theta_2 + I_3\theta_3 + I_4\theta_4 = 0 \tag{5.27}$$

The governing equation then reduces to

$$\begin{bmatrix} 100 & -100 & 0 \\ -400 & 800 & -400 \\ 400 & -100 & 600 \end{bmatrix} \begin{bmatrix} \theta_1 \\ \theta_2 \\ \theta_3 \end{bmatrix}_{(k)} = \omega^2 \begin{bmatrix} \theta_1 \\ \theta_2 \\ \theta_3 \end{bmatrix}_{(k+1)} \tag{5.28}$$

giving, for an initial displacement vector of $\{1, -1, 1\}$,

$$\omega^2 \begin{bmatrix} \theta_1 \\ \theta_2 \\ \theta_3 \end{bmatrix}_{(2)} = 1100 \begin{bmatrix} 0{\cdot}182 \\ -1{\cdot}45 \\ 1 \end{bmatrix} \tag{5.29}$$

After ten iterations the solution converges to

$$\omega^2 \begin{bmatrix} \theta_1 \\ \theta_2 \\ \theta_3 \end{bmatrix} = 994 \begin{bmatrix} 0{\cdot}305 \\ -2{\cdot}72 \\ 1 \end{bmatrix} \tag{5.30}$$

θ_4 is then given by equation (5.27) as $-0{\cdot}25$.

The lowest eigenvalue of S may be obtained as the highest eigenvalue of S^{-1}, that is

$$\frac{1}{\omega_1{}^2} x_{(k+1)} = \lambda x_{(k+1)} = S^{-1} x_{(k)} \tag{5.31}$$

where λ is the highest eigenvalue of S^{-1}. This is known as inverse iteration.

We shall find the lowest eigenvalue of the system shown in Fig. 5.2. In this case it is not only convenient to eliminate the rigid body degree of freedom, it is essential, otherwise the matrix S will be singular, making inversion impossible. Equation (5.31) becomes

$$\frac{1}{3600} \begin{bmatrix} 44 & 6 & 4 \\ 8 & 6 & 4 \\ -28 & -3 & 4 \end{bmatrix} \begin{bmatrix} \theta_1 \\ \theta_2 \\ \theta_3 \end{bmatrix}_{(k)} = \frac{1}{\omega_1{}^2} \begin{bmatrix} \theta_1 \\ \theta_2 \\ \theta_3 \end{bmatrix}_{(k+1)} \tag{5.32}$$

Taking the initial vector {1, 1, −1} gives

$$\frac{1}{\omega_1^2}\begin{bmatrix}\theta_1\\ \theta_2\\ \theta_3\end{bmatrix} = \frac{1}{103}\begin{bmatrix}1{\cdot}315\\ 0{\cdot}286\\ -1\end{bmatrix} \tag{5.33}$$

After five iterations the solution converges to

$$\frac{1}{\omega_1^2}\begin{bmatrix}\theta_1\\ \theta_2\\ \theta_3\end{bmatrix} = \frac{1}{85{\cdot}8}\begin{bmatrix}1{\cdot}34\\ 0{\cdot}186\\ -1\end{bmatrix} \tag{5.34}$$

That the above method iterates towards the value of the highest eigenvalue of the matrix concerned can be demonstrated as generally true for matrices associated with undamped linear systems. If we express the vector x in terms of normal coordinates A, that is

$$\mathrm{x} = \mathrm{XA} \tag{5.35}$$

where X is the complete set of normal modes for the system, the equation

$$[\mathrm{S} - \omega^2\mathrm{I}]\mathrm{x} = 0 \tag{5.36}$$

may be transformed by methods discussed in chapters 2 and 3 to give

$$[\mathrm{X^TSX} - \omega^2\mathrm{X^TIX}]\mathrm{A} = 0 \tag{5.37}$$

$[\mathrm{X^TSX}] = \lambda$, the diagonal matrix of eigenvalues and $[\mathrm{X^TIX}] = \mathrm{I}$ hence

$$\omega^2\mathrm{A} = \lambda\mathrm{A} \tag{5.38}$$

In the matrix iteration process we have

$$\omega^2\mathrm{A}^{(k+1)} = \lambda_n[\lambda/\lambda_n]\mathrm{A}^{(k)} \tag{5.39}$$

where λ_n is the highest eigenvalue. Hence $\lambda_r/\lambda_n < 1$ for $r \neq n$. In terms of the initial vector $\mathrm{A}^{(1)}$ equation (5.39) becomes

$$\omega^2\mathrm{A}^{(k+1)} = \lambda_n[\lambda/\lambda_n]^k\mathrm{A}^{(1)} \tag{5.40}$$

As $k \to \infty$ this approaches

$$\omega^2\mathrm{A}^{(k+1)} = \lambda_n\begin{bmatrix}0 & 0 & 0 & \dots & . & . & .\\ 0 & 0 & 0 & \dots & . & . & .\\ \vdots & & & & & & \vdots\\ . & . & . & \dots & 0 & 0 & 0\\ . & & . & \dots & 0 & 0 & 1\end{bmatrix}\mathrm{A}^{(1)} \tag{5.41}$$

therefore, ω^2 approaches λ_n and x approaches the component of the n'th eigenvector in the initial x vector. Obviously the rate of convergence depends upon the values of the ratios λ_r/λ_n; if the higher eigenvalues are bunched in the region of λ_n convergence will be slow.

A method for finding eigenvalues other than the highest or lowest for the system now becomes obvious. We may shift the origin on the eigenvalue scale to the region in which we think an eigenvalue exists and solve for the smallest eigenvalue in the new coordinates. For example, consider again the system of Fig. 5.2, if we wish to look for the eigenvalue λ nearest to $\omega^2 = 400$ we transform to λ' coordinates where $\lambda = 400 + \lambda'$, giving

$$[\mathrm{S} - (400 + \lambda')\mathrm{I}]\theta = 0 \qquad (5.42)$$

To find the lowest characteristic value of λ' we use

$$[\mathrm{S} - 400\mathrm{I}]^{-1}\theta_{(k)} = \frac{1}{\lambda'}\,\theta_{(k+1)} \qquad (5.43)$$

or

$$-\frac{1}{400}\begin{bmatrix} 4 & 2 & 4 \\ -8 & -6 & -12 \\ -12 & -7 & -16 \end{bmatrix}\begin{bmatrix} \theta_1 \\ \theta_2 \\ \theta_3 \end{bmatrix}_{(k)} = \frac{1}{\lambda'}\begin{bmatrix} \theta_1 \\ \theta_2 \\ \theta_3 \end{bmatrix}_{(k+1)} \qquad (5.44)$$

Starting from $\{1, -1, -2{\cdot}5\}$, the solution converges to $(1/21{\cdot}6)\{1, -3{\cdot}22, -4{\cdot}05\}$ after three iterations. Hence

$$\lambda = 421{\cdot}6 \text{ rad}^2/\text{s}^2$$

Exercises

1. Find the highest and lowest natural frequencies of torsional vibration and the corresponding amplitude ratios for the systems shown in Figs. 5.4(*a*) and (*b*). Check your answers by the Holzer method.
2. By changing origin on the eigenvalue scale, is it possible to find the lowest natural frequency of a system by direct iteration? If so, find the lowest natural frequencies called for in exercise 1 above, by this method.
3. Find the natural frequencies of the beam system shown in Fig. 5.8.

5.2.6 The number of natural frequencies in a frequency range

The determinantal form of the frequency equation for the system shown in Fig. 5.2 has coefficients on the leading diagonal and on each of the adjacent diagonals, all others being zero. This tridiagonal form is convenient for the evaluation of the determinant, as it may be performed by means of a simple recurrence relationship. Before proceeding it will be more convenient if we reduce the equations of motion to the form

$$[\mathrm{K} - \lambda\mathrm{I}]\theta = 0$$

where K is a symmetrical matrix and I is the unit matrix. This was shown possible in chapter 2, thus

$$\mathrm{K} = \begin{bmatrix} \dfrac{k_1}{I_1} & -\dfrac{k_1}{\sqrt{I_1 I_2}} & 0 & 0 \\ \dfrac{-k_1}{\sqrt{I_1 I_2}} & \dfrac{k_1 + k_2}{I_2} & -\dfrac{k_2}{\sqrt{I_2 I_3}} & 0 \\ 0 & -\dfrac{k_2}{\sqrt{I_2 I_3}} & \dfrac{k_2 + k_3}{I_3} & -\dfrac{k_3}{\sqrt{I_3 I_4}} \\ 0 & 0 & -\dfrac{k_3}{\sqrt{I_3 I_4}} & \dfrac{k_3}{I_4} \end{bmatrix} \tag{5.45}$$

In the present case the frequency equation becomes

$$\begin{vmatrix} 100 - \lambda & -200 & 0 & 0 \\ -200 & 800 - \lambda & -283 & 0 \\ 0 & -283 & 400 - \lambda & -200 \\ 0 & 0 & -200 & 200 - \lambda \end{vmatrix} = 0 \tag{5.46}$$

or in general of the form

$$\begin{vmatrix} a_{11} - \lambda & a_{12} & 0 & 0 \\ a_{21} & a_{22} - \lambda & a_{23} & 0 \\ 0 & a_{32} & a_{33} - \lambda & a_{34} \\ 0 & 0 & a_{43} & a_{44} - \lambda \end{vmatrix} = 0 \tag{5.47}$$

(with det_1, det_2, det_3, det_4 marked off by dotted lines over the leading principal submatrices)

where

$$a_{ij} = a_{ji}$$

Let $det_0 = 1$ and consider the determinants det_1, det_2, etc., marked off by the dotted lines, then

$$det_0 = 1; \quad det_1 = a_{11} - \lambda$$

and

$$det_r = (a_{rr} - \lambda) det_{r-1} - a^2_{r,\, r-1} det_{r-2} \tag{5.48}$$

For a given value of λ (say $\lambda = b$) the sequence $det_0, det_1, \ldots, det_n$ may be evaluated easily by this simple recurrence relationship: it is known as a Sturm sequence (Givens, 1953), and has the property that the number of distinct real roots of det_n with an algebraic value less than b is equal to the number of changes of sign in it. For example, the sequence of signs of det_0 to det_4 for frequency equation (5.46) is $+ + - - +$ when $\lambda = 90$. Hence, below 90 rad^2/s^2 there are two natural frequencies. When one of the determinants has a value of zero, it is given the sign of the previous determinant in the sequence.

For a tridiagonal symmetric matrix it is thus possible to determine in a precise and rational manner the number of distinct real eigenvalues which lie in a given range. Given the number, they may be located by a systematic search procedure. For example, if only one occurred in an interval it could be located by successive bisection of the appropriate interval. Obviously, for a given accuracy the number of bisections may be determined.

Exercises

1. Repeat the exercise relating to Fig. 5.2 using the properties of the Sturm sequence to locate the roots.
2. Show that equation (5.48) is true.

5.2.7 Reduction to tridiagonal form

The method for finding the eigenvalues in a given frequency range described in section 5.2.6 applies to symmetrical tridiagonal matrices. All systems studied in chapters 2 and 4 could be described in terms of symmetrical matrices, but in general they are not tridiagonal. Several numerical analysts have devised methods for reducing a symmetrical matrix to tridiagonal form. One, known as Givens' method, may be described as follows; initially we have

$$[\mathrm{A} - \lambda \mathrm{I}]\mathrm{x} = 0 \tag{5.49}$$

and we require a coordinate transformation

$$\mathrm{Q}^{\mathrm{T}}[\mathrm{A} - \lambda \mathrm{I}]\mathrm{QX} = 0 \tag{5.50}$$

giving

$$[\mathrm{B} - \lambda \mathrm{I}]\mathrm{X} = 0 \tag{5.51}$$

where B is of tridiagonal form. Obviously Q must be an orthogonal matrix: it is derived in stages, at each a factor (say Q_i) is defined as a unitary matrix of the form

$$\mathrm{Q}_i = \begin{bmatrix} 1 & 0 & 0 & 0 \\ 0 & C & S & 0 \\ 0 & -S & C & 0 \\ 0 & 0 & 0 & 1 \end{bmatrix} \tag{5.52}$$

where $C = \cos\theta$ and $S = \sin\theta$.

For a 3×3 matrix A with coefficients a_{ij}, a tridiagonal matrix B with $b_{13} = b_{31} = 0$ could be formed by the transformation

$$\begin{bmatrix} 1 & 0 & 0 \\ 0 & C & S \\ 0 & -S & C \end{bmatrix} \mathrm{A} \begin{bmatrix} 1 & 0 & 0 \\ 0 & C & -S \\ 0 & S & C \end{bmatrix} = \mathrm{B} \tag{5.53}$$

where $$\tan\theta = a_{13}/a_{12} = a_{31}/a_{21}$$

Similarly a 4 × 4 matrix A could be reduced to tridiagonal form in the following three stages:

1.
$$\begin{bmatrix} 1 & 0 & 0 & 0 \\ 0 & C & S & 0 \\ 0 & -S & C & 0 \\ 0 & 0 & 0 & 1 \end{bmatrix} \mathrm{A} \begin{bmatrix} 1 & 0 & 0 & 0 \\ 0 & C & -S & 0 \\ 0 & S & C & 0 \\ 0 & 0 & 0 & 1 \end{bmatrix} = \mathrm{B}^{(1)} \quad (5.54)$$

where $$\tan\theta = \frac{a_{13}}{a_{12}} = \frac{a_{31}}{a_{21}}$$

gives $$b_{13}^{(1)} = b_{31}^{(1)} = 0$$

2.
$$\begin{bmatrix} 1 & 0 & 0 & 0 \\ 0 & C & 0 & S \\ 0 & 0 & 1 & 0 \\ 0 & -S & 0 & C \end{bmatrix} \mathrm{B}^{(1)} \begin{bmatrix} 1 & 0 & 0 & 0 \\ 0 & C & 0 & -S \\ 0 & 0 & 1 & 0 \\ 0 & S & 0 & C \end{bmatrix} = \mathrm{B}^{(2)} \quad (5.55)$$

$$\tan\theta = b_{14}^{(1)}/b_{12}^{(1)} = b_{41}^{(1)}/b_{21}^{(1)}$$

gives $$b_{14}^{(2)} = b_{41}^{(2)} = 0$$

also $$b_{13}^{(2)} = b_{31}^{(2)} = 0$$

3.
$$\begin{bmatrix} 1 & 0 & 0 & 0 \\ 0 & 1 & 0 & 0 \\ 0 & 0 & C & S \\ 0 & 0 & -S & C \end{bmatrix} \mathrm{B}^{(2)} \begin{bmatrix} 1 & 0 & 0 & 0 \\ 0 & 1 & 0 & 0 \\ 0 & 0 & C & -S \\ 0 & 0 & S & C \end{bmatrix} = \mathrm{B}^{(3)} \quad (5.56)$$

where $$\tan\theta = b_{24}^{(2)}/b_{23}^{(2)} = b_{42}^{(2)}/b_{32}^{(2)}$$

gives $$b_{24}^{(3)} = b_{42}^{(3)} = 0$$

also $$b_{13}^{(3)} = b_{31}^{(3)} = b_{14}^{(3)} = b_{41}^{(3)} = 0$$

Note that the zeros established at an earlier stage are not affected by a current transformation. Consequently a symmetrical matrix may be reduced to tridiagonal form in a finite number of stages.

In another method, known as Householder's method, the reduction is again achieved by orthogonal transformations but in this case it reduces to zero all the required coefficients in a row or column at one stage. It is computationally more efficient than Givens' method. The transformation matrices Q_i in Householder's method are of the form

$$Q_i = \begin{bmatrix} 1 & 0 & 0 & 0 \\ 0 & 1 - 2\omega_2^2 & -2\omega_2\omega_3 & -2\omega_2\omega_4 \\ 0 & -2\omega_3\omega_2 & 1 - 2\omega_3^2 & -2\omega_3\omega_4 \\ 0 & -2\omega_4\omega_2 & -2\omega_4\omega_3 & 1 - 2\omega_4^2 \end{bmatrix} \tag{5.57}$$

where $$Q_i^T = Q_i^{-1} = Q_i$$

this implies that $$\omega_2^2 + \omega_3^2 + \omega_4^2 = 1$$

The above matrix could be used to reduce the coefficients b_{13}, b_{14}, b_{31}, b_{41} to zero, of a 4×4 matrix.

Wilkinson (1965) details and discusses the relative merits of the Givens and Householder methods. He also discusses other methods. Standard library programs are available for the reduction of a general symmetric matrix to tridiagonal form and subsequent solution for the eigenvalues in a given frequency range, on most large scientific computers.

5.2.8 The Q-R transformation

If it were possible to take the Givens method, outlined in section 5.2.7, further and reduce a general symmetrical matrix to a diagonal matrix by unitary transformations, the eigenvalues would be given directly. As would be expected from the nature of the problem, because root finding of high order polynomials involves iteration, in this case it is not possible to reduce to zero all the desired coefficients without affecting zero coefficients already established. However, Francis (1961), who discovered the method and named it, as he says 'somewhat arbitrarily', the Q-R transformation, proved that convergence to diagonal form exists if we repeatedly reduce the non-zero off-diagonal elements to zero.

Numerical analysts claim this to be one of the most effective methods known currently for solving eigenvalue problems, further details are given by Wilkinson (1965). Standard library programs utilising the method are available on many large scientific computers.

5.3 Frequency equations of transcendental form

In chapter 3 it was shown that the frequency equation for a system with distributed mass and elasticity has, in determinantal form, elements which are transcendental functions, involving frequency implicitly. The eigenvalues may be found by a systematic scanning procedure, the determinant being evaluated at each stage. In the vicinity of a sign change in the determinant a more accurate value of the eigenvalue may be found by an interpolation procedure or by refining the scanning step. Once an eigenvalue is found, the corresponding eigenvector is calculated by matrix inversion, with one of the amplitudes set.

Evaluation of each transcendental function at each frequency in the above scanning procedure is expensive in computer time. Particularly at the stage where we are improving the accuracy of an eigenvalue, having found an upper and a lower bound, linearisation and inverse matrix iteration bring considerable economies in computer time, see Cohen and McCallion (1967). Expressed symbolically we have at an approximate value of a natural frequency Ω_n

$$\mathrm{YA} \neq 0 \tag{5.58}$$

where Y is the dynamic stiffness matrix and A is the vector describing the vibratory form. Assume that Ω_n is sufficiently close to the true natural frequency ω_n for linearisation of the transcendental functions to involve little or no error, then

$$\{\mathrm{Y}_{\Omega_n} - \lambda \mathrm{B}\}\mathrm{A} = 0 \tag{5.59}$$

where

$$\mathrm{B} = -\frac{\partial \mathrm{Y}_{\Omega_n}}{\partial \omega^2} = -\frac{1}{2\Omega_n}\frac{\partial \mathrm{Y}_{\Omega_n}}{\partial \omega} \tag{5.60}$$

and

$$\lambda = \omega_n^{\ 2} - \Omega_n^{\ 2} \tag{5.61}$$

As Y_{Ω_n} is non-singular λ may be found by inverse iteration, see section 5.2.5, namely from

$$\frac{1}{\lambda^{(k)}}\,\mathrm{A}_{(k+1)} = \mathrm{Y}_{\Omega_n}^{-1}\mathrm{BA}_{(k)} \tag{5.62}$$

A then gives the eigenvector, and the required natural frequency is

$$\omega_n = (\lambda + \Omega_n^{\ 2})^{1/2}$$

Cohen and McCallion compared results found using the above method with those found directly from *det* Y, for the first seven natural frequencies of a five-legged transformer core: the frequencies were within 0·1 per cent and the mode shapes indistinguishable, even when the point of linearisation was about 6 Hz away from the natural frequency.

Miscellaneous exercises (each involves the use of a digital computer)

1. Referring to Fig. 5.9 and using each appropriate method, find the natural frequencies of the following:
   (a) vibration in the x-direction of the system shown at (a),
   (b) flexural vibration of the light uniform cantilever shown in (b), where m = 50 kg, L = 1 m, and $EI = 5 \times 10^6$ N m^2,
   (c) torsional vibration of the system shown at (c).

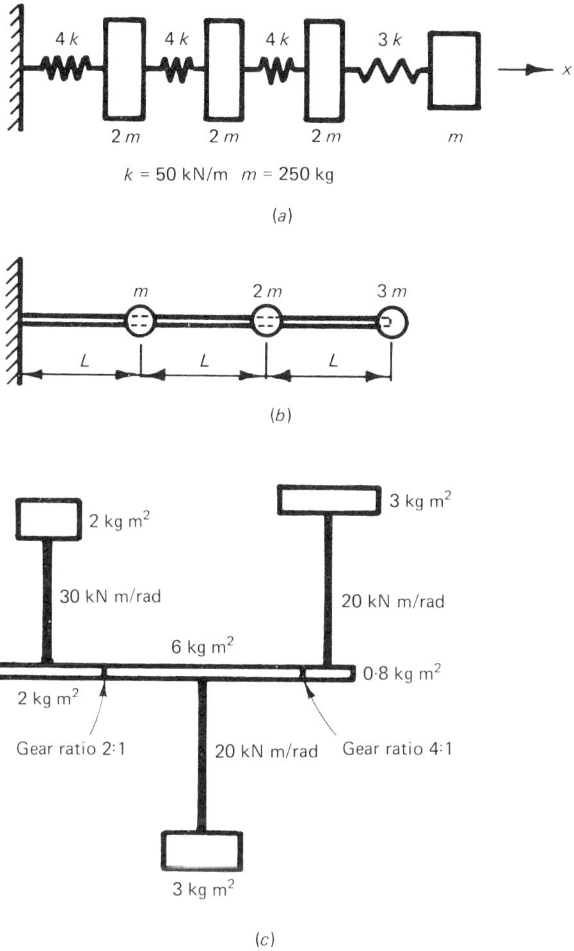

Fig. 5.9(a), (b), (c)

2. Find the natural frequencies of the system shown in Fig. 4.4 and described in exercise 2 of section 4.1.2.
3. For torsional vibration, a six-cylinder diesel engine and generator may be represented as shown in Fig. 5.10. The stiffness k_3 of the shaft between the flywheel I_1 and the generator I_2 is to be designed to give the torsional natural frequency of the one-node mode at 1100 vibrations per minute.

   With the above value for k_3, find the next two higher natural frequencies and then, assuming each shaft to be torsionally equivalent to a plain steel shaft of 0·36 m diameter, estimate the maximum shear stress in each shaft, per radian amplitude of the crank at the free end, when the system is excited at each of its natural frequencies.

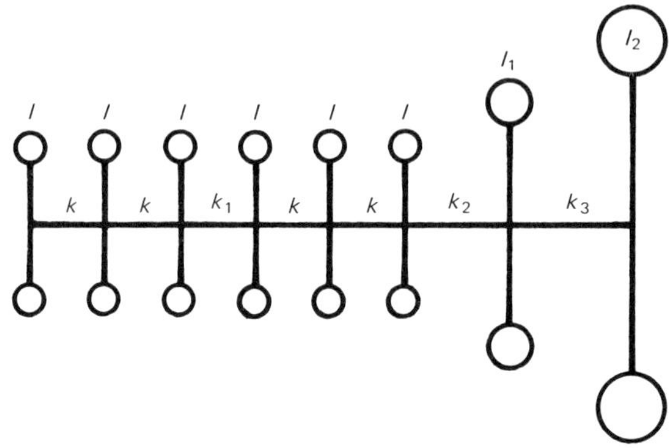

Fig. 5.10 (I = 166 kg m^2; I_1 = 392 kg m^2; I_2 = 883 kg m^2; k = 144 MN m/rad; k_1 = 136 MN m/rad; k_2 = 142 MN m/rad; k_3 = is to be found)

4. Define a set of functions for use with the Rayleigh-Ritz method and estimate the first two natural frequencies of the solid-steel roll shown in Fig. 5.11. The ends are simply supported in short bearings.

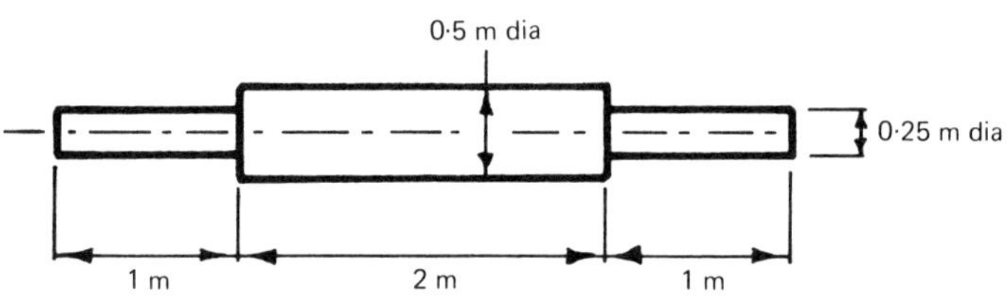

Fig. 5.11

Check your results against those you find by setting up and solving the frequency equation on

(*a*) thin beam theory, and
(*b*) thick beam theory,

as given in chapter 3.

Also estimate the first two natural frequencies using

(*a*) a lumped parameter model,
(*b*) a finite element model, and
(*c*) a finite difference model of the system,

assuming the beams to be thin.

5. For a thin square plate, clamped at all edges find values for the first six natural frequencies by

(*a*) the Rayleigh-Ritz method, using characteristic beam functions,
(*b*) the finite element method, and
(*c*) the finite difference method.

References

COHEN, E. and McCALLION, H.
'Economical methods for finding eigenvalues and eigenvectors', *J. Sound Vib.*, **5**, No. 3, 1967, 397–406.

CRANDALL, S. H. and STRANG, W. G.
'An improvement of the Holzer table based on a suggestion of Rayleigh's, *J. Appl. Mech.*, **24**, 1957, 228–30.

FRANCIS, J. G. F.
'The Q-R transformation, Parts I and II', *Computer J.*, **4**, 1961, 265–71, 332–45.

GIVENS, W.
'A method of computing eigenvalues and eigenvectors suggested by classical results on symmetric matrices', *Appl. Math. Ser. Nat. Bur. Stand.*, **29**, 1953, 117–22.

MAHALINGHAM, S.
'An improvement of the Holzer method', *J. Appl. Mech.*, **25**, 1958, 618–20.

MYKLESTAD, N. O.
'A new method of calculating natural modes of uncoupled bending vibration of airplane wings and other types of beams', *J. Aeronaut. Sc.*, **11**, No. 2, April 1944, 153–62.

RAYLEIGH, LORD
Theory of sound, Macmillan, 1894, p. 110.

THOMSON, W. T.
'Matrix solution for the vibration of non-uniform beams', *J. Appl. Mech.*, **17**, 1950, 337–9.

WARBURTON, G. B.
'The vibration of rectangular plates', *Proc. Inst. Mech. Engrs.*, **168**, No. 12, 1954, 371–84.

YOUNG, D.
'Vibration of rectangular plates by the Ritz method', *J. Appl. Mech. Trans. Am. Soc. Mech. Engrs.*, **72**, 1950, 448–53.

Further reading

FOX, L.
An introduction to numerical linear algebra, Oxford University Press, 1964, chaps. 9 and 10.

WILKINSON, J. H.
The algebraic eigenvalue problem, Oxford University Press, 1965.

Chapter 6

Uniform systems

A uniform system will be defined as one comprised of an assemblage of identical sub-systems. Many examples occur in practice, for instance packets of turbine blades, wave filters for electronic circuits, uniform grillages, plane frameworks, reinforced grillages, and reinforced shells. They also arise when studying the accuracy of lumped parameter representations of continuous systems. Often it is easier to calculate the vibratory behaviour of these systems by taking into account this uniformity when setting up the frequency equation. This is because, except at the ends of the system, the equations of motion relating to each subsystem take the same form: they may, therefore, be represented by equations called recurrence or finite difference equations.

As examples of the method employed we shall examine the accuracy with which the frequencies of continuous systems may be predicted using some of the piecewise representations discussed in chapter 4. Examples of such studies are Karman and Biot (1940) and Ellington (1956).

6.1 Lumped parameter representation of continuous systems

6.1.1 Longitudinal vibration of bars

Suppose we wish to investigate the accuracy with which the frequencies of longitudinal vibration of a continuous bar may be estimated by representing it as a series of identical masses and springs. Let the bar under consideration be rigidly held at one end and free at the other. The discrete mass and spring representation is illustrated in Fig. 6.1,

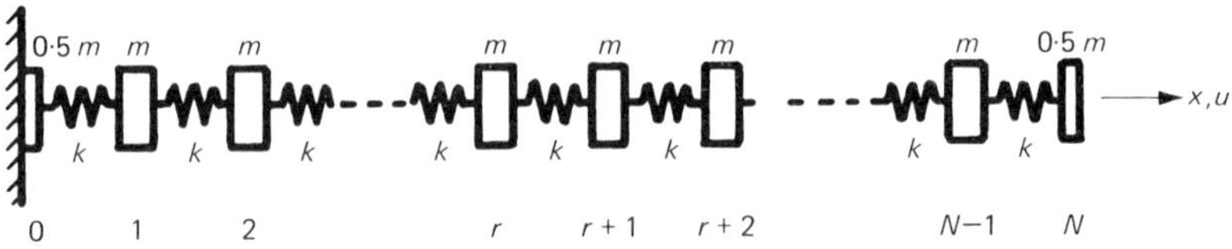

Fig. 6.1 A lumped parameter representation for studying longitudinal vibrations of a continuous uniform bar

where $m = \rho AL/N$
$k = EAN/L$ (6.1)
ρ = mass density of the bar material
E = Young's modulus
A = cross-sectional area of the bar
L = length of bar
N = number of segments used in discrete mass representation of the bar
u_r = instantaneous displacement in the x-direction of the r'th mass, from its equilibrium position

Applying Newton's second law at the $r + 1$'th mass point, that is any mass point with the exception of those at either end, gives

$$m\ddot{u}_{r+1} + k(2u_{r+1} - u_r - u_{r+2}) = 0 \tag{6.2}$$

As the system is linear we may assume a solution of the form

$$u_r = \sum_n U_{r,n}(C \sin \omega_n t + D \cos \omega_n t) \tag{6.3}$$

where for each of the normal modes $U_{r,n}$ is a function of r only. When the value of u_r given by expression (6.3) is substituted into equation (6.2) the following relationship between the amplitudes of adjacent mass points is found:

$$U_{r+2,n} + 2\left(\frac{m\omega_n^2}{2k} - 1\right) U_{r+1,n} + U_{r,n} = 0 \tag{6.4}$$

This type of relationship is known as a difference equation or as a recurrence relationship. By analogy with differential equations it is a second-order linear homogeneous difference equation which may be satisfied by a solution of the form.

$$U_{r,n} = B_n \alpha_n^r \tag{6.5}$$

Substituting this value into equation (6.4) gives

$$(\alpha_n^2 + 2\beta_n\alpha_n + 1)B_n\alpha_n^r = 0 \tag{6.6}$$

where $\beta_n = (m\omega_n^2/2k) - 1$

Neither B_n nor α_n is zero or there would be no vibration and, therefore, the term in parentheses must be zero. Hence, excluding the special case of

$$\beta_n^2 = 1,$$

$$\alpha_n = -\beta_n + \sqrt{(\beta_n^2 - 1)} \quad \text{or} \quad -\beta_n - \sqrt{(\beta_n^2 - 1)} \tag{6.7}$$

Now, for convenience, let

$$-\beta_n + \sqrt{(\beta_n^2 - 1)} = \exp(\theta_n)$$

then $$-\beta_n - \sqrt{(\beta_n^2 - 1)} = \exp(-\theta_n) \tag{6.8}$$

The complete solutions of equation (6.4) may then be written as

$$U_{r,n} = {}_1B_n \exp(r\theta_n) + {}_2B_n \exp(-r\theta_n) \tag{6.9}$$

or as $$U_{r,n} = G_n \sinh r\theta_n + H_n \cosh r\theta_n \tag{6.10}$$

the relative values of ${}_1B_n$ and ${}_2B_n$ or of G_n and H_n being fixed to satisfy the boundary conditions. In the case being considered we have a kinematic constraint at $r = 0$, namely $U_{0,n} = 0$ giving $H_n = 0$ and at $r = N$ the boundary condition states that the end mass is free from external force, hence Newton's second law gives

$$-0{\cdot}5m\omega_n^2 U_{N,n} + k(U_{N,n} - U_{N-1,n}) = 0 \tag{6.11}$$

or

$$\left(1 - \frac{m\omega_n^2}{2k}\right) G_n \sinh N\theta_n - G_n \sinh (N - 1)\theta_n = 0 \tag{6.12}$$

Obviously $G_n \neq 0$ or there would be no vibration, also from equations (6.6) and (6.8)

$$1 - \frac{m\omega_n^2}{2k} = \cosh \theta_n \tag{6.13}$$

These observations may be used to reduce equation (6.12) to

$$\cosh \theta_n \sinh N\theta_n - \sinh (N - 1)\theta_n = 0 \tag{6.14}$$

or

$$\sinh \theta_n \cosh N\theta_n = 0 \tag{6.15}$$

which is the frequency equation: it is satisfied if

$$\theta_n = \frac{2n - 1}{2N} i\pi$$

where

$$i = \sqrt{-1} \tag{6.16}$$

Using these values expression (6.13) gives

$$\omega_n^2 = \frac{2k}{m}\left(1 - \cos\left(\frac{2n - 1}{2N}\right)\pi\right) \tag{6.17}$$

where n takes each integral value in the range 1 to N inclusive. By inspection, it is seen that the range of values for ω_n^2 is between zero and $4k/m$. Hence β lies in the range $-1 < \beta < +1$, giving α in the range $+1 > \alpha > -1$. Obviously no new values of ω_n^2 are given if n lies outside the above range; the values within it are merely repeated.

The mode shapes are given by equation (6.10) which becomes

$$U_{r,n} = G_n \sin\left(\frac{2n - 1}{2N} \pi r\right) \tag{6.18}$$

The restriction $\beta_n^2 \neq 1$ introduced above is not in this case a real restriction, as we shall now prove, because with the present system β_n^2 cannot be equal to unity. If it were possible the following solutions would apply

(*i*) for $\beta = -1$

$$U_r = A + Br \tag{6.19}$$

The boundary conditions are

(*a*) at $r = 0$, $U_r = 0$, hence $A = 0$

(*b*) at $r = N$

$$-\beta BN - B(N - 1) = 0 \tag{6.20}$$

giving $B = 0$, hence $\beta = -1$ is not associated with a natural frequency.

(*ii*) for $\beta = 1$

$$U_r = (-1)^r(A + Br) \tag{6.21}$$

Again, fitting the boundary conditions gives $A = B = 0$, therefore $\beta_n \neq 1$. Hence equation (6.17) gives all the natural frequencies and (6.18) gives the corresponding mode shapes, for the system illustrated in Fig. 6.1.

We are now equipped to discuss the accuracy with which the natural frequencies of a uniform continuous bar may be found from a discrete mass representation. Comparing the mode shape given by equation (6.18) with that found in chapter 3 for the continuous bar, we find that at the equally spaced mass points the mode shapes agree for the first N natural frequencies. That is

$$U_{r,n} = G_n \sin\left(\frac{2n - 1}{2N}\pi r\right) \tag{6.18}$$

and

$$U_n(x) = C_n \sin\left(\frac{2n - 1}{2L}\pi x\right) \tag{3.29b}$$

where at the mass points $r/N = x/L$. Obviously the discrete mass representation does not predict natural frequencies or mode shapes for higher than the N'th mode.

Agreement on the values of the natural frequencies is not so good as that for the mode shapes, but as equations (6.17) and (3.29*a*) differ in form we modify equation (6.17) before making the comparison.

$$\omega_n^2 = \frac{2k}{m}\left(1 - \cos\frac{2n - 1}{2N}\pi\right) \qquad \text{from (6.17)}$$

$$\simeq \left\{\frac{(2n - 1)\pi}{2L}\right\}^2 \frac{E}{\rho}\left\{1 - \frac{[(2n - 1)\pi]^2}{48N^2} + \cdots\right\}$$

whereas

$$\omega_n^2 = \left\{\frac{(2n - 1)\pi}{2L}\right\}^2 \frac{E}{\rho} \qquad \text{from (3.29}a\text{)}$$

Hence for the first few natural frequencies, n being much less than N, the discrete mass approximation underestimates the true natural frequency. Equation (6.17) shows that the error is of order $1/N^2$ in this case.

Exercises

1. How many equally spaced mass points would be required for the above representation of a uniform bar, held rigidly at one end and free at the other, if the lowest natural frequency of longitudinal oscillations had to be predicted to within one per cent of the exact value?

How many would be required to predict the first five natural frequencies to within one per cent?

2. Derive expressions giving the accuracy with which the torsional natural frequencies of a uniform shaft may be found from a discrete mass representation. Assume it to be divided into N sections with the mass of each section concentrated (*a*) at the centre of each section and (*b*) half at each end of a section. About its axis the shaft is free from external constraints.

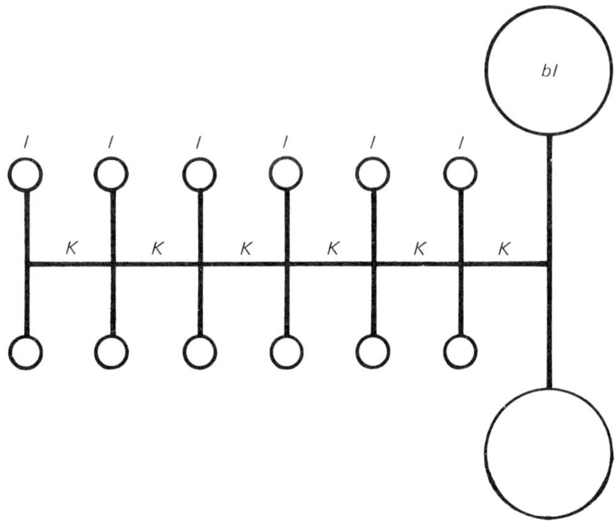

Fig. 6.2

3. A six-cylinder engine is used to drive a centrifugal blower of very large moment of inertia. The combined engine and blower may be represented by the equivalent system shown in Fig. 6.2 for the prediction of torsional natural frequencies. Show that the natural frequencies are given by

$$\omega_n{}^2 = \frac{2K}{I}\left\{1 - \cos\frac{2n-1}{13}\pi\right\}$$

where I has units of moment of inertia
K is the torsional stiffness of the shafts
b is a very large number
and n is a positive integer the range of which is to be defined.

6.1.2 Transverse vibration of beams

Lumped-parameter models for representing flexural vibrations of slender beams lead to fourth-order difference equations. The solutions are, therefore, slightly more complicated than those of the second-order equations relating to longitudinal or torsional oscillations.

In the following discussion we shall consider the free oscillations of a system, the elasticity of which is supplied by a uniform massless beam of modulus of elasticity E and second moment of area I, about the principal axis in the neutral plane perpendicular to the longitudinal axis of the beam. Rigidly attached at equal distances, ℓ, the beam carries a number of equal concentrated masses. Rotary inertia and shearing deformation will be neglected.

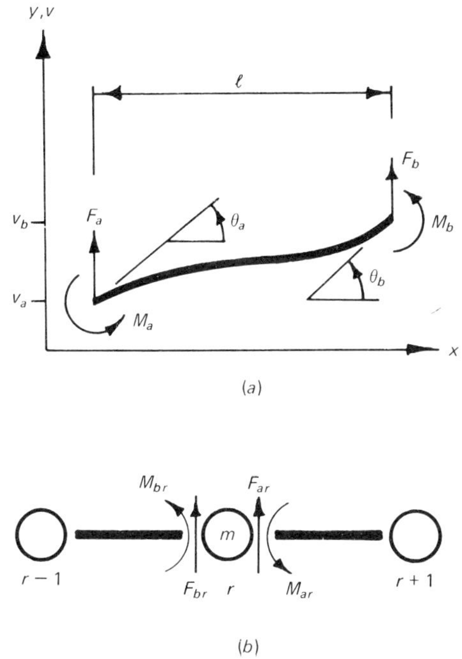

Fig. 6.3(a), (b)

The positive sign convention adopted for applied moment, applied force, deflection and rotation at the ends of a beam section is given in Fig. 6.3. In this convention, the applied end forces and moments are related to the end deflections and rotations by

$$\begin{bmatrix} F_a \\ F_b \\ M_a \\ M_b \end{bmatrix} = \frac{EI}{\ell^3}\begin{bmatrix} 12 & -12 & 6\ell & 6\ell \\ -12 & 12 & -6\ell & -6\ell \\ 6\ell & -6\ell & 4\ell^2 & 2\ell^2 \\ 6\ell & -6\ell & 2\ell^2 & 4\ell^2 \end{bmatrix}\begin{bmatrix} v_a \\ v_b \\ \theta_a \\ \theta_b \end{bmatrix} \tag{6.22}$$

Applying Newton's second law, the equation of motion in the y-direction for the mass m at a typical mass point r, as shown in Fig. 6.36(b), may be written

$$m\ddot{v}_r + F_{ar} + F_{br} = 0 \tag{6.23}$$

or $$m\ddot{v}_r + \frac{EI}{\ell^3}\{-12v_{r-1} + 24v_r - 12v_{r+1} + 6\ell(\theta_{r+1} - \theta_{r-1})\} = 0 \tag{6.24}$$

Neglecting rotary inertia, the moments at each mass point must be in equilibrium, hence

$$M_{ar} + M_{br} = 0 \tag{6.25}$$

or

$$6\ell(v_{r-1} - v_{r+1}) + 2\ell^2(\theta_{r-1} + 4\theta_r + \theta_{r+1}) = 0 \tag{6.26}$$

Manipulation of equations (6.24) and (6.26) is facilitated by the introduction of the shifting operator **E**, which by definition gives $\mathbf{E}v_r = v_{r+1}$, $\mathbf{E}^2 v_r = v_{r+2}$, and so on. Also as the system is linear we may expect solutions of the form

$$v_{r,n} = \sum_{n=1}^{N} V_{r,n}\{A_n \sin \omega_n t + B_n \cos \omega_n t\} \tag{6.27}$$

where n is 1, 2, 3, . . . for the first, second, third, . . . mode of vibration respectively. Similarly,

$$\ell\theta_{r,n} = \sum_{n=1}^{N} \Theta_{r,n}\{A_n \sin \omega_n t + B_n \cos \omega_n t\} \tag{6.28}$$

With these substitutions equation (6.24) becomes

$$2\{\mathbf{E} - (2 - \gamma_n) + \mathbf{E}^{-1}\}V_{r,n} - (\mathbf{E} - \mathbf{E}^{-1})\Theta_{r,n} = 0 \tag{6.29}$$

and equation (6.26) becomes

$$-3(\mathbf{E} - \mathbf{E}^{-1})V_{r,n} + (\mathbf{E} + 4 + \mathbf{E}^{-1})\Theta_{r,n} = 0 \tag{6.30}$$

where

$$\gamma_n = \frac{m\omega_n{}^2\ell^3}{12EI}$$

Elimination of $\Theta_{r,n}$ gives

$$[2(\mathbf{E} + 4 + \mathbf{E}^{-1})\{\mathbf{E} - (2 - \gamma_n) + \mathbf{E}^{-1}\} - 3(\mathbf{E} - \mathbf{E}^{-1})^2]V_{r,n} = 0 \tag{6.31}$$

or

$$[(\mathbf{E} + \mathbf{E}^{-1})^2 - 2(2 + \gamma_n)(\mathbf{E} + \mathbf{E}^{-1}) + 4(1 - 2\gamma_n)]V_{r,n} = 0 \tag{6.32}$$

This is an homogeneous fourth-order linear difference equation which has solutions of the form

$$V_{r,n} = \sum_{i=1}^{4} A_{i,n} \exp(\phi_{i,n} r) \tag{6.33}$$

These solutions may be used in equation (6.32) to give the characteristic equation

$$\cosh^2 \phi_{i,n} - (2 + \gamma_n) \cosh \phi_{i,n} + (1 - 2\gamma_n) = 0 \tag{6.34}$$

from which

$$\cosh \phi_{i,n} = (1 + 0{\cdot}5\gamma_n) \pm \sqrt{\{0{\cdot}25\gamma_n(12 + \gamma_n)\}} \tag{6.35}$$

The maximum possible value of ω_n occurs when each mass is oscillating in antiphase with its neighbours. By inspection γ_n lies in the range $0 \leqslant \gamma_n \leqslant 2$, so one value of $\cosh \phi$, say $\cosh \phi_{1,n}$, will be less than or equal to unity, the other, $\cosh \phi_{2,n}$, being greater than unity. The former gives two imaginary values for ϕ, say $\pm i\psi_{1,n}$, and the latter gives two real values $\pm\psi_{2,n}$. Hence the solutions, equation (6.33), may be written in the form

$$V_{r,n} = B_{1,n}\cos\psi_{1,n}r + B_{2,n}\sin\psi_{1,n}r + B_{3,n}\cosh\psi_{2,n}r + B_{4,n}\sinh\psi_{2,n}r \tag{6.36}$$

From equation (6.30)

$$\Theta_{r,n} = \frac{3(\mathbf{E} - \mathbf{E}^{-1})}{(\mathbf{E} + 4 + \mathbf{E}^{-1})} V_{r,n}$$

and noting that for the functions used here the **E** operator has the property that

$$f(\mathbf{E})a^r = a^r f(a)$$

so that

$$\begin{aligned}\frac{(\mathbf{E} - \mathbf{E}^{-1})}{(\mathbf{E} + 4 + \mathbf{E}^{-1})}\cos\psi r &= f(\mathbf{E})\,\mathrm{Re}\exp(i\psi r)\\ &= \mathrm{Re}\,\frac{\exp(i\psi r)[\exp(i\psi) - \exp(-i\psi)]}{[\exp(i\psi) + \exp(-i\psi) + 4]}\\ &= -\frac{\sin\psi\sin\psi r}{2 + \cos\psi}, \text{ etc.,}\end{aligned}$$

gives

$$\begin{aligned}\Theta_{r,n} = \mu_{1,n}\{-B_{1,n}\sin\psi_{1,n}r + B_{2,n}\cos\psi_{1,n}r\}\\ + \mu_{2,n}\{B_{3,n}\sinh\psi_{2,n}r + B_{4,n}\cosh\psi_{2,n}r\}\end{aligned} \tag{6.37}$$

where
$$\mu_{1,n} = \frac{3\sin\psi_{1,n}}{2 + \cos\psi_{1,n}}$$

and
$$\mu_{2,n} = \frac{3\sinh\psi_{2,n}}{2 + \cosh\psi_{2,n}}$$

To simplify the notation we shall omit the subscript n and understand that the expressions apply to each principal mode.

The relative values of the constants B_i are determined from the end conditions. They may be eliminated from the four equations expressing the end constraints, to give the frequency equation, which is satisfied by the characteristic values of ω.

Before considering particular end conditions the relationship between the end forces and moments applied to a beam and the coefficients B_i will be presented. This relationship eases the task of deriving the frequency equation for particular end conditions. It is derived by substituting into equations (6.22) the values for V and Θ given by equations (6.36) and (6.37). After reduction it is found that if

$$\nu_1 = 2\sigma_1\sin\frac{\psi_1}{2} \qquad \nu_2 = 2\sigma_2\sinh\frac{\psi_2}{2}$$

$$\sigma_1 = \frac{6(1 - \cos\psi_1)}{2 + \cos\psi_1} \qquad \sigma_2 = \frac{6(1 - \cosh\psi_2)}{2 + \cosh\psi_2}$$

then

$$\begin{bmatrix} \dfrac{\ell^3 F_{ar}}{EI} \\ \dfrac{\ell^3 F_{br}}{EI} \\ \dfrac{\ell^2 M_{ar}}{EI} \\ \dfrac{\ell^2 M_{br}}{EI} \end{bmatrix} = \begin{bmatrix} \nu_1 \sin (r + \tfrac{1}{2})\psi_1 & -\nu_1 \cos (r + \tfrac{1}{2})\psi_1 & -\nu_2 \sinh (r + \tfrac{1}{2})\psi_2 & -\nu_2 \cosh (r + \tfrac{1}{2})\psi_2 \\ -\nu_1 \sin (r - \tfrac{1}{2})\psi_1 & \nu_1 \cos (r - \tfrac{1}{2})\psi_1 & \nu_2 \sinh (r - \tfrac{1}{2})\psi_2 & \nu_2 \cosh (r - \tfrac{1}{2})\psi_2 \\ \boldsymbol{\sigma}_1 \cos r\psi_1 & \boldsymbol{\sigma}_1 \sin r\psi_1 & \boldsymbol{\sigma}_2 \cosh r\psi_2 & \boldsymbol{\sigma}_2 \sinh r\psi_2 \\ -\boldsymbol{\sigma}_1 \cos r\psi_1 & -\boldsymbol{\sigma}_1 \sin r\psi_1 & -\boldsymbol{\sigma}_2 \cosh r\psi_2 & -\boldsymbol{\sigma}_2 \sinh r\psi_2 \end{bmatrix} [\mathrm{B}]$$

where

$$[\mathrm{B}] = \begin{bmatrix} B_1 \\ B_2 \\ B_3 \\ B_4 \end{bmatrix} (A \sin \omega t + B \cos \omega t) \tag{6.38}$$

We shall examine the vibrations of a system of length $N\ell$ with masses $0{\cdot}5m$ at the ends and $N - 1$ masses m equally spaced at a distance ℓ apart along its length. Both ends of the beam are assumed to be pinned.

At end $r = 0$, $V_0 = 0$ and $M_{a0} = 0$
At end $r = N$, $V_N = 0$ and $M_{bN} = 0$

Therefore, the frequency equation is

$$\begin{vmatrix} 1 & 0 & 1 & 0 \\ \sigma_1 & 0 & \sigma_2 & 0 \\ \cos\psi_1 N & \sin\psi_1 N & \cosh\psi_2 N & \sinh\psi_2 N \\ \sigma_1\cos\psi_1 N & \sigma_1\sin\psi_1 N & \sigma_2\cosh\psi_2 N & \sigma_2\sinh\psi_2 N \end{vmatrix} = 0 \tag{6.39}$$

or

$$(\sigma_1 - \sigma_2)^2 \sin\psi_1 N \sinh\psi_2 N = 0 \tag{6.40}$$

which is satisfied if

$$\psi_1 = \frac{n\pi}{N} \tag{6.41}$$

Substituting for ψ_1 in equation (6.34) gives

$$\omega_n^{\ 2} = \frac{12EI}{m\ell^3} \cdot \frac{\left(\cos\dfrac{n\pi}{N} - 1\right)^2}{2 + \cos\dfrac{n\pi}{N}} \tag{6.42}$$

When the above system is used as a discrete mass representation of a uniform continuous beam the errors in the predicted natural frequencies may be estimated as follows. Consider the lower natural frequencies, that is when n is much smaller than N. The total mass of the beam will be $mN = \rho A\ell N = \rho AL$. Hence

$$\frac{\rho A\omega_n^{\ 2}}{EI} L^4 \simeq (n\pi)^4 \left(1 - \frac{1}{720}\left(\frac{n\pi}{N}\right)^4\right) \tag{6.43}$$

For the continuous beam its value is $(n\pi)^4$, therefore, the error in the estimation of the lower natural frequencies by this lumped mass representation is of order $1/N^4$.

Gladwell (1962) showed that for beams with clamped, pinned, or sliding ends the above discrete mass representation of a continuous beam leads to errors in natural frequencies which are proportional to $1/N^4$. With one or both ends of the beam free the error is proportional to $1/N^2$.

Exercises

1. Show that the frequency equation for a lumped mass representation of a uniform continuous beam clamped at each end is

$$2(1 - \cos\psi_1 N \cosh\psi_2 N) + \left(\frac{\mu_2}{\mu_1} - \frac{\mu_1}{\mu_2}\right) \sin\psi_1 N \sinh\psi_2 N = 0$$

Hence show, for the lower natural frequencies, that the values predicted from this model are in error compared with the exact values for the continuous beam, the error being proportional to $1/N^4$.

N is the number of beam segments used and half the mass of a segment is to be concentrated at each of its ends.

2. When a uniform continuous cantilever is represented by N beam segments, half the mass of a segment being concentrated at each of its ends, show that the error in the lower natural frequencies is given by

$$-\frac{(\lambda L)^2 \sin \lambda L \sinh \lambda L}{3N^2(\cos \lambda L \sinh \lambda L - \sin \lambda L \cosh \lambda L)}$$

where λL is a root of the equation

$$1 + \cos \lambda L \cosh \lambda L = 0$$

6.2 Finite difference representations of continuous systems

Finite difference approximations were introduced in chapter 4. Here we shall examine the accuracy with which they allow natural frequencies to be predicted; longitudinal vibrations of a uniform bar will be examined, other cases will be set as exercises.

6.2.1 Longitudinal vibration of bars

The finite difference form for the equation of motion was given as equation (4.52), that is

$$U_{i+1} - (2 - \alpha^2 h^2)U_i + U_{i-1} = 0 \qquad (4.52)$$

where h is the distance between adjacent nodes and $\alpha^2 = \omega^2 \rho/E$.

Equation (4.52) may be satisfied by solutions of the form

$$U_r = A \exp(\phi r) \qquad (6.44)$$

That is
$$\cosh \phi = 1 - 0{\cdot}5\alpha^2 h^2 \qquad (6.45)$$

Considering the lowest natural frequency of a bar of length h together with that of one length L it is seen that $\cosh \phi$ lies in the range $-1 \leqslant \cosh \phi \leqslant 1$. Thus there are two imaginary values for ϕ, namely $\pm i\psi$. Therefore,

$$U_r = A_1 \cos \psi r + A_2 \sin \psi r \qquad (6.46)$$

where the relative magnitude of A_1 and A_2 depends upon the boundary conditions.

Consider a bar of total length L to be represented at $N - 1$ nodes, that is 0 to N inclusive, it being held rigidly at end 0 and free from stress at end N; then the boundary conditions in finite difference form become,

(*i*) at $r = 0$, $\quad U_0 = 0$

(*ii*) at $r = N$, $\quad (U_{N+1} - U_{N-1})/2h = 0 \qquad (6.47)$

where U_{N+1} is a fictitious node, as discussed in section 4.3.1. The frequency equation is

$$\cos N\psi \sin \psi = 0 \tag{6.48}$$

which is satisfied if

$$N\psi = \frac{2n - 1}{2} \pi \tag{6.49}$$

Hence
$$\omega_n^2 = \frac{2E}{\rho h^2}\left\{1 - \cos \frac{2n - 1}{2N} \pi\right\} \tag{6.50}$$

which for the lower natural frequencies, that is n small compared with N, may be expressed as

$$\omega_n^2 \simeq \left(\frac{2n - 1}{2L} \pi\right)^2 \frac{E}{\rho}\left\{1 - \frac{(2n - 1)^2\pi^4}{48N^2}\right\} \tag{6.51}$$

The error in the frequency of this finite difference approximation is of order $1/N^2$, compared with the exact value.

The above analysis, using a central difference expression for the gradient boundary condition which involved the introduction of a fictitious node, has given the same error as the lumped mass representation examined in section 6.1.1. If the gradient boundary condition were represented by a one-sided difference approximation, namely

$$(U_N - U_{N-1})/h = 0 \tag{6.52}$$

the frequency equation would be

$$(1 - \cos \psi) \sin N\psi + \sin \psi \cos N\psi = 0 \tag{6.53}$$

We expect the solution to differ very little from that of the previous case, therefore let

$$N\psi = \frac{2n - 1}{2} \pi + \beta \tag{6.54}$$

Neglecting second and higher order terms equation (6.53) becomes

$$\beta = \frac{2n - 1}{4N} \pi \tag{6.55}$$

Hence

$$\omega_n^2 \simeq \left(\frac{2n - 1}{2L} \pi\right)^2 \frac{E}{\rho}\left\{1 + \frac{1}{N}\right\} \tag{6.56}$$

Thus this one-sided approximation to the first derivative leads to an error in the lower natural frequencies proportional to $1/N$, which is less accurate than the central difference form used in the previous case.

Exercises

1. Using the central difference form for the equation of motion of a uniform beam given in equation (4.62), show that the vibratory form of the beam is given by

$$V_r = A_1 \cos \phi_1 r + A_2 \sin \phi_1 r + A_3 \cosh \phi_2 r + A_4 \sinh \phi_2 r$$

where $$\cos \phi_1 = 1 - 0{\cdot}5\lambda^2 h^2$$

and $$\cosh \phi_2 = 1 + 0{\cdot}5\lambda^2 h^2$$

2. Show for a uniform cantilever, clamped at $r = 0$, free at $r = N$, that the frequency equation is,

$$\sin \phi_1 \sinh \phi_1(\mu_1{}^2 + \mu_2{}^2 + 2\mu_1\mu_2 \cos N\phi_1 \cosh N\phi_2) + \mu_1\mu_2(\sin^2 \phi_1 - \sinh^2 \phi_2) \sin N\phi_1 \sinh N\phi_2 = 0$$

where $\mu_1 = 1 - \cos \phi_1$ and $\mu_2 = \cosh \phi_2 - 1$

when the boundary conditions are represented by central difference expressions.

3. The frequency equation for a continuous uniform cantilever has the form

$$1 + \cos \psi \cosh \psi = 0$$

the characteristic values, ψ_n, of which should satisfy approximately the frequency equation given in exercise 2. Hence or otherwise, show that the errors in the lower natural frequencies involved by representing the equations of motion and boundary conditions for a cantilever by central finite difference forms are of order $1/N^2$.

4. Finite difference forms for the equations of motion and boundary conditions of a simply supported uniform plate were discussed in chapter 4 (section 4.3.3). Show that those equations, for a simply supported rectangular plate, are satisfied by solutions of the form

$$W_{r,s} = A \sin \frac{rp\pi}{M} \sin \frac{sq\pi}{N}$$

where p and q are integers, and M and N are the number of mesh lengths in the r and s directions respectively.

Hence show that the errors involved in predicting the lower natural frequencies by this finite difference representation are of order $1/N^2$ or $1/M^2$.

6.3 Miscellaneous systems

The methods outlined above may be used to study the accuracy of finite element solutions but in this case the analysis even for a simply supported beam becomes tedious. However this does not exhaust the usefulness of the method as will be seen from the bibliography. Here we shall use it to illustrate a well-known phenomenon associated with the vibrations of elastically-coupled identical subsystems. When vibrated, many different modes with virtually the same frequency may be executed. The phenomenon is utilised in the narrow band-pass filters for electrical circuits; the designers of turbines are aware of it in turbine blade vibration.

We may illustrate it by considering the system shown in Fig. 6.4. Consider the system to be comprised of a set of $N + 1$ identical light cantilevers, each of stiffness k, and carrying a mass m at their free ends. The masses are coupled elastically

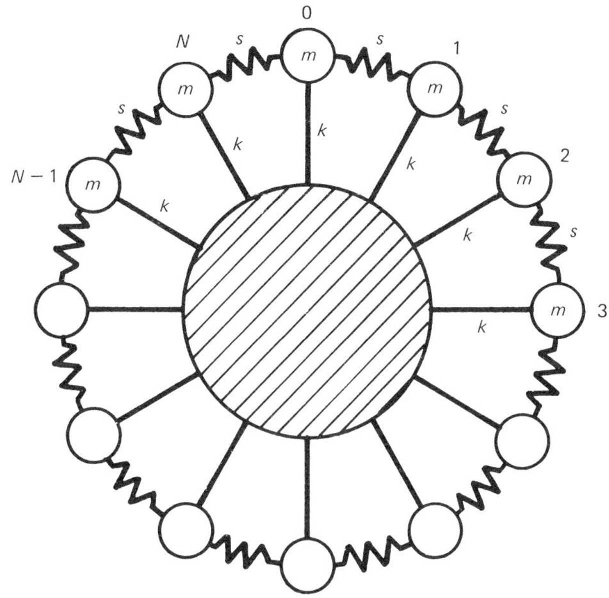

Fig. 6.4

by springs of stiffness s, as shown. The fixed ends of the cantilevers are clamped to a rigid stationary foundation.

The equation of motion for a typical mass is

$$m\ddot{x}_r + kx_r + s(2x_r - x_{r+1} - x_{r-1}) = 0 \tag{6.57}$$

the solutions of which have the form

$$x_r = X_r(A \sin \omega t + B \cos \omega t) \tag{6.58}$$

On substitution of this expression for x_r, equation (6.57) becomes

$$\left[\mathbf{E} - 2\left(1 + \frac{k - m\omega^2}{2s}\right) + \mathbf{E}^{-1}\right] X_r = 0 \tag{6.59}$$

The solutions to equation (6.59) have the form

$$X_r = A \sinh \phi r + B \cosh \phi r \tag{6.60}$$

where $$\cosh \phi = 1 + \frac{k - m\omega^2}{2s}$$

but as $1 - (k - m\omega^2)/2s$ lies in the range -1 to $1 + k/2s$, imaginary values of ϕ might be found. The boundary conditions are in this case

$$X_0 = X_{N+1}$$

and

$$X_{-1} = X_N \tag{6.61}$$

giving the frequency equation

$$\sinh \phi \, [1 - \cosh (N + 1)\phi] = 0 \tag{6.62}$$

which is satisfied if

$$\phi = \frac{i2p\pi}{N + 1} \tag{6.63}$$

Hence

$$\omega^2 = \frac{k}{m} + \frac{2s}{m}\left(1 - \cos \frac{2p\pi}{N + 1}\right) \tag{6.64}$$

and the principal modes are given by

$$X_r = B \cos \frac{2p\pi r}{N + 1} \tag{6.65}$$

Equation (6.64) shows that when s is small compared with k, that is when the coupling is weak, the $N + 1$ natural frequencies lie in a narrow band about the natural frequency of a single cantilever. Equation (6.65) shows that the normal modes cover the range from all cantilevers vibrating in phase to each being 180 degrees out of phase with its neighbours. Obviously the phase of one mode relative to any other is arbitrary, depending in the case of free vibrations upon the way the motion starts.

Some systems, instead of leading to homogeneous difference equations, give rise to a non-homogeneous form. By analogy with linear differential equations, the complete solution then requires a primitive or complementary solution satisfying

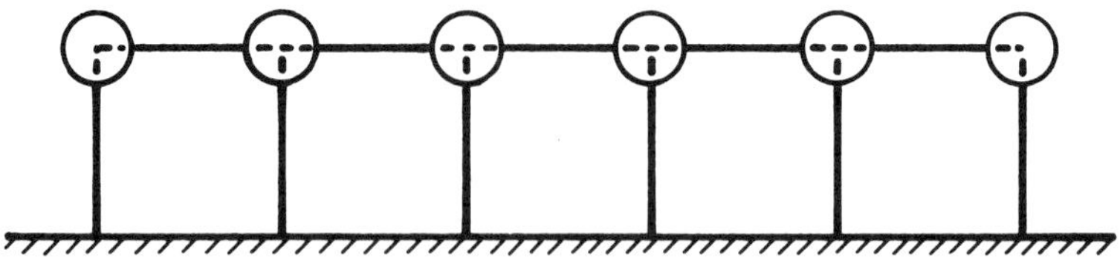

Fig. 6.5

the homogeneous equation together with a particular solution satisfying the whole equation. The simple system shown in Fig. 6.5 will be used to illustrate the method of solution. The system is comprised of a light elastic framework with a disc of moment of inertia J and mass m attached to each joint. Each beam is of length L, it has a bending rigidity EI, and an infinite longitudinal stiffness.

Utilising the expressions given in equation (6.22), the equations of motion become

$$J\ddot{\theta}_r + \frac{2EI}{L}\{\theta_{r+1} + 6\theta_r + \theta_{r-1}\} = \frac{6EI}{L^2}v_r \tag{6.66}$$

and

$$\sum_r \left\{\frac{12EI}{L^3}v_r - \frac{6EI}{L^2}\theta_r + m\ddot{v}_r\right\} = 0 \tag{6.67}$$

Since the beams have been assumed inextensible v_r is independent of r and equations (6.66) and (6.67) are satisfied by solutions of the form

$$\begin{aligned}\theta_r &= \Theta_r(A \sin \omega t + B \cos \omega t)\\ v_r &= V(A \sin \omega t + B \cos \omega t)\end{aligned} \tag{6.68}$$

which on substitution in equation (6.66) gives

$$(\mathbf{E} + 2\gamma + \mathbf{E}^{-1})\Theta_r = \beta \tag{6.69}$$

where $\beta = 3V/L$

and $\gamma = 3 - J\omega^2 L/4EI$

Equation (6.69) is a non-homogeneous second-order linear difference equation. It will be remembered that the complete solution to non-homogeneous linear differential equations is comprised of a complementary solution and a particular solution, also that the form of the solution depends upon the roots of the characteristic equation; similar findings apply in the case of linear difference equations.

To find the complementary solution we assume a solution of the form

$$\Theta_r = P\alpha^r \tag{6.70}$$

where P and α are constants. The characteristic equation is formed by substituting for Θ_r in equation (6.69) with β set to zero, that is

$$\alpha + 2\gamma + \alpha^{-1} = 0 \tag{6.71}$$

Hence

$$\alpha = -\gamma \pm \sqrt{(\gamma^2 - 1)} \tag{6.72}$$

Now we may estimate bounds for the value of γ as follows: the lowest natural frequency must be greater than zero so an upper bound for γ is 3; the highest natural frequency of the system will be in the vicinity of that when V is zero and Θ_r is a constant for a system with an infinite number of mass points, for this case $\gamma = -1$. To be certain we shall assume that γ may be less than -1.

When $3 \geqslant \gamma > -1$ the two values of α given by equation (6.72) are both negative; let them be $-\nu$ and $(-\nu)^{-1}$. Then the complementary solution from equation (6.70) is

$$\begin{aligned}{}_c\Theta_r &= P_1(-\nu)^r + P_2(-\nu)^{-r}\\ &= (-1)^r\{P_1\nu^r + P_2\nu^{-r}\}\end{aligned} \tag{6.73}$$

or

$${}_c\Theta_r = (-1)^r\{P \sinh \phi r + Q \cosh \phi r\} \tag{6.74}$$

where $$\cosh \phi = \gamma$$

The particular solution for this range of γ is

$$_p\Theta_r = \frac{\beta}{2(1 + \cosh \phi)} \tag{6.75}$$

found by letting Θ_r be a constant in equation (6.69). Hence the complete solution is

$$\Theta_r = \frac{\beta}{2(1 + \cosh \phi)} + (-1)^r \{P \sinh \phi r + Q \cosh \phi r\} \tag{6.76}$$

In the range $3 \geqslant \gamma > -1$, the case when $\gamma = 1$ requires investigation as the assumed solution, equation (6.70), is not then complete. When $\gamma = 1$, it is easily shown that the complete solution is

$$\Theta_r = \frac{\beta}{4} + (-1)^r \{P + Qr\} \tag{6.77}$$

Another special solution is required when $\gamma = -1$, the complete solution being

$$\Theta_r = \frac{\beta}{2} r^2 + P + Qr \tag{6.78}$$

When $\gamma < -1$ the complete solution may be found by the procedure outlined in deriving equation (6.76). It is

$$\Theta_r = \frac{\beta}{2(1 - \cosh \phi)} + P \sinh \phi r + Q \cosh \phi r \tag{6.79}$$

where $\cosh \phi = -\gamma$.

For a system with $N + 1$ mass points the moment balance at each end together with the overall force balance, represented by equation (6.67), allow the three unknowns P, Q, and V to be eliminated to form the frequency equation for each of the solutions (6.76) to (6.79).

As an example we shall derive the frequency equation for a system with $N + 1$ mass points, r taking the values 0 to N. We shall restrict consideration to the frequency range associated with the solution (6.76).

At end $r = 0$ moment balance requirements give

$$\left(-J\omega^2 + \frac{8EI}{L}\right) \Theta_0 + 2\frac{EI}{L} \Theta_1 = \frac{6EI}{L^2} V \tag{6.80}$$

or

$$2(\cosh \phi - 1)\Theta_0 + \Theta_1 = \beta \tag{6.81}$$

therefore

$$\frac{-9V}{2L(1 + \cosh \phi)} - (\sinh \phi)P + (\cosh \phi - 2)Q = 0 \tag{6.82}$$

At $r = N$ moment balance requirements give

$$2(\cosh\phi - 1)\Theta_N + \Theta_{N-1} = \beta$$

or

$$\frac{-9}{2L(1+\cosh\phi)}V + (-1)^N\{2(\cosh\phi - 1)\sinh\phi N - \sinh\phi(N-1)\}P$$
$$+ (-1)^N\{2(\cosh\phi - 1)\cosh\phi N - \cosh\phi(N-1)\}Q = 0 \quad (6.83)$$

the overall force balance gives

$$(N+1)\left\{\frac{12EI}{L^3} - m\omega^2\right\}V - \frac{6EI}{L^2}\sum_{r=0}^{N}\Theta_r = 0$$

or

$$(N+1)\left\{\frac{3EI(1+4\cosh\phi)}{L^3(1+\cosh\phi)} - m\omega^2\right\}V + \left(\sum_{r=0}^{N}(-1)^r\sinh\phi r\right)P$$
$$+ \left(\sum_{r=0}^{N}(-1)^r\cosh\phi r\right)Q = 0 \quad (6.84)$$

From equations (6.82), (6.83), and (6.84) the frequency equation may be formed. The two summations may be performed analytically; the characteristic values of ω may then be found by a step-by-step search procedure using a digital computer.

Exercises

1. The essential components of a narrow band-pass mechanical wave filter, which is used in connection with electronic circuits, are shown in Fig. 6.6. They consist of a series of identical subsystems, each with a natural frequency ω, which

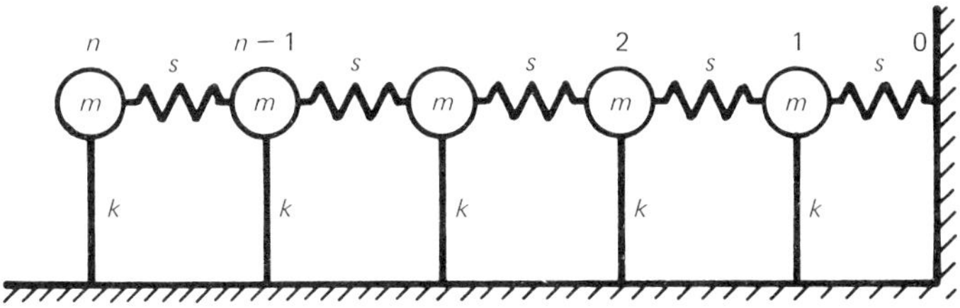

Fig. 6.6

are elastically coupled by springs of stiffness s. Show that the natural circular frequencies p_q of the system are given by

$$p_q^2 = \omega^2\left[\frac{2s}{k}\left\{1 - \cos\left(\frac{2q+1}{2n+1}\right)\pi\right\} + 1\right]$$

where n = the number of identical elements
k = the stiffness of each cantilever spring
q = an integer, the range of which must be defined.

2. Show that the natural frequencies p_q of the system shown in Fig. 6.7 are given by

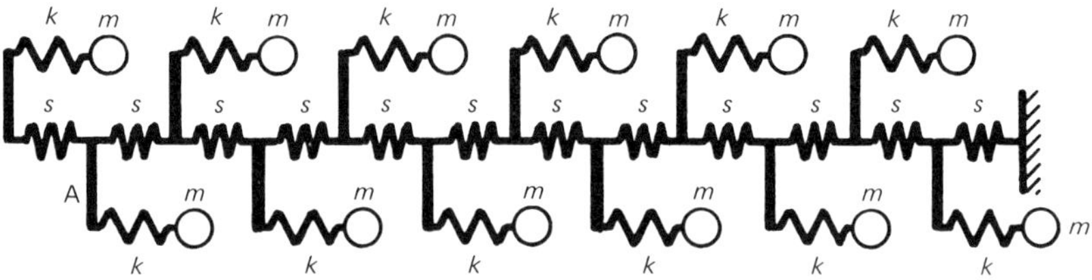

Fig. 6.7 (**Note**: Bars of type 'A' are massless, rigid, and remain perpendicular to the axes of the springs)

$$p_q^2 = \frac{\omega^2\left[1 - \cos\left(\frac{2q+1}{2n+1}\right)\pi\right]}{\frac{k}{2s} + 1 - \cos\left(\frac{2q+1}{2n+1}\right)\pi}$$

where $\omega^2 = k/m$
n = the number of masses
q = an integer, the range of which must be defined.

3. The system illustrated in Fig. 6.8 is comprised of a number of identical beams of length L, cross-sectional area A, bending rigidity EI, and mass density ρ. Each beam is free at one end and rigidly fastened to a light elastic beam at the other. The light beam has a bending rigidity E_1I_1 and the distance between each of the pinned supports is L_1. If Θ_r is the rotational amplitude of the light beam at the r'th support show that

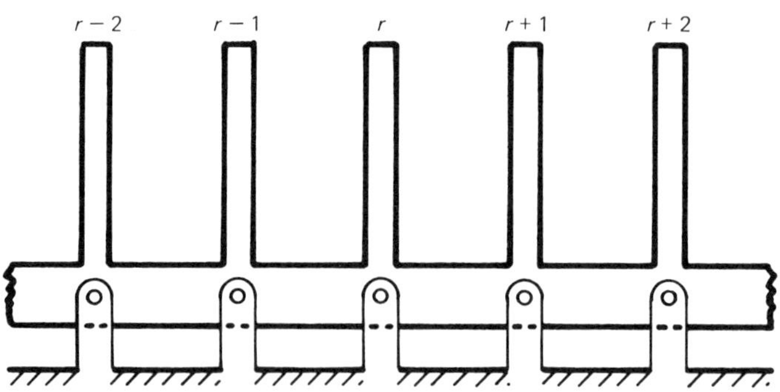

Fig. 6.8

$$\Theta_{r+1} + \left(4 + K\frac{\lambda L}{2}\frac{F_5}{F_4}\right)\Theta_r + \Theta_{r-1} = 0$$

where $K = \dfrac{EI}{L}\dfrac{L_1}{E_1 I_1}$

$\lambda^4 = \rho A \omega^2 / EI$

F_4 and F_5 = functions of λL defined in section 3.3.1.

If the light elastic beam is in the form of a ring with $N + 1$ beams cantilevered from it, show that the frequency equation has the form

$$\lambda L \frac{F_5}{F_4} = -\frac{4}{K}\left\{2 + \cos\left[\frac{2p\pi}{(N+1)}\right]\right\}$$

Hence show that an infinite number of bands of natural frequencies occur.

Further reading

(*a*) *Books*

MILNE-THOMSON, L. M.
The calculus of finite differences, Macmillan, London, 1951.

WAH, T. and CALCOTE, L. R.
Structural analysis by finite difference calculus, Van Nostrand Reinhold, New York, 1970.

(*b*) *Papers*

ELLINGTON, J. P. and MCCALLION, H.
'Moments and deflections of a simply-supported beam grillage', *Aeronaut. Quart.*, **8**, 1957, 360–8.

ELLINGTON, J. P. and MCCALLION, H.
'Blade frequency of turbines—effect of disc elasticity', *Engng.*, **187**, 1959, 645–6.

ELLINGTON, J. P. and MCCALLION, H.
'The free vibration of grillages', *J. Appl. Mech.—Trans. A.S.M.E.*, **26**, 1959, 603–7.

RIEGER, N. F. and MCCALLION, H.
'The natural frequencies of portal frames—II', *Int. J. Mech. Sci.*, **7**, 1965, 263–76.

References

ELLINGTON, J. P.
'The vibration of segmented beams', *Br. J. Appl. Phys.*, **7**, 1956, 299–303.

GLADWELL, G. M. L.
'The approximation of uniform beams in transverse vibration by sets of masses elastically connected', *Proc. 4th U.S. Nat. Congress of Applied Mechanics, Amer. Soc. Mech. Engrs.*, 1962, 169–76.

KARMAN, T. V. and BIOT, M. A.
Mathematical methods in engineering, McGraw-Hill, New York, 1940.

Chapter 7

Rotating shafts

The motion of systems with quickly rotating components, turbomachinery for example, is strongly dependent upon the complexity of the system. For instance, it is influenced and often markedly changed by whether or not a rotating shaft is symmetrical, by the type of bearings it runs in, and by whether or not the bearing support stiffnesses are symmetrical. The results of a comprehensive investigation were given by Smith (1933), this chapter has been modelled on that classic paper.

In this chapter, the influence of a number of the above variables, taken separately, will be discussed, but before doing so it is necessary to understand the behaviour of a very simple system. This understanding will form a basis from which many of the other investigations of the chapter will commence.

7.1 A concentrated mass on a light elastic shaft

7.1.1 Basic principles

Imagine a straight uniform light circular elastic rod to be held horizontally, with one end clamped in the chuck of a lathe and the other firmly fixed to a concentrated mass; the centre of mass coincides with the geometric and elastic axes of the rod. As the rod is horizontal, its axis will deflect from its unloaded straight horizontal position, under the end load. The rod is uniform and circular, its geometrical and elastic properties are, therefore, the same about all diameters. As the geometrical and elastic axes of the rod coincide in this case, we shall refer to both as the axis.

When the chuck is given a rotation the position of the mass will remain unmoved, by symmetry considerations, but each cross-section of the rod normal to the axis (in its deflected position) rotates through the same angle as the chuck. Thus when the chuck is set in motion at a slow constant angular velocity Ω, each cross-section normal to the axis (in the deflected position) rotates about the axis with the same constant angular velocity Ω. This may be demonstrated easily, but, so that we may continue to reason out the shaft motion for the case when the centre of mass does not lie on the axis, we shall discuss this case further.

Consider diagram (*a*) of Fig. 7.1, which represents the motion of the end of a light cantilevered rod when acted upon by a constant amplitude force. The direction of the force is rotating at an angular velocity $-\Omega_a$, the counter-clockwise direction being positive. The rod is not rotating about its axis (in the deformed state). A mark made on the end of the shaft at A would not move relative to the centre of the end cross-section, indicating that the fibres of the rod do not rotate relative to its axis. Turn now to diagram (*b*) of Fig. 7.1, here both the constant amplitude force vector and the rod rotate at a constant angular velocity $+\Omega_b$ (i.e., counter-clockwise). Note in this case that the force is always directed along the same diameter of the rod cross-section. The condition which applies to the slowly rotating shaft is a combination of these. That is, we wish to rotate the shaft and hold the force constant in magnitude and direction. This may be achieved by combining a rotation of the force and shaft together at an angular velocity $+\Omega$ and a rotation of the force at an angular velocity $-\Omega$. That is the resultant motion may be built up by rotating the force and shaft anticlockwise through an angle Ωt as indicated in Fig. 7.1(*b*) and then, from this displaced position, add motion of the type indicated in Fig. 7.1(*a*) corresponding to a clockwise rotation of the force vector through an angle Ωt. The resultant positions of the shaft for $\Omega t = 0, \pi/2, \pi$, and $3\pi/2$ are shown in Fig. 7.1(*c*).

It is now clear that an initially straight circular shaft carrying a perfectly balanced mass will rotate about its static equilibrium axis, when rotated slowly. What is more we could correctly argue that as the mass is perfectly balanced no other forces act to disturb this state, even when the shaft rotates quickly.

If the shaft is not perfectly balanced, the mass centre being a distance h from the axis, a rotating force of amplitude $m\Omega^2 h$ will be generated, where m is the value of the mass and Ω the angular velocity of the shaft. This force will rotate at an angular velocity Ω. Thus a rotating force will act upon an otherwise quiescent system, giving a state very like those discussed in chapter 2. Obviously, then, an initially straight circular shaft which carries an imperfectly balanced mass will vibrate about its static equilibrium position, when it rotates. The system considered here has two degrees of freedom, x and y in the plane in which the mass vibrates, the two natural frequencies being identical. In common with all undamped linear systems, when the angular velocity of the exciting force is coincident with a system natural frequency the amplitude of the oscillations becomes very large. For rotating systems this angular velocity is called a critical speed. It is a resonant state. Also in the case of a rotating system the vibratory behaviour is referred to as whirling.

The above phenomena may be expressed mathematically.

7.1.2 An initially straight shaft

Let the system be comprised as follows: A torsionally-rigid light elastic circular shaft is held firmly at one of its ends in the chuck of a lathe. To its other end is attached an imperfectly balanced mass m, the distance from the centre of mass to the axis of the shaft being h. Again the geometrical and elastic axes are assumed to

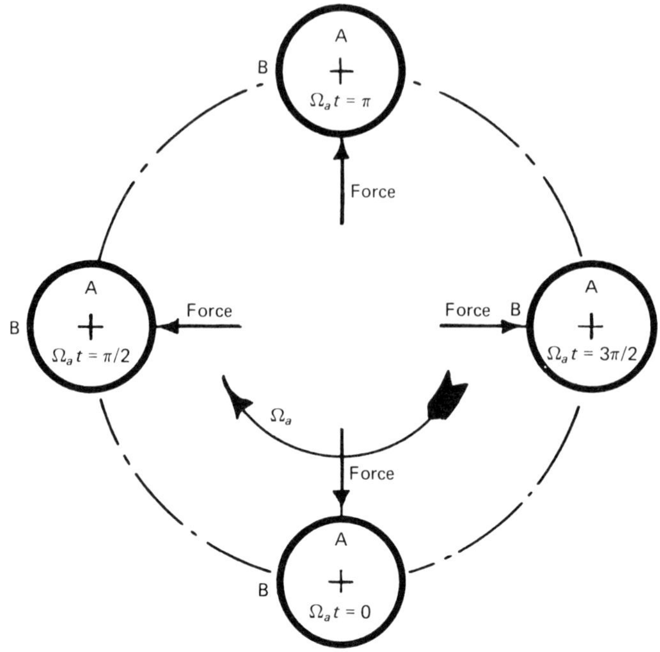
A
B
$\Omega_a t = \pi$
Force
A
B
$\Omega_a t = \pi/2$
Force
Force
B
A
$\Omega_a t = 3\pi/2$
Ω_a
Force
A
B
$\Omega_a t = 0$

(*a*)

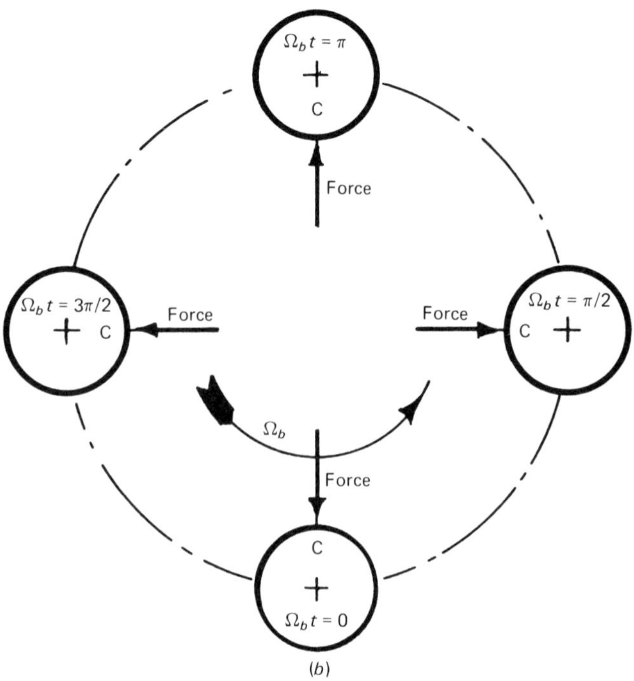
$\Omega_b t = \pi$
C
Force
$\Omega_b t = 3\pi/2$
C
Force
Force
$\Omega_b t = \pi/2$
C
Ω_b
Force
C
$\Omega_b t = 0$

(*b*)

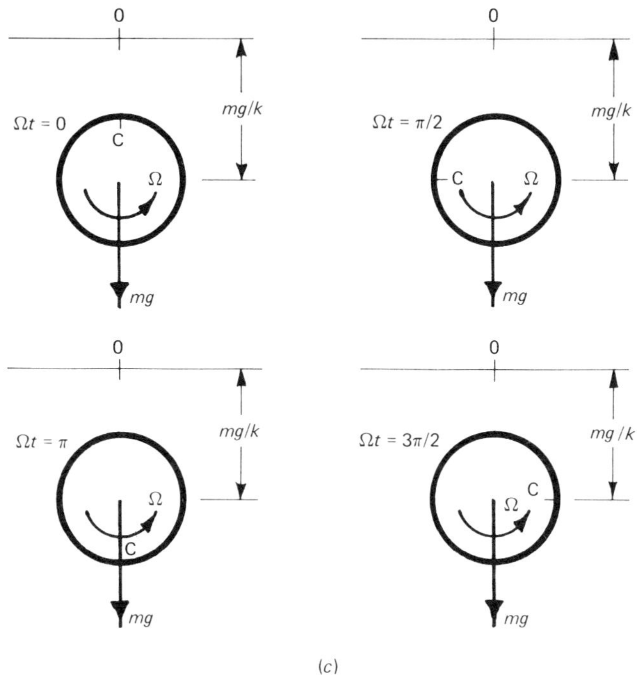

Fig. 7.1(*a*) Force vector rotates clockwise; shaft does not rotate about its own centre
(*b*) Force vector rotates anticlockwise; shaft rotates anticlockwise about its own centre
(*c*) Motions (*a*) and (*b*) have been summed to give the motion of a fibre of a rotating shaft which carries a perfectly balanced mass. The shaft is rotating anticlockwise about its own axis

coincide, the elastic axis being the locus of the points of intersection of the principal axes of elasticity. Resulting from the bending rigidity of the shaft, a horizontal deflection of the mass, x, requires a force kx and a vertical deflection y requires a force ky. Instead of depicting the system as being three-dimensional we may represent it diagrammatically, as shown in Fig. 7.2, where the springs are equivalent to the elasticity of the shaft. The lathe chuck, which is represented by a rigid ring, is made to rotate with a constant angular velocity Ω. The rollers represent the support bearings. 0 is the equilibrium position of the shaft centre, c is the displaced position of the shaft centre, and m is at the mass centre a distance h from c.

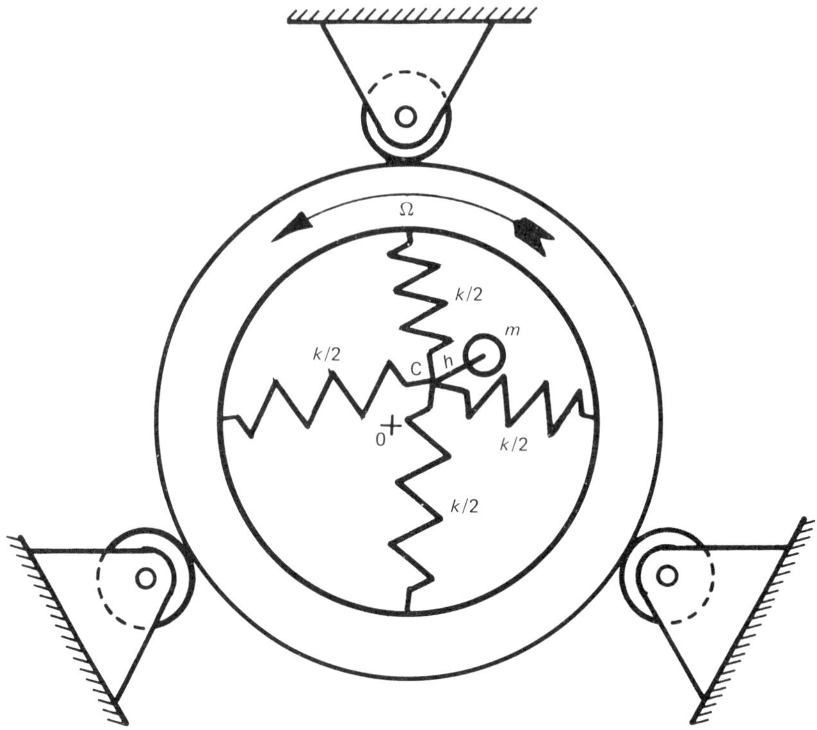

(a)

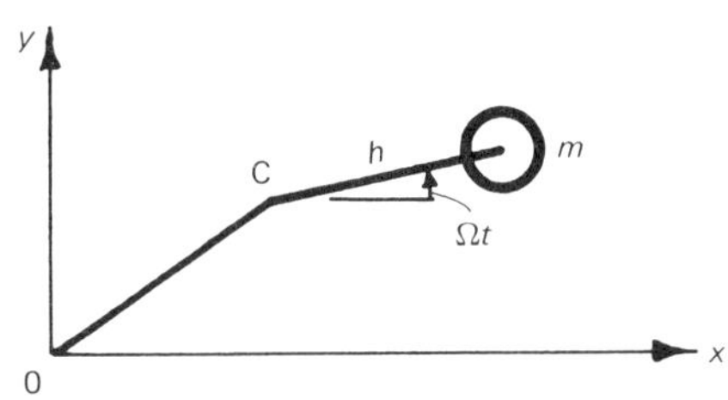

(b)

Fig. 7.2(a), (b)

The equations of motion

$$m \frac{d^2}{dt^2}\{x + h\cos\Omega t\} = -kx$$

$$m \frac{d^2}{dt^2}\{y + h\sin\Omega t\} = -ky$$

or
$$m(\ddot{x} - h\Omega^2\cos\Omega t) + kx = 0 \tag{7.1}$$

$$m(\ddot{y} - h\Omega^2\sin\Omega t) + ky = 0 \tag{7.2}$$

may be solved, giving

$$x = A \sin \omega t + B \cos \omega t + \frac{h\Omega^2 \cos \Omega t}{\omega^2 - \Omega^2} \tag{7.3}$$

$$y = C \sin \omega t + D \cos \omega t + \frac{h\Omega^2 \sin \Omega t}{\omega^2 - \Omega^2} \tag{7.4}$$

where $$\omega^2 = k/m$$

The first two terms in the expressions for x and y obviously represent the free vibratory motion; the third in each case is the forced motion resulting from the out-of-balance. Note that the amplitude of the forced motion depends upon the value of h, the radius of the mass centre, and that the resonant or critical speed is at $\Omega = \omega$. The natural angular frequency ω of the system is unaffected by the rotation. That is, at any rotational speed the free vibrations have the same frequency as for the non-rotating system. The modes of free vibration are also unaffected. For an imperfectly balanced system a critical speed occurs when the rotational velocity of the shaft coincides with the natural angular frequency of the system.

A shaft of an inherently stable system may be run through critical speeds of this type. That is, a shaft may be operated safely above its resonance-type critical speeds provided the system is not dynamically unstable. Dynamic instability will be discussed in sections 7.4 to 7.8.

It is interesting to follow the change in the position of the centre of mass relative to the axis of the shaft from below to above its critical speed. When the rotational velocity of the shaft Ω is lower than its natural angular frequency ω the rotating displacement vector, of amplitude $h\Omega^2/(\omega^2 - \Omega^2)$, is in phase with the rotating force vector, of amplitude $mh\Omega^2$. So the centre of mass rotates at a radius greater than that of the shaft axis. When the shaft velocity Ω is above ω, the displacement and force vectors are in antiphase, that is the mass centre lies between the axis at the end of the shaft and its equilibrium position 0. The mass centre approaches 0 as Ω becomes very large, the end of the shaft then vibrates with an amplitude equal to h.

With what is termed stationary damping, the change in phase between the exciting force and displacement is gradual, as for the oscillatory motion of systems discussed in chapters 1 and 2. An examination of the influence of damping on rotating systems will be deferred until we have discussed the direction in which the shaft whirls, the motion of a bent shaft and the balancing and vibration of a shaft with mass distributed along its length.

7.1.3 Directions of whirls

In the discussion associated with Fig. 7.1(*b*) we saw that a constant amplitude force rotating anticlockwise at an angular velocity Ω caused the shaft centre to orbit its equilibrium position with a constant amplitude at the same angular velocity, Ω. The shaft was also considered to be rotating anticlockwise and the resulting motion of the shaft centre is referred to as a forward whirl. That is, the shaft centre whirls in the same direction as the shaft rotates about its own axis. If the direction of the whirl were opposite to that of rotation of the shaft about its own axis the motion would be referred to as a reverse whirl. The directions of the whirls which may arise in rotating systems are easily found from the solutions of the equations of motion, as follows.

Consider the complex quantity $z = x + iy$, where x and y are given by equations (7.3) and (7.4), and $i = \sqrt{-1}$. Substituting for x and y gives

$$z = E\,\mathrm{e}^{i\omega t} + F\,\mathrm{e}^{-i\omega t} + \frac{h\Omega^2}{\omega^2 - \Omega^2}\,\mathrm{e}^{j\Omega t} \tag{7.5}$$

where

$$E = 0{\cdot}5\{B + C + i(-A + D)\}$$
$$F = 0{\cdot}5\{B - C + i(A + D)\}$$

Obviously, the free response is a combination of a forward and a reverse whirl; the amplitudes E and F depend upon the initial conditions. Whereas, for this type of excitation, the forced motion is a forward whirl.

As equations (7.1) and (7.2) are uncoupled, we could have solved for z in the following manner. Combining equations (7.1) and (7.2) gives

$$\ddot{z} + \omega^2 z = h\Omega^2\,\mathrm{e}^{i\Omega t} \tag{7.6}$$

the solution of which is again equation (7.5).

Exercises

1. An initially straight uniform steel shaft of length 1 m and diameter 3 mm is supported at its ends in self-aligning ball bearings. At the centre of the span a concentrated mass of 0·5 kg is rigidly attached such that its mass centre is 0·1 mm from the centre of the circular cross-section of the shaft. Calculate the critical speed of the system, plot the whirl amplitude as a function of rotational speed and state the direction of this whirl.

   Neglect the mass of the shaft.
2. If the system described in the previous exercise were perfectly balanced and subjected to an exciting force $A \sin \omega t$ acting horizontally on the mass, plot the response at the mass in the x and y directions as a function of ω when the shaft rotates at a constant speed of (*a*) 0·5 and (*b*) 1·5 times the critical speed. Does the shaft whirl?

7.1.4 An initially bent shaft

When, in the unloaded condition, the circular shaft considered previously is slightly bent its free end describes a circular path of radius r_0 about a centre c, as the lathe rotates slowly. The gravitational force, due to a mass attached to the free end, now causes a constant vertical displacement, relative to the circular path traced out in the unloaded condition, when rotated slowly. That is, the path traced

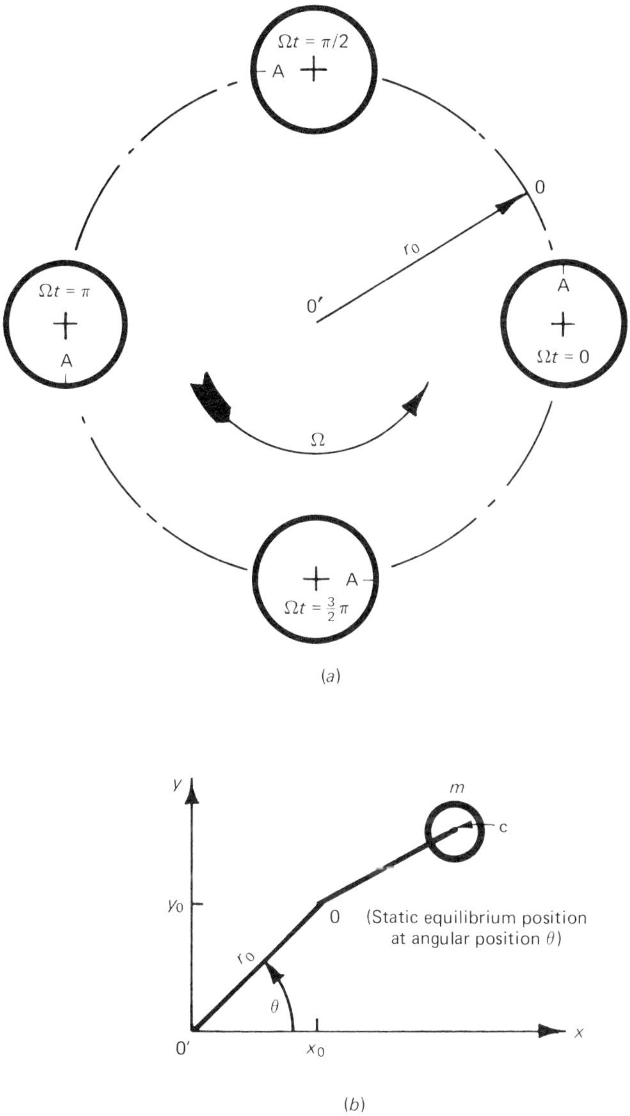

Fig. 7.3 Motion of a rotating initially bent shaft (*a*) and (*b*)

out by the end when carrying a mass is a circle of the same radius as that in the unloaded condition but with centre c displaced vertically downwards a distance equal to the static elastic deflection of the shaft. Thus the initial lack of straightness may now be taken relative to the static deflection curve of the initially straight shaft. To examine the motion of the system when rotating quickly, we set up the equations of motion. Figure 7.3 illustrates the state; the mass centre m and the geometrical/elastic axis at the end of the shaft in this instance are assumed coincident. $0'$ would be the static equilibrium position of an initially straight shaft. r_0 is due to the initial lack of straightness. The equations of motion are

$$m\ddot{x} + k(x - x_0) = 0 \tag{7.7}$$

and

$$m\ddot{y} + k(y - y_0) = 0 \tag{7.8}$$

where k is the bending stiffness of the shaft,

$$x_0 = r_0 \cos \Omega t$$

and

$$y_0 = r_0 \sin \Omega t$$

The solutions to equations (7.7) and (7.8) are

$$x = A \sin \omega t + B \cos \omega t + \frac{\omega^2 r_0 \cos \Omega t}{(\omega^2 - \Omega^2)} \tag{7.9}$$

and

$$y = C \sin \omega t + D \cos \omega t + \frac{\omega^2 r_0 \sin \Omega t}{(\omega^2 - \Omega^2)} \tag{7.10}$$

where

$$\omega^2 = k/m$$

This means that the radius of the mass centre is greater than the initial radius r_0 when the rotational velocity Ω is between zero and the natural angular frequency ω, and in phase with it. Whereas when Ω is greater than ω the radius to the mass centre is 180 degrees out of phase with the initial radius r_0. Note also that when Ω^2 is greater than $2\omega^2$ the radius of the mass centre is less than r_0. Thus the whirl radius of the mass approaches zero as the rotational speed of the shaft becomes very high. Therefore, in this case, when there is no out-of-balance relative to the shaft axis, the end of the shaft will run with very little amplitude at speeds high compared to the natural frequency.

Exercises

1. An initially bent cantilevered shaft of circular cross-section has attached to it a mass m, the centre of which is offset a distance h from the centre of the end of the shaft. Show that the whirl amplitude of the mass approaches h at very high rotational speeds.
2. Discuss the whirling behaviour of the following system. Each end of a light straight circular shaft is rigidly fastened to a disc. The discs are held in bearings driven so that they rotate about a common axis, but, due to manufacturing imperfections the shaft axis is a distance h from it, both axes lying in one plane. A mass, perfectly balanced relative to the shaft axis, is attached at one third the way along the shaft.
3. In the previous sections it has been assumed that the elastic and geometric axes of the shaft coincided. Now assume that this is not so but it is possible for the stiffness to remain symmetrical. Discuss the motion of an initially straight

cantilevered shaft carrying an end mass; in the unstressed state the geometrical and elastic axes are parallel. Consider the following combinations:

(*a*) mass centre and elastic axis coinciding, imposed rotation at clamped end being about the elastic axis;
(*b*) mass centre and elastic axis coinciding, imposed rotation at clamped end being about the geometrical axis; and
(*c*) mass centre and geometrical axis coinciding, imposed rotation at clamped end being about the elastic axis.

7.2 Balancing of shafts with distributed mass

In the previous sections it has been shown that the critical speeds of shafts carrying concentrated masses coincide with the resonant frequencies of the same shafts in a non-rotating state. For two or more concentrated masses the additional critical speeds will also be unaltered by shaft rotation and, therefore, their prediction raises no new problems. Similarly, the critical speeds of shafts with distributed mass and elasticity may be found as the natural frequencies of the non-rotating shafts by the methods given in chapter 3.

The previous section has also shown that the whirl amplitude of the rotating shaft depends upon the amount of out-of-balance, that is, the distance between the mass centre and the axis about which shaft rotation is imposed. If the out-of-balance were zero no forced vibration would be present and the shaft would run smoothly at the critical speed. Similarly, for two or more concentrated masses on a light straight shaft, if all the masses were perfectly balanced the shaft would run smoothly throughout the speed range. This may be possible for a small number of masses, but when the number is large or the mass is continuously distributed, as in the case of large turbo-alternator rotors, it becomes impractical if not impossible. What is more, only the critical speeds which lie below or near the normal running speed are important enough to require balancing out completely. Another approach has, therefore, been made (Bishop and Gladwell, 1959).

Essentially, the whirling motion considered here is the resultant of forced vibrations in two mutually perpendicular planes. In these planes, the exciting forces arise from out-of-balance. This being so, we may investigate the motion of a shaft with distributed mass and elasticity by the method described in section 3.4.2. That is we may describe its motion in terms of principal coordinates and find the associated generalised forces. The unwanted, that is deleterious, critical speeds may then be eliminated by placing balance weights which reduce the associated generalised forces to zero. Mathematically, this may be expressed as follows.

Figure 7.4 illustrates the cross-section of a thin circular shaft, the fibres of which are rotating about its elastic axis at a constant angular velocity Ω. x, y, z is a fixed coordinate system and u, v, z is a coordinate system which rotates about the z axis with an angular velocity Ω. At the section considered the mass centre of the elemental length is a distance r from the elastic axis, as shown. The components of this radius vector in the u and v directions are r_u and r_v respectively, and the corresponding components of the resulting exciting force are F_u and F_v, where

$$F_u = \rho a \Omega^2 r_u \, dz \tag{7.11}$$

$$F_v = \rho a \Omega^2 r_v \, dz \tag{7.12}$$

ρ is the mass density of shaft material and a is the cross-sectional area of the shaft.

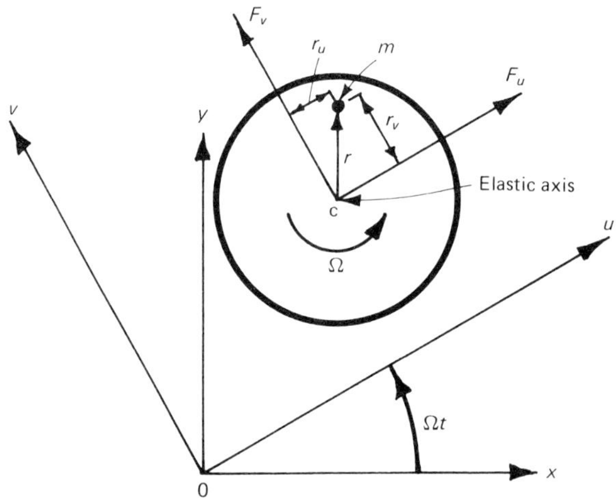

Fig. 7.4

In the fixed coordinate system the exciting forces become

$$\begin{bmatrix} F_x \\ F_y \end{bmatrix} = \begin{bmatrix} \cos \Omega t & -\sin \Omega t \\ \sin \Omega t & \cos \Omega t \end{bmatrix} \begin{bmatrix} F_u \\ F_v \end{bmatrix} \tag{7.13}$$

Therefore, the equations of motion for the element are

$$EI \frac{\partial^4 x}{\partial z^4} + \rho a \frac{\partial^2 x}{\partial t^2} = \rho a \Omega^2 (r_u \cos \Omega t - r_v \sin \Omega t) \tag{7.14}$$

and

$$EI \frac{\partial^4 y}{\partial z^4} + \rho a \frac{\partial^2 y}{\partial t^2} = \rho a \Omega^2 (r_u \sin \Omega t + r_v \cos \Omega t) \tag{7.15}$$

where EI is the bending rigidity of the shaft. The deflections x and y may be expressed in terms of normal coordinates that is

$$x = \sum_k A_k(t) p_k(z); \qquad y = \sum_k B_k(t) q_k(z) \tag{7.16}$$

where $p_k(z)$ and $q_k(z)$ are the normalised eigenfunctions of the k'th modes.

$A_k(t)$ and $B_k(t)$ are the normal coordinates which are trigonometric functions of the argument Ωt so that

$$\frac{\partial^2 A_k}{\partial t^2} = -\Omega^2 A_k \quad \text{and} \quad \frac{\partial^2 B_k}{\partial t^2} = -\Omega^2 B_k.$$

As we are considering a uniform circular shaft $p_k = q_k$.
Substituting for x in equation 7.14 gives

$$\sum_k A_k \left\{ EI \frac{d^4 p_k}{dz^4} - \rho a \Omega^2 p_k \right\} = \rho a \Omega^2 (r_u \cos \Omega t - r_v \sin \Omega t)$$

or

$$\sum_k A_k \{(\omega_k{}^2 - \Omega^2) p_k\} = \Omega^2 (r_u \cos \Omega t - r_v \sin \Omega t) \tag{7.17}$$

where ω_k is the natural angular frequency of the k'th mode of vibration.
We may find the normal coordinates A_k separately by using the orthogonal properties of the normal modes, that is

$$\int_0^L p_k p_\ell \, dz = 1 \quad \text{for } k = \ell$$
$$= 0 \quad \text{for } k \neq \ell$$

Therefore,

$$A_k = \frac{\Omega^2}{\omega_k{}^2 - \Omega^2} \int_0^L p_k (r_u \cos \Omega t - r_v \sin \Omega t) \, dz \tag{7.18}$$

similarly,

$$B_k = \frac{\Omega^2}{\omega_k{}^2 - \Omega^2} \int_0^L p_k (r_u \sin \Omega t + r_v \cos \Omega t) \, dz \tag{7.19}$$

To avoid a large amplitude at the k'th critical speed would require that

$$\int_0^L p_k r_u \, dz = 0 \quad \text{and} \quad \int_0^L p_k r_v \, dz = 0 \tag{7.20}$$

In general this is not practicable. Normally shafts are balanced by the addition of concentrated masses in planes which are referred to as balancing planes. These concentrated masses are generally considered to make a negligible change to the normal modes but, as intended, they modify the generalised forces associated with each mode. If the additional forces in the u and v directions due to the j'th balancing mass M_j are

$$F_{ju} = M_j R_{ju} \Omega^2$$

and

$$F_{jv} = M_j R_{jv} \Omega^2 \tag{7.21}$$

where R_{ju} and R_{jv} are the u and v components of the out of balance radius of the j'th mass, then

$$A_k = \frac{\Omega^2}{\omega_k{}^2 - \Omega^2} \left\{ \int_0^L p_k (r_u \cos \Omega t - r_v \sin \Omega t) \, dz + \frac{1}{\rho a} \sum_j ({}_j p_k)(M_j R_{ju} \cos \Omega t - M_j R_{jv} \sin \Omega t) \right\} \tag{7.22}$$

$({}_jp_k)$ is the amplitude of the normal mode at the j'th mass point. For the whirl amplitude of the k'th mode to be zero

$$\int_0^L p_k r_u \, \mathrm{d}z + \frac{1}{\rho a} \sum_j {}_jp_k M_j R_{ju} = 0$$

and

$$\int_0^L p_k r_v \, \mathrm{d}z + \frac{1}{\rho a} \sum_j {}_jp_k M_j R_{jv} = 0 \tag{7.23}$$

The minimum number of balancing planes required is obviously equal to the number of critical speeds to be balanced out, say n. The associated $2n$ equations of the above form give the $2n$ unknowns, M_jR_{ju} and M_jR_{jv}.

Consider now a very simple example, the balancing of a free-free shaft in its rigid body modes. Let r_u and r_v be given functions of z, the coordinate along the axis of the shaft, the normal mode in translation be $p_1 = 1/\sqrt{L}$ and the normal mode in rotation be

$$p_2 = \sqrt{\left(\frac{12}{L}\right)}\left\{\frac{z}{L} - \frac{1}{2}\right\}.$$

Then in the u direction we have

$$M_1R_{1u} + M_2R_{2u} = -\rho a \int_0^L r_u \, \mathrm{d}z \tag{7.24}$$

for the first mode and

$$M_1R_{1u}z_1 + M_2R_{2u}z_2 - 0{\cdot}5L(M_1R_{1u} + M_2R_{2u}) = -\rho a \int_0^L \{r_u z - 0{\cdot}5Lr_u\} \, \mathrm{d}z \tag{7.25}$$

for the second mode where z_1 and z_2 are the distances of the balancing planes from the origin. Obviously if the first mode were balanced the additional constraint necessary for balancing out the second would be

$$M_1R_{1u}z_1 + M_2R_{2u}z_2 = -\rho a \int_0^L r_u z \, \mathrm{d}z \tag{7.26}$$

Similar expressions have to be satisfied for balance in the v-direction.

Note, the above expressions are those developed in courses on elementary dynamics for the balancing of non-coplanar masses rotating about a common axis.

As a further example, consider the case of a uniform, initially straight, circular steel shaft which is required to run in short bearings at a speed in excess of its third critical speed, the fourth critical being sufficiently far above the running speed to be of no importance. The shaft is 2 m long and has a total mass of 500 kg.

Relative to the axis of the shaft, the components in two mutually perpendicular directions of the radius to the mass axis are given by

$$r_u = [1{\cdot}1 - 2(z/L)] \times 10^{-4}\ \text{m}$$

$$r_v = 4[(z/L) - 1] \times 10^{-5}\ \text{m} \tag{7.27}$$

where L is the length of the shaft and z the distance from one end. The out-of-balance associated with unconstrained rigid body motions has already been balanced in the planes of the bearings.

We are required to find the magnitudes and angular positions of the three masses needed to balance the shaft in its first three modes of vibration, when the balancing planes are at 0·5, 1, and 1·5 m from one end of the shaft and the radius to each mass centre is 0·1 m.

The critical speeds of a shaft supported at its ends in short bearings, which restrain its position only, are given by

$$\omega_n = \frac{n^2\pi^2}{L^2}\left(\frac{EI}{\rho a}\right)^{1/2} \tag{7.28}$$

and its normal modes may be expressed as

$$p_n = \sqrt{\left(\frac{2}{L}\right)} \sin \frac{n\pi z}{L} \tag{7.29}$$

Let $\qquad M_i R_{iu} = C_i \quad \text{and} \quad M_i R_{iv} = D_i$

then equations (7.23) give

$$\begin{bmatrix} 0{\cdot}707 & 1 & 0{\cdot}707 \\ 1 & 0 & -1 \\ 0{\cdot}707 & -1 & 0{\cdot}707 \end{bmatrix} \begin{bmatrix} C_1 \\ C_2 \\ C_3 \end{bmatrix} = -\begin{bmatrix} 31{\cdot}8 \\ 159 \\ 10{\cdot}6 \end{bmatrix} \times 10^{-4}$$

and

$$\begin{bmatrix} 0{\cdot}707 & 1 & 0{\cdot}707 \\ 1 & 0 & -1 \\ 0{\cdot}707 & -1 & 0{\cdot}707 \end{bmatrix} \begin{bmatrix} D_1 \\ D_2 \\ D_3 \end{bmatrix} = \begin{bmatrix} 63{\cdot}7 \\ 31{\cdot}8 \\ 21{\cdot}2 \end{bmatrix} \times 10^{-4} \tag{7.30}$$

which may be solved easily. Therefore,

$$\begin{bmatrix} C_1 \\ C_2 \\ C_3 \end{bmatrix} = \begin{bmatrix} -94{\cdot}5 \\ -10{\cdot}6 \\ 64{\cdot}5 \end{bmatrix} \times 10^{-4}\ \text{kg m}; \qquad \begin{bmatrix} D_1 \\ D_2 \\ D_3 \end{bmatrix} = \begin{bmatrix} 45{\cdot}9 \\ 21{\cdot}2 \\ 14{\cdot}1 \end{bmatrix} \times 10^{-4}\ \text{kg m}$$

If θ_i is the angle of mass M_i relative to the $+u$ axis rotating towards $+v$ axis, then

$$\theta_i = \tan^{-1}(D_i/C_i)$$

therefore

$$\theta_1 = 153 \text{ degrees}, \quad \theta_2 = 116 \text{ degrees}, \quad \text{and} \quad \theta_3 = 12 \text{ degrees}$$

each relative to the $+u$-axis rotating towards the $+v$-axis.

$$M_i = (C_i^2 + D_i^2)^{1/2} \times 10 \text{ kg}$$

giving $$M_1 = 0{\cdot}105 \text{ kg}; \quad M_2 = 0{\cdot}024 \text{ kg}; \quad M_3 = 0{\cdot}067 \text{ kg}$$

Exercises

1. Find the magnitudes and angular positions of three masses to balance the above shaft in its first three modes of vibration, when the balancing planes are located at one-third, one-half, and two-thirds of the shaft length from one end.
2. What restrictions are there on the positions selected for the balancing planes?
3. If the lowest critical speed of a uniform circular steel shaft is p rad/s, how would you expect it to change for each of the following separate modifications:
   1. the attachment of masses at a number of points along the length of the shaft?
   2. the repositioning of the bearings to positions distant a from each end?
   3. the subjection of the shaft to an end thrust?

7.3 The influence of damping

Rotating systems respond in two different ways to friction or damping forces depending upon whether the forces rotate with the shaft or not. When the positions at which the forces act rotate in space with the shaft, for example forces due to the internal friction of the shaft material, the damping is referred to as rotary damping. When the positions at which they act remain fixed in space, for example forces causing energy losses in the bearing support structure, the damping is called stationary damping.

7.3.1 Stationary damping

We shall return now to the system comprised of a light circular cantilevered shaft carrying a mass at its free end, and consider the mass point to be subjected to an additional force arising from external or stationary damping. Let us assume that these forces are of the linear viscous type and are symmetrical with respect to the axis of the shaft. That is, they are proportional to and oppose the velocities of the mass, relative to a fixed reference frame. The equations of motion [(7.1) and (7.2)] require an additional term taking the form

$$m\ddot{x} + c_s\dot{x} + kx = mh\Omega^2 \cos \Omega t \tag{7.31}$$

and $$m\ddot{y} + c_s\dot{y} + ky = mh\Omega^2 \sin \Omega t \tag{7.32}$$

where c_s is the stationary damping coefficient. Again, as in equations (7.1) and (7.2) these equations are uncoupled; therefore, we may introduce the complex quantity $z(=x + iy)$ to combine them, but before doing so it is worth replacing c_s/m by $2\mu_s\omega$, to simplify the form of the solution. μ_s is a non-dimensional stationary damping coefficient. Performing this transformation gives

$$\ddot{z} + 2\mu_s\omega\dot{z} + \omega^2 z = h\Omega^2 e^{i\Omega t} \tag{7.33}$$

the solution of which is

$$z = \exp(-\mu_s\omega t)\{E \exp(i\omega\sqrt{(1 - \mu_s^2)}t) + F\exp(-i\omega\sqrt{(1 - \mu_s^2)}t)\} + \frac{h\Omega^2 \exp(i\Omega t)}{\omega^2 + 2i\mu_s\omega\Omega - \Omega^2} \tag{7.34}$$

From this it may be seen that both the forward and reverse whirls of the free response decay exponentially with time and that their angular velocities are, in magnitude, $(1 - \mu_s^2)^{1/2}$ times that of the undamped case.

The forced response is more easily interpreted if it is expressed as

$$\frac{h\Omega^2 \exp\{i(\Omega t - \xi)\}}{\{(\omega^2 - \Omega^2)^2 + 4\mu_s^2\omega^2\Omega^2\}^{1/2}} \tag{7.35}$$

where $$\xi = \tan^{-1}\left(\frac{2\mu_s\omega\Omega}{\omega^2 - \Omega^2}\right)$$

$\xi = \pi/2$, that is the mass centre lags 90 degrees behind the geometrical centre of the shaft, when the rotational velocity of the shaft Ω equals the undamped natural angular frequency of the system ω. ξ varies gradually from zero at $\Omega = 0$ to π at very high values of Ω.

In other words, with stationary damping the system behaves in very much the same way as a non-rotating system having the same physical parameters, when subjected to out-of-balance type forced vibration.

7.3.2 Rotary damping

Rotary damping forces arise from the varying strain distribution in the rotating members. For example, the fibres at C of the shaft depicted in Fig. 7.1(*c*) are stretched at $\Omega t = 0$, unstrained at $\Omega t = \pi/2$, compressed at $\Omega t = \pi$ and again unstrained at $\Omega t = 3\pi/2$. The rate of straining is, of course, greatest at $\Omega t = \pi/2$ and at $\Omega t = 3\pi/2$. In general, materials exhibit inelastic behaviour; they dissipate energy on straining. This is often referred to as internal friction, and for metals the proportion of energy dissipated to elastic energy stored is small. We shall assume that the dissipative forces generated are of the linear viscous type, proportional to the strain rate in a fibre and opposed to it.

Let us follow, again, the fibres at C of Fig. 7.1(*c*). When $\Omega t = \pi/2$ the fibres are changing from a state of tension to a state of compression, the rotary damping force would therefore generate a moment about the vertical diameter of the shaft cross-section which would result in a horizontal force to the right acting on the mass. Similarly, when $\Omega t = 3\pi/2$ the fibres are lengthening and the rotary damping forces would again cause a horizontal force to the right to act on the mass.

Obviously the forces due to the rate of straining of the fibres at C when $\Omega t = 0$ or π are zero, for momentarily the strain rate is zero in both positions. In intermediate positions, a horizontal force to the right proportional to the strain rate will result. The corresponding elastic forces are proportional to the strain. As in the undamped case previously studied, instead of depicting the system as three-dimensional we may represent it diagrammatically as illustrated in Fig. 7.5,

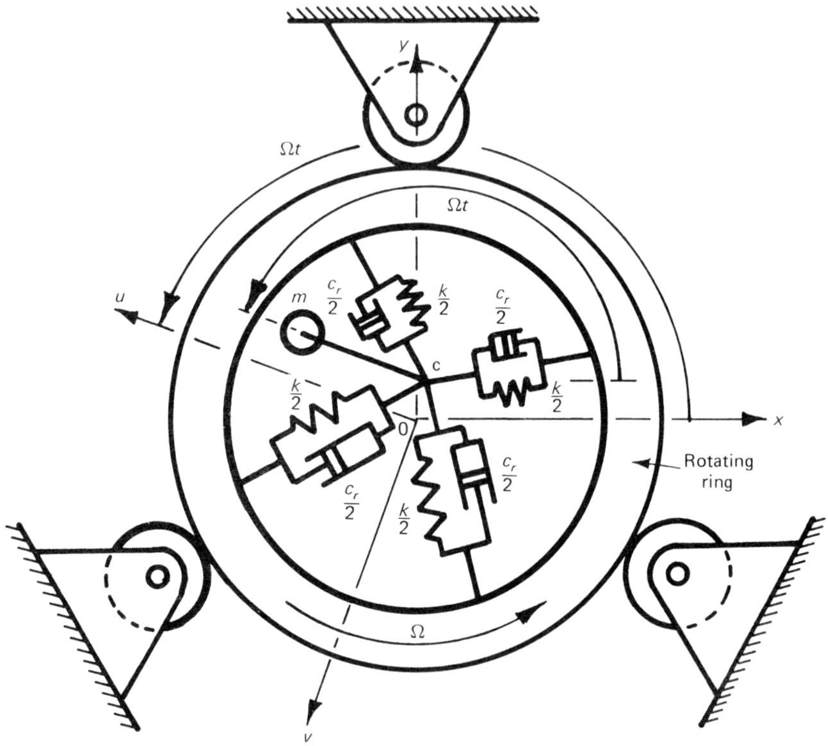

Fig. 7.5

where the springs are equivalent to the shaft elasticity and the dash pots are equivalent to the damping effects of the shaft material. Here m is the mass point. The springs, for this system all of equal stiffness $k/2$, represent the elastic bending stiffnesses of the shaft about two mutually perpendicular axes. The dash pots, for this system each of damping coefficients $c_r/2$, represent the internal friction and other dissipative forces which rotate with the shaft, again about two mutually perpendicular axes. Note, in the diagram 0c should be small enough for the angles between the four visco-elastic elements to remain right angles. It is exaggerated for clarity only.

Exercise

Show that the directions of the forces due to the dashpots of Fig. 7.5 agree with those found in discussing the behaviour of a set of fibres of a shaft rotating under steady load. Derive an expression for the equilibrium position when the shaft is rotating slowly.

7.4 Equations of motion with rotary damping

Often the analysis of the motion of a rotating system is more easily accomplished, or the results more clearly envisaged, if the investigation is conducted in a rotating coordinate system. This is certainly so for some of the systems considered later. Therefore, to familiarise the reader with the considerations involved, we shall examine the motion of the light circular cantilevered shaft, carrying an end mass and allowing for rotary damping, in a rotating coordinate system and in a fixed coordinate system.

7.4.1 In rotating coordinates

Refer to Fig. 7.5 and consider motion relative to the equilibrium position 0 for the shaft when perfectly balanced and rotating with the angular velocity Ω. In the u, v coordinate system of Fig. 7.5, the elastic and damping forces acting on the mass are

$$F_u = -c_r\dot{u} - ku \tag{7.36}$$

$$F_v = -c_r\dot{v} - kv \tag{7.37}$$

and the forces required to cause accelerations along fixed axes momentarily coinciding with the u and v axes are

$$F_u = m(\ddot{u} - \Omega^2 u - 2\dot{v}\Omega) \tag{7.38}$$

$$F_v = m(\ddot{v} - \Omega^2 v + 2\dot{u}\Omega) \tag{7.39}$$

giving the equations of motion for a perfectly balanced shaft as

$$m\ddot{u} + c_r\dot{u} + (k - m\Omega^2)u - 2m\Omega\dot{v} = 0 \tag{7.40}$$

and

$$2m\Omega\dot{u} + m\ddot{v} + c_r\dot{v} + (k - m\Omega^2)v = 0 \tag{7.41}$$

These equations are coupled. Consequently, it is not permissible to combine them in terms of the single variable used in the previous cases, sections 7.1.3 and 7.3.1. Eliminating v by the normal procedure gives the following fourth-order differential equation in u,

$$\{m\mathrm{D}^2 + (c_r + i2m\Omega)\mathrm{D} + k - m\Omega^2\}\{m\mathrm{D}^2 + (c_r - i2m\Omega)\mathrm{D} + k - m\Omega^2\}u = 0 \tag{7.42}$$

where

$$\mathrm{D} \equiv \frac{\mathrm{d}}{\mathrm{d}t}$$

which has solutions of the form

$$u = A\,\mathrm{e}^{\lambda t} \tag{7.43}$$

The equation has been expressed in this factorised form for ease of solution. From the first term it may be shown that if $c_r \ll 2\sqrt{(km)}$

$$\lambda = -\mu_r(\omega - \Omega) + i(\omega - \Omega) \quad \text{or} \quad -\mu_r(\omega + \Omega) - i(\omega + \Omega) \tag{7.44}$$

and from the second term

$$\lambda = -\mu_r(\omega + \Omega) + i(\omega + \Omega) \quad \text{or} \quad -\mu_r(\omega - \Omega) - i(\omega - \Omega) \tag{7.45}$$

where $\mu_r = c_r/2m\omega$ and $\omega^2 = k/m$

That is

$$u = \exp[-\mu_r(\omega - \Omega)t]\{A_1 \exp[i(\omega - \Omega)t] + A_2 \exp[-i(\omega - \Omega)t]\}$$
$$+ \exp[-\mu_r(\omega + \Omega)t]\{A_3 \exp[i(\omega + \Omega)t] + A_4 \exp[-i(\omega + \Omega)t]\} \tag{7.46}$$

A similar expression could be derived for v.

The amplitudes of the second pair of whirls obviously decay with time, whereas the time variation of the amplitudes of the first pair of whirls depends upon the value of the imposed rotational velocity of the shaft Ω relative to the value of the natural angular frequency of the system when not rotating (ω). When Ω is less than ω the whirls decay and when Ω is greater than ω they grow exponentially with time. That is, when the imposed rotational speed of the shaft exceeds the natural angular frequency of the system ω, the system is unstable. This is known as a dynamic instability; in this case, it is not possible to run through it.

Let us now examine the magnitudes and directions of the whirl velocities. The first pair of whirls consists of a forward and a reverse whirl of magnitude $\omega - \Omega$ relative to the moving coordinate system. Relative to the fixed coordinate system (x, y) they are forward whirls of magnitude ω and $2\Omega - \omega$ respectively. The second pair consists of forward and reverse whirls respectively, of magnitude $\omega + \Omega$ relative to the moving coordinate system, and of magnitude $2\Omega + \omega$ and ω relative to the fixed coordinate system. The whirl velocities p, that is the natural angular frequencies when rotating, are shown as functions of Ω for each coordinate system in Fig. 7.6.

At first sight this is a surprising result. As we have assumed light damping, the natural frequencies do not depend upon the magnitude of damping. Therefore, we may regard the damping factor μ_r as equal to zero and expect the result given in section 7.1.2. However, it is seen that this is not so: Fig. 7.6(*b*) indicates two extra natural frequencies which are speed dependent. As a similar result will arise later in connection with the whirling of unsymmetrical shafts we shall proceed to study the forced vibration of the present system. This will show that for the perfectly symmetrical shaft it is not possible to excite these speed-dependent natural frequencies. Therefore, in practice they will not exist.

7.4.2 Forced vibration

Let the system described by equations (7.40) and (7.41) be subjected to a constant amplitude force which rotates at an angular velocity p relative to the moving coordinate system. The equations of motion will then be

$$m\ddot{u} + c_r\dot{u} + (k - m\Omega^2)u - 2m\Omega\dot{v} = F\cos pt \tag{7.47}$$

$$2m\Omega\dot{u} + m\ddot{v} + c_r\dot{v} + (k - m\Omega^2)v = F\sin pt \tag{7.48}$$

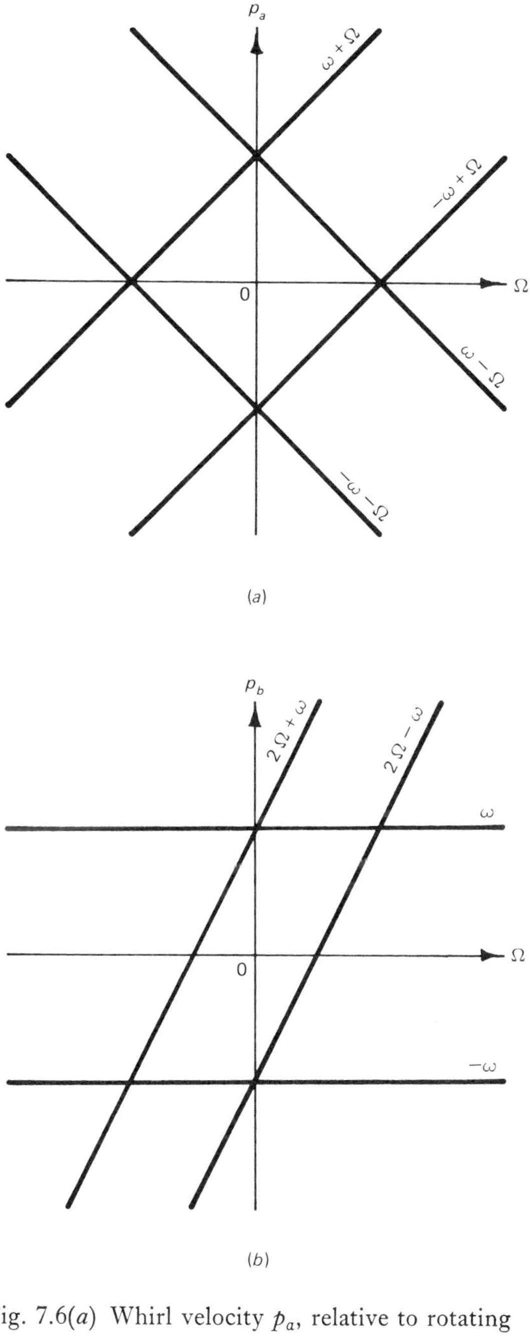

Fig. 7.6(*a*) Whirl velocity p_a, relative to rotating coordinate system
(*b*) Whirl velocity p_b, relative to fixed coordinate system

Eliminating v gives

$$u = \text{Re}\left\{\frac{F\{m\text{D}^2 + (c_r - i2m\Omega)\text{D} + k - m\Omega^2\}\,\text{e}^{ipt}}{\{m\text{D}^2 + (c_r + i2m\Omega)\text{D} + k - m\Omega^2\}\{m\text{D}^2 + (c_r - i2m\Omega)\text{D} + k - m\Omega^2\}}\right\}$$

$$= \text{Re}\left\{\frac{F\,\text{e}^{ipt}}{m\text{D}^2 + (c_r + i2m\Omega)\text{D} + k - m\Omega^2}\right\} \tag{7.49}$$

Therefore, this type of excitation does not excite whirls associated with the second term of equation (7.46). What is more, they cannot be excited by any other means. That is, for a perfectly symmetrical system only two natural angular frequencies may be excited, being, in the fixed coordinate system, $\pm\,\omega$.

In this section, we have studied the motion of a perfectly-balanced symmetrical single-mass system and found that rotary damping causes it to have a dynamically unstable forward whirl when the shaft velocity exceeds the critical speed.

Exercises

1. Show, with vector diagrams, for a number of positions in a whirl, that the viscous forces associated with rotary damping urge the mass forward when the shaft is driven at above its critical speed. Under these conditions, what is the source of the energy which is fed into the whirl?
2. Derive expressions to show how the whirl velocities are modified by the presence of rotary damping when $\mu_r{}^2$ is not very small, compared with unity.

7.4.3 In fixed coordinates

In the x, y coordinate system, fixed relative to earth, the forces F_x and F_y may be found from the values given in equations (7.36) and (7.37) for F_u and F_v, thus

$$\begin{bmatrix} F_x \\ F_y \end{bmatrix} = \begin{bmatrix} \cos\Omega t & -\sin\Omega t \\ \sin\Omega t & \cos\Omega t \end{bmatrix} \begin{bmatrix} F_u \\ F_v \end{bmatrix} \tag{7.50}$$

Substituting for u and v and for their derivatives with respect to time, in terms of x and y and their derivatives, in the above, equations (7.36) and (7.37) give

$$F_x = -c_r\dot{x} - c_r\Omega y - kx \tag{7.51}$$

$$F_y = -c_r\dot{y} + c_r\Omega x - ky \tag{7.52}$$

The equations of motion for a perfectly balanced system thus become

$$m\ddot{x} + c_r\dot{x} + kx + c_r\Omega y = 0 \tag{7.53}$$

$$-c_r\Omega x + m\ddot{y} + c_r\dot{y} + ky = 0 \tag{7.54}$$

We could proceed as follows: eliminating x, substituting $\omega^2 = k/m$, and $2\mu_r\omega = c_r/m$ gives

$$(\text{D}^2 + 2\mu_r\omega\text{D} + \omega^2 - i2\mu_r\omega\Omega)(\text{D}^2 + 2\mu_r\omega\text{D} + \omega^2 + i2\mu_r\omega\Omega)x = 0 \tag{7.55}$$

The roots of which are, for $\mu_r \ll 1$

$$\lambda = -\mu_r(\omega + \Omega) \pm i\omega \quad \text{or} \quad -\mu_r(\omega - \Omega) \pm i\omega \tag{7.56}$$

However, this analysis masks the importance of the direction of Ω. For Ω in the positive direction some of these roots do not apply. We may decide which do by considering equations in a polar form. Each of the solutions represented may be put into the form

$$x + iy = R\,e^{ipt} = A\,e^{\lambda t} \tag{7.57}$$

where $$R = A\,e^{\alpha t} \quad \text{and} \quad \lambda = \alpha + ip$$

Combining equations (7.53) and (7.54) and substituting the above form for the solution gives

$$m(\ddot{R} + 2i\dot{R}p - Rp^2) + c_r(\dot{R} + iRp) + kR - ic_r\Omega R = 0 \tag{7.58}$$

Equating the imaginary parts to zero we find

$$\dot{R} = \frac{c_r R(\Omega - p)}{2mp} = \left(\frac{-\mu_r\omega(p - \Omega)R}{p}\right) \tag{7.59}$$

If the terms involving c_r^2 are negligible, the real parts together with the expression for $\dot{R}$ give $p^2 = \omega^2$, namely, $p = \pm\omega$.

Therefore, when $\quad p = \omega, \quad \lambda = -\mu_r(\omega - \Omega) + i\omega$

and when $\quad p = -\omega, \quad \lambda = -\mu_r(\omega + \Omega) - i\omega \qquad (7.60)$

therefore,

$$x + iy = A\exp\left[-\mu_r(\omega - \Omega)t\right]\exp(i\omega t) + B\exp\left[-\mu_r(\omega + \Omega)t\right]\exp(-i\omega t) \tag{7.61}$$

which is exactly the same motion as found previously from the rotating coordinate system.

Exercise

Show that if symmetrical stationary damping, μ_s, is applied to the mass of the system considered above, the limit of stable motion is given by

$$\Omega = \frac{\mu_s + \mu_r}{\mu_r}\,\omega.$$

7.5 Stability of linear systems

In section 7.4.1 we found that rotary damping gives rise to dynamically unstable motion at speeds above the first critical. The instability was such that after an initial disturbance the amplitude of the free response would grow exponentially with time. Before examining more complex systems which exhibit a similar phenomenon we need to be empowered to decide whether or not a system is stable without having to solve its characteristic equation.

The equations governing the free motion of linear systems reduce to the general form

$$[a_0 D^n + a_1 D^{n-1} + a_2 D^{n-2} + \cdots + a_{n-1} D + a_n]x = 0 \qquad (7.62)$$

which is satisfied by a solution of the form

$$x = A\,e^{st} \qquad (7.63)$$

On substituting this value for x, equation (7.62) gives the characteristic equation

$$a_0 s^n + a_1 s^{n-1} + \cdots + a_{n-1} s + a_n = 0 \qquad (7.64)$$

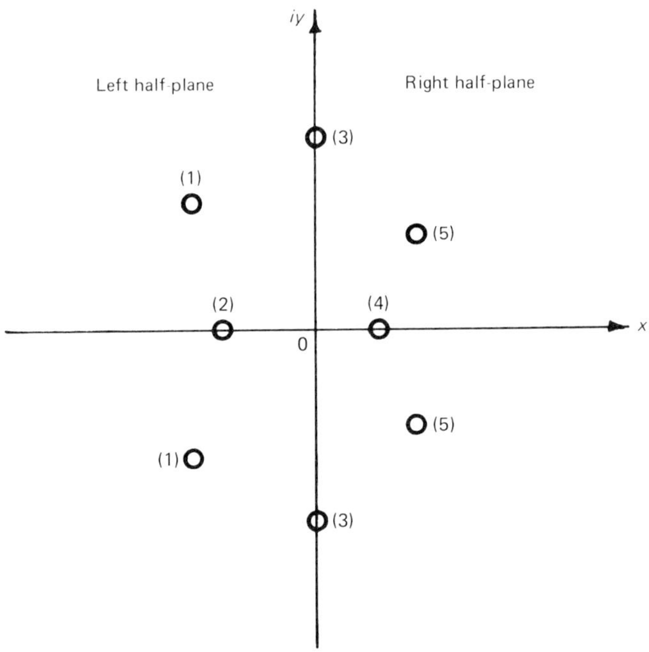

Position of root	Type of motion
1	Stable; damped oscillatory
2	Stable; damped aperiodic
3	Oscillatory
4	Unstable; divergent (distance increases exponentially with time)
5	Unstable; oscillatory with increasing amplitude, referred to as dynamically unstable

Fig. 7.7

which has n roots, s_i. That is

$$x = \sum_{i=1}^{n} A_i \exp(s_i t) \qquad (7.65)$$

excluding the possibility of repeated roots.

In this section we are not attempting to make a comprehensive study of stability, but merely to cover it in sufficient detail to proceed with the examination of the remaining systems of this chapter and those in the next. Hence the study of instability arising from repeated roots will not be required.

From equation (7.65), it is obvious that if any s_i has a positive real part the system will be unstable. The types of motion associated with the position of a root are summarised in Fig. 7.7.

For higher order systems than those examined so far, it would be convenient if we could establish whether or not a system was unstable without solving for the roots of its characteristic equation. This problem exercised the minds of a number of nineteenth century mathematicians of whom Routh, in 1877, established an algorithm for determining the number of roots of a real polynomial which lie in the right half-plane. Independently, Hurwitz, in 1895, gave a second solution: the determinant inequalities obtained by him are known as Routh-Hurwitz conditions. They have been used in practical investigations associated with many different manifestations of dynamic instability in linear systems, for example, instability of rotating shafts, flutter of aircraft parts, control of engines, hunting of de-aerators, chatter of machine tools and many more. In determinantal form the stability conditions are as follows.

Given an algebraic equation of the form

$$F(s) = a_0 s^n + a_1 s^{n-1} + \cdots + a_{n-1} s + a_n = 0 \qquad (7.66)$$

where the coefficients a_r are real, construct the following n determinants

$$D_1 = a_1$$

$$D_2 = \begin{vmatrix} a_1 & a_0 \\ a_3 & a_2 \end{vmatrix}$$

$$D_3 = \begin{vmatrix} a_1 & a_0 & 0 \\ a_3 & a_2 & a_1 \\ a_5 & a_4 & a_3 \end{vmatrix}$$

$$D_n = \begin{vmatrix} a_1 & a_0 & 0 & 0 & \cdots & \cdot \\ a_3 & a_2 & a_1 & a_0 & \cdots & \\ a_5 & a_4 & a_3 & a_2 & \cdots & \\ \vdots & & & & & \vdots \\ a_{2n-1} & a_{2n-2} & \cdot & \cdot & \cdots & a_n \end{vmatrix} \qquad (7.67)$$

If $r > n$ or $r < 0$ then $a_r = 0$. A necessary and sufficient condition for all roots of $F(s) = 0$ to lie in the left half-plane is that $D_i > 0$.

It will be noted that if $a_n > 0$ it is only necessary to test the determinants from D_1 to D_{n-1} as $D_n = a_n D_{n-1}$.

The coefficients a_r may be expressed in terms of the roots s_i of equation (7.66), for example

$$a_1 = -a_0 \sum_{i=1}^{n} s_i, \qquad a_2 = a_0 \sum_{i=1}^{n} \sum_{\substack{j=1 \\ i \neq j}}^{n} s_i s_j$$

and so on. By considering the forms of these expressions it is easy to show that all a_i have the same sign as a_0 if all the roots have negative real parts. The only way in which any of the coefficients may be zero or take the opposite signs to a_0 is for one or more of the roots to have positive real parts.

Hence, a necessary, but not sufficient, condition for all roots of $F(s) = 0$ to lie in the left half-plane is that all a_r be positive. This very easily tested condition will be employed in the next section.

Exercise

Have any of the following equations roots which lie in the right half-plane?

$$\begin{aligned} Z^3 + 8Z^2 + 23Z + 22 &= 0 \\ -Z^3 + 4Z^2 + 17Z + 10 &= 0 \\ Z^5 + 8Z^4 + 33Z^3 + 82Z^2 + 110Z &= 0 \\ Z^4 + 6Z^3 + 15Z^2 + 24Z + 44 &= 0 \\ Z^4 + 5{\cdot}9Z^3 + 14{\cdot}4Z^2 + 22{\cdot}9Z + 44 &= 0 \end{aligned}$$

7.6 An unsymmetrical shaft supported in rigid bearings

A shaft may have unsymmetrical elastic properties due to slots being cut in it, such as keyways and armature winding slots, or due to manufacturing imperfections. It will be shown that if the stiffnesses about the two principal axes of the shaft cross-section are different a range of speeds exists in which the shaft motion is unstable, even in the absence of rotary damping. Initially, damping will be neglected.

7.6.1 Equations of motion for a vertical shaft

In this study it is more convenient to work in a coordinate system which rotates with the principal axes of the shaft cross-section, thus avoiding time varying coefficients. The motion of a light cantilevered shaft with a vertical axis of rotation, carrying a perfectly balanced mass at the free end, is to be studied. Referring to Fig. 7.8, let the bending stiffness about the principal axis in the u-direction be k_u and about that in the v-direction be k_v. Then in the u-v coordinate system we have, for a shaft rotating about its elastic axis at a constant angular velocity Ω,

$$m(\ddot{u} - \Omega^2 u - 2\dot{v}\Omega) + k_v u = 0 \qquad (7.68)$$

$$m(\ddot{v} - \Omega^2 v + 2\dot{u}\Omega) + k_u v = 0 \tag{7.69}$$

Eliminating v from equations (7.68) and (7.69) gives

$$\{D^4 + (2\Omega^2 + \omega_u{}^2 + \omega_v{}^2)D^2 + (\Omega^2 - \omega_u{}^2)(\Omega^2 - \omega_v{}^2)\}u = 0 \tag{7.70}$$

where
$$D \equiv d/dt$$
$$\omega_u{}^2 = k_u/m$$
and
$$\omega_v{}^2 = k_v/m$$

As this is a fourth-order linear homogeneous differential equation we may assume that the solution is of the form

$$u = A \exp(st) \tag{7.71}$$

Thus

$$s^2 = -0{\cdot}5(2\Omega^2 + \omega_u{}^2 + \omega_v{}^2) \pm \sqrt{\{0{\cdot}25(2\Omega^2 + \omega_u{}^2 + \omega_v{}^2)^2 - (\Omega^2 - \omega_u{}^2)(\Omega^2 - \omega_v{}^2)\}} \tag{7.72}$$

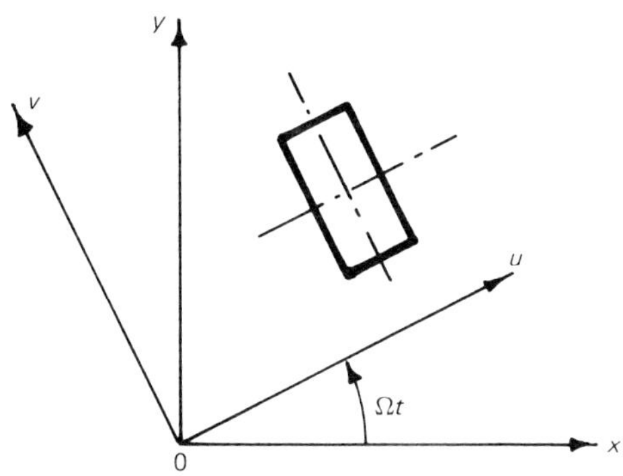

Fig. 7.8

For this discussion, let us take $k_u > k_v$, hence $\omega_u > \omega_v$. When $\Omega < \omega_v$ both values of s^2 are negative, the corresponding values of s being $\pm is_1$, $\pm is_2$. The motion will therefore, have the form

$$u = A \sin s_1 t + B \cos s_1 t + C \sin s_2 t + D \cos s_2 t \tag{7.73}$$

Similarly,

$$v = A' \sin s_1 t + B' \cos s_1 t + C' \sin s_2 t + D' \cos s_2 t \tag{7.74}$$

Let us examine now the nature of the motion when $\omega_v < \Omega < \omega_u$. In this case the sign of the second term under the square root will change. It will be a positive quantity, making one value of s^2 positive and the other negative. Thus the four roots will be $\pm is_a$, $\pm s_b$, and the motion will take the form

$$u = A \sin s_a t + B \cos s_a t + C \sinh s_b t + D \cosh s_b t \tag{7.75}$$

Motion in the v direction takes a similar form. This means that when the imposed rotational velocity of the shaft has a value between the two angular frequencies

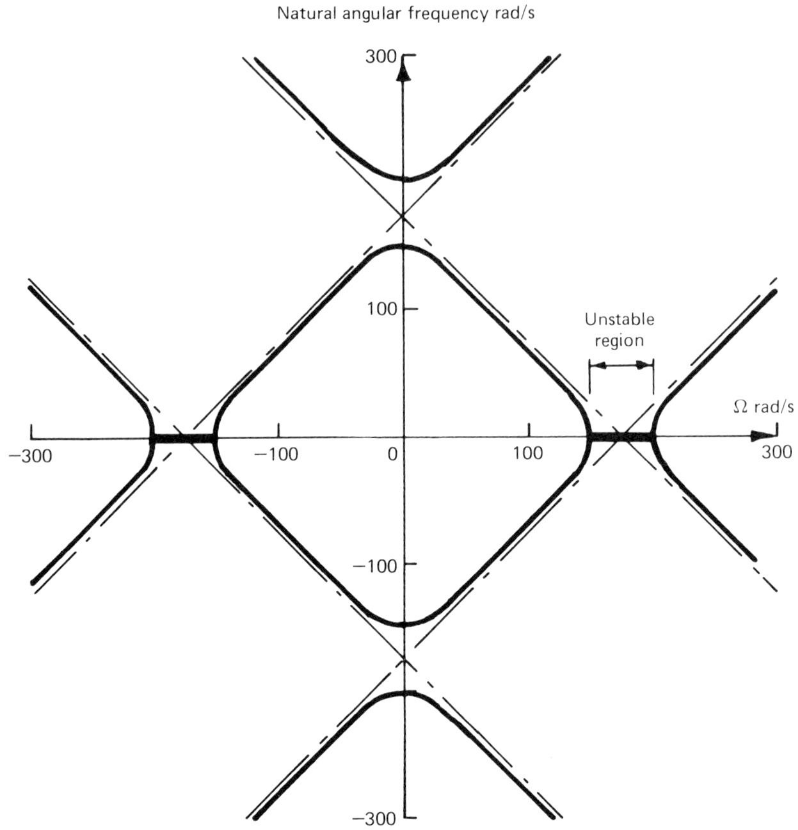

Fig. 7.9(*a*) Variation of natural angular frequencies, relative to a coordinate system which rotates with the shaft at an angular velocity Ω

ω_u and ω_v of the system the motion is unstable; the amplitude grows with time. The unstable motion in this speed range could have been predicted from equation (7.70) by observing that the coefficient a_4 is negative.

Stable running is again achieved when $\Omega > \omega_u$, for then the second term under the square root is again a negative quantity and it cannot exceed the first term in the braces regardless of how great Ω becomes.

This is the first case we have investigated in which the natural frequencies of the free motion depend upon the rotational speed of the shaft. Figure 7.9(*a*) shows graphically how the natural angular frequencies of a system, in which $\omega_u = 200$ rad/s and $\omega_v = 150$ rad/s, vary with rotational speed; the natural frequencies are relative to the rotating coordinate system. Figure 7.9(*b*) shows a similar plot relative to a stationary coordinate system.

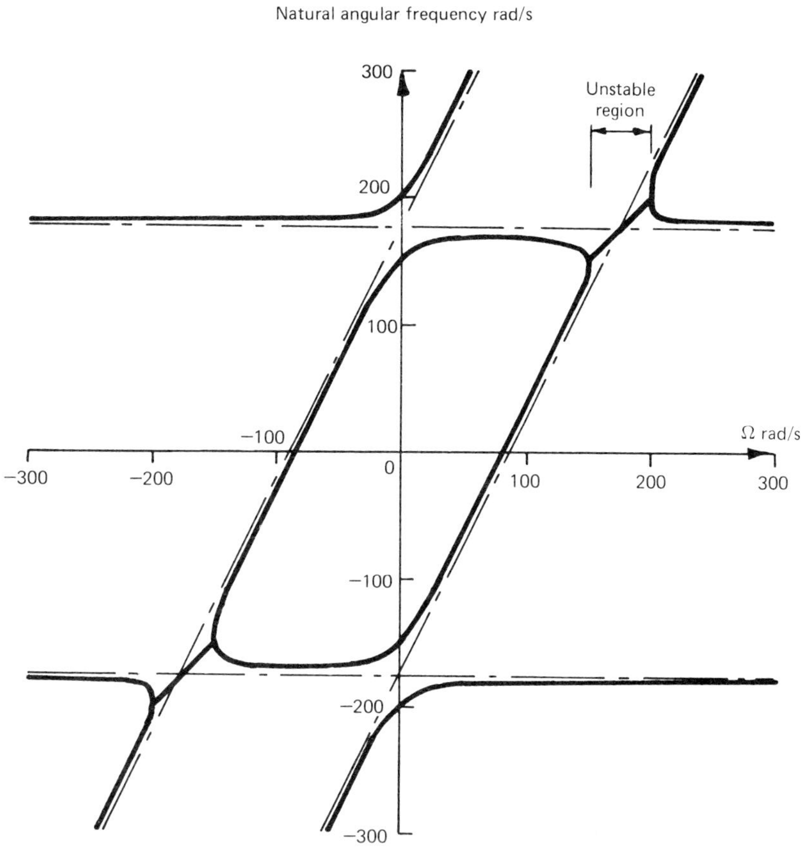

Fig. 7.9(*b*) Variation of natural angular frequencies, relative to the stationary coordinate system, with shaft rotational speed Ω

Exercise

With the aid of Fig. 7.9 explain how you would expect the above system to respond to the action of a transverse force which remains stationary in space.

7.6.2 Equations of motion for a horizontal shaft

The motion of a perfectly balanced rotating unsymmetrical shaft exhibits an additional peculiarity when its axis is horizontal. The gravitational force induces a secondary resonance. This may be seen from natural frequency plots or, as we shall now do, by solving the equations of motion for a horizontal shaft.

Including the gravitational force mg acting on the mass, equations (7.68) and (7.69) become

$$m(\ddot{u} - \Omega^2 u - 2\dot{v}\Omega) + k_v u = -mg \sin \Omega t \tag{7.76}$$

$$m(\ddot{v} - \Omega^2 v + 2\dot{u}\Omega) + k_u v = -mg \cos \Omega t \tag{7.77}$$

Thus the free vibration of a horizontal shaft, that is, the motion given by the complementary function of the equations, is the same as for the same shaft in the vertical position. However, even for a perfectly balanced horizontal shaft its motion has forced components given by

$$u = \frac{(4\Omega^2 - \omega_u{}^2)g \sin \Omega t}{\omega_u{}^2\omega_v{}^2 - 2\Omega^2(\omega_u{}^2 + \omega_v{}^2)} \tag{7.78}$$

$$v = \frac{(4\Omega^2 - \omega_v{}^2)g \cos \Omega t}{\omega_u{}^2\omega_v{}^2 - 2\Omega^2(\omega_u{}^2 + \omega_v{}^2)} \tag{7.79}$$

A resonance will be observed when

$$\Omega^2 = \frac{1}{2}\left(\frac{\omega_u{}^2\omega_v{}^2}{\omega_u{}^2 + \omega_v{}^2}\right) = \Omega_s{}^2 \tag{7.80}$$

that is, at approximately one half the mean of the two critical speeds. It is known as a secondary critical speed, and is caused entirely by the gravitational forces. It will be noted that in the neighbourhood of the secondary critical speed, both the numerator and denominator of the gravity forced motion are of the same order of magnitude. That is, the steady state amplitude depends upon the ratio $(\omega_u{}^2 - \omega_v{}^2)/(\Omega_s{}^2 - \Omega^2)$. Therefore, the rate of amplitude build-up at this speed is slow. Obviously any transverse disturbing forces which remain stationary in space would cause secondary resonance.

Exercises

1. For the above horizontal shaft set up and solve the equations of motion allowing for out-of-balance and initial lack of straightness. From the solution, what do you conclude about the relative severity of the primary and secondary resonances?
2. Show that if a small amount of stationary damping acts on a mass attached to a perfectly balanced unsymmetrical vertical shaft which rotates in rigidly supported bearings the width of the unstable speed band is reduced. For small values of c_s show that the unstable speed band is given approximately by

$$\omega_v(1 + \delta_v) \leqslant \Omega \leqslant \omega_u(1 - \delta_u)$$

where

$$\delta_u = \delta_v = \frac{(c_s/m)^2}{2(\omega_u{}^2 - \omega_v{}^2)}$$

3. If rotary damping only is present in the system described in question 2 above, show that the shaft motion is unstable at all speeds greater than ω_v, where $\omega_v < \omega_u$.

   Discuss the relative intensities of the instabilities, that is the rate of growth of amplitude of the free motion due to (*i*) unsymmetrical shaft flexibility and (*ii*) rotary damping.

4. When both rotary and stationary damping are present in the system described in question 2 show that for stability in addition to Ω lying outside the interval given in question 2 it has to be less than a transition speed Ω_t given by

$$\Omega_t^2 = \frac{1}{8}\left(\frac{c_s + c_r}{c_r}\right)^2 \left\{\left[\left(2(\omega_u^2 + \omega_v^2) + \left(\frac{c_r}{m}\right)^2\right)^2 + 4\left(\frac{c_r}{c_s + c_r}\right)^2 (\omega_u^2 - \omega_v^2)^2\right]^{1/2} + 2(\omega_u^2 + \omega_v^2) - \left(\frac{c_r}{m}\right)^2\right\}$$

7.7 Unsymmetrical shaft with distributed mass in rigidly supported perfect bearings

It has been shown earlier in this chapter that a symmetrical shaft with distributed mass and elasticity has an infinite number of critical speeds. In the case of an unsymmetrical shaft we would, therefore, expect an infinite number of pairs of critical speeds. Also the single mass representation of the previous section indicated that the motion of the shaft was unstable when its speed of rotation lay in a band between the pair of critical speeds. Therefore, at speeds between the pairs of critical speeds of a shaft with distributed mass and elasticity we would expect its motion to be unstable, Ariaratnam (1965) discussed this case. The above deductions will be validated.

7.7.1 Equations of motion

In rotating coordinates, for an element of a thin unbalanced uniform shaft, of unsymmetrical cross-section, rotating about its longitudinal elastic axis at a constant angular velocity Ω, and supported in bearings which impose the same degree of restraint in each principal direction at the ends of the shaft, we have

$$\rho a\left(\frac{\partial^2 u}{\partial t^2} - \Omega^2 u - 2\Omega\frac{\partial v}{\partial t}\right) = -EI_{vv}\frac{\partial^4 u}{\partial z^4} - \rho a(g\sin\Omega t - \Omega^2 r_u) \tag{7.81}$$

$$\rho a\left(\frac{\partial^2 v}{\partial t^2} - \Omega^2 v + 2\Omega\frac{\partial u}{\partial t}\right) = -EI_{uu}\frac{\partial^4 v}{\partial z^4} - \rho a(g\cos\Omega t - \Omega^2 r_v) \tag{7.82}$$

where ρ is the mass density of the shaft material
a is the cross-sectional area of the shaft
E is Young's modulus
I_{uu}, I_{vv} are the principal second moments of area of the shaft cross-section
g is the gravitational acceleration
r_u, r_v are the coordinates of the mass centre of the cross-section relative to its elastic axis.

The deflections u and v may be expressed in terms of the characteristic functions P_k and Q_k of the shaft when not rotating. That is,

$$u(z, t) = \sum_k P_k(z)F_k(t) \tag{7.83}$$

$$v(z, t) = \sum_k Q_k(z)H_k(t) \tag{7.84}$$

Previously we have specified that the shaft is uniform and that the bearings apply the same degree of restraint in each principal direction at its ends, therefore, P_k and Q_k will be equal.

We may now substitute from equations (7.83) and (7.84) in equations (7.81) and (7.82), let

$$\frac{EI_{vv}\alpha_k^{\ 4}}{\rho a} = \omega_{uk}^{\ 2} \quad \text{and} \quad \frac{EI_{uu}\alpha_k^{\ 4}}{\rho a} = \omega_{vk}^{\ 2} \tag{7.85}$$

multiply through by P_k dz and integrate with respect to z to give

$$\{\mathrm{D}^2 + (\omega_{uk}^{\ 2} - \Omega^2)\}F_k - 2\Omega \mathrm{D} H_k = -G_k \sin \Omega t + \Omega^2 R_{uk} \tag{7.86}$$

$$2\Omega \mathrm{D} F_k + \{\mathrm{D}^2 + (\omega_{vk}^{\ 2} - \Omega^2)\}H_k = -G_k \cos \Omega t + \Omega^2 R_{vk} \tag{7.87}$$

where
$$\mathrm{D} \equiv \frac{\mathrm{d}}{\mathrm{d}t}$$

$$G_k = g\int_0^L P_k \, \mathrm{d}z$$

$$R_{uk} = \int_0^L r_u P_k \, \mathrm{d}z$$

$$R_{vk} = \int_0^L r_v P_k \, \mathrm{d}z$$

7.7.2 Forced vibration

The particular solutions which satisfy equations (7.86) and (7.87) are

$$F_k = \frac{\Omega^2 R_{uk}}{\omega_{uk}^{\ 2} - \Omega^2} - \frac{(\omega_{vk}^{\ 2} - 4\Omega^2)G_k \sin \Omega t}{\omega_{uk}^{\ 2}\omega_{vk}^{\ 2} - 2\Omega^2(\omega_{uk}^{\ 2} + \omega_{vk}^{\ 2})} \tag{7.88}$$

$$H_k = \frac{\Omega^2 R_{vk}}{\omega_{vk}^{\ 2} - \Omega^2} - \frac{(\omega_{uk}^{\ 2} - 4\Omega^2)G_k \cos \Omega t}{\omega_{uk}^{\ 2}\omega_{vk}^{\ 2} - 2\Omega^2(\omega_{uk}^{\ 2} + \omega_{vk}^{\ 2})} \tag{7.89}$$

$$k = 1, 2, 3, \ldots$$

7.7.3 Free vibration

The solutions of the homogeneous equations are found, by eliminating one of the unknowns, from the left-hand sides of equations (7.86) and (7.87). That is from

$$\{D^4 + (\omega_{uk}^2 + \omega_{vk}^2 + 2\Omega^2)D^2 + (\omega_{uk}^2 - \Omega^2)(\omega_{vk}^2 - \Omega^2)\}F_k = 0 \qquad (7.90)$$

This equation has the same form as equation (7.70) for the light shaft carrying a single mass. For each mode pair of the system presently under investigation, the behaviour found for the shaft with the concentrated mass will be repeated. That is, when the shaft's rotational speed lies in the range between its n'th bending natural frequency in the u direction and its n'th bending natural frequency in the v direction its whirl amplitude is unbounded so that the free motion is unstable.

7.7.4 The total motion

The system will have primary resonances when the rotational speed of the shaft coincides with a natural frequency of bending vibrations, the motion at speeds between a pair of n'th critical speeds being unstable. Due to out-of-balance forces the shaft rotates with the rotating coordinate system; the forced whirls in a stationary coordinate system are, therefore, in the forward direction and of velocity Ω. Away from resonant and unstable conditions the amplitudes depend upon the amount of out-of-balance.

The system will also have secondary resonances (secondary critical speeds), due to the gravitational force acting on the shaft in a transverse direction. One secondary resonance may occur for each pair of n'th primary resonances, but the rate of amplitude build-up is low. At a secondary resonance the shaft moves in a forward whirl, of velocity Ω, relative to the rotating coordinate system. Therefore, in a stationary coordinate system it moves in a forward whirl of velocity 2Ω.

Exercises

1. For a thin uniform shaft, running in bearings which act as position-fixed, direction-free restraints, plot the variations in the first two pairs of natural frequencies with shaft rotational-speed. Take the shaft as 3 m long, $I_{uu} = 7 \times 10^{-6}$ m^4, $I_{vv} = 4 \times 10^{-6}$ m^4, $E = 207$ GN/m^2, $\rho = 7{\cdot}8$ Mg/m^3, and $a = 0{\cdot}008$ m^2. From the graphs find the first four critical speeds and the first two secondary critical speeds.
2. Show that the forced vibration due to initial lack of straightness has amplitudes

$$F_k = \frac{\omega_{uk}^2 \Delta_{uk}}{\omega_{uk}^2 - \Omega^2}$$

$$H_k = \frac{\omega_{vk}^2 \Delta_{vk}}{\omega_{vk}^2 - \Omega^2}$$

In a coordinate system which rotates at an angular velocity Ω the shaft is stationary.

$$\Delta_{uk} = \int_0^L \frac{EI_{vv}}{\rho a} \frac{\partial^4 u_0}{\partial z^4} P_k \, dz$$

$$\Delta_{vk} = \int_0^L \frac{EI_{uu}}{\rho a} \frac{\partial^4 v_0}{\partial z^4} P_k \, dz$$

where u_0 and v_0 are initial values of u and v, representing the initial lack of straightness of the elastic axis of the shaft.

3. For the shaft described in question 1 above, show that the width of the unstable regions increases as k, the mode number, increases. Hence, show that above a certain frequency the unstable regions overlap, and write down the conditions which must be imposed upon the shaft speed Ω for the free vibrations to be stable.
4. How are the conditions found in question 3 modified by the presence of stationary and rotary damping?

7.8 The influence of flexibility in the bearings

The simple model of a rotating shaft, shown in Fig. 7.5, may be extended as shown in Fig. 7.10 to represent elasticity and damping in the bearings or their supports. It is necessary to allow for bearing flexibility, particularly when very stiff shafts are used, to predict accurately the system natural frequencies. In some instances unstable motion may arise due to the bearing characteristics; such behaviour may be induced by the characteristics of oil films in journal bearings.

A symmetrical shaft without damping, running in bearings with symmetrical elastic characteristics and no damping, is a simple case, and it will be left to the reader to show that equations (7.1) and (7.2) apply where k is now the combined stiffness of shaft and bearing. That is

$$k = \frac{k_1 k_2}{k_1 + k_2} \tag{7.91}$$

where k_1 is the shaft stiffness and k_2 the bearing stiffness. Thus, when the shaft and bearings are each symmetrically elastic, the natural frequency of the rotating shaft is independent of rotational speed, and decreases with increased bearing flexibility.

We shall now examine a number of shaft-bearing combinations.

7.8.1 Symmetrical shaft, asymmetrical bearings

Often, in practice, the bearings of rotating machinery are not equally flexible in all directions. For instance, in turbo-alternator sets the bearing supports are more flexible in the lateral direction than in the vertical direction. The system we shall investigate is shown diagrammatically in Fig. 7.10, from which it will be seen that both the elasticity and the damping are to be taken as asymmetrical. The rotating ring represents the bearing journals. The principal axes of both bearing stiffness and bearing damping effect will be assumed coincident with $0x$ and $0y$ axes, for

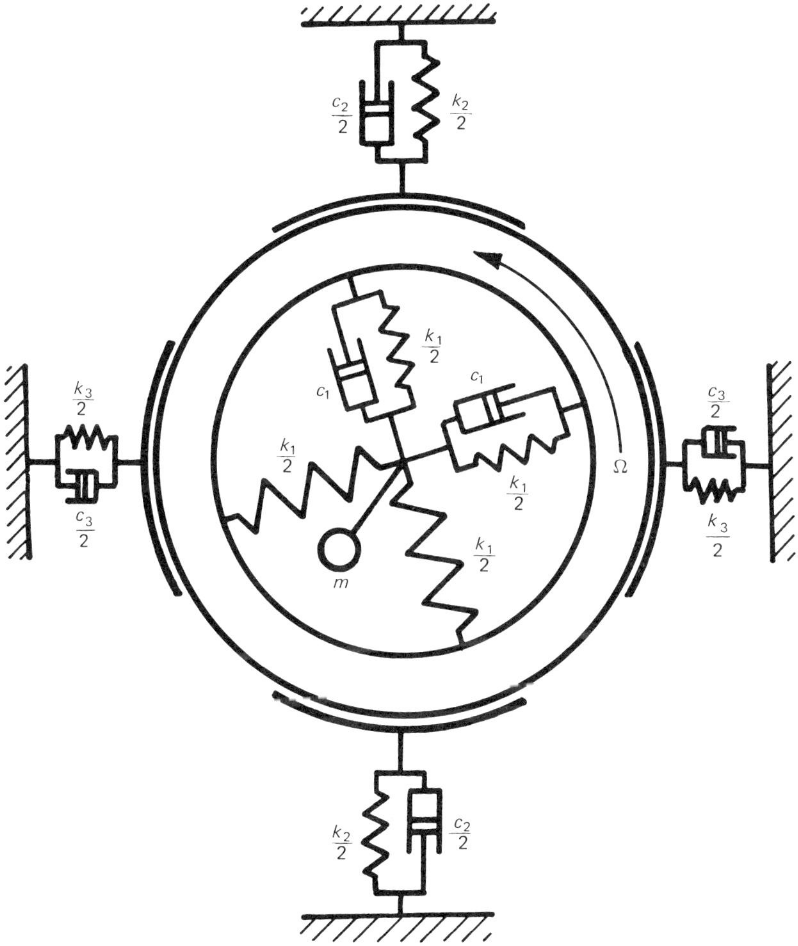

Fig. 7.10

each bearing. Let the absolute displacements of the mass m from its equilibrium position be x, y in the fixed coordinate system. Also let the displacements of the mass m from its equilibrium position be x_1, y_1 relative to the rotating ring and let the absolute displacements of the rotating ring from its equilibrium position be x_2, y_2.

That is $$x = x_1 + x_2$$

and $$y = y_1 + y_2 \tag{7.92}$$

We found previously, see section 7.4.3, that the force on the mass due to a shaft with rotary damping is

$$F_x = -k_1x_1 - c_1\dot{x}_1 - c_1\Omega y_1 \qquad [(7.51)]$$

and

$$F_y = -k_1y_1 - c_1\dot{y}_1 + c_1\Omega x_1 \qquad [(7.52)]$$

Therefore the equations of motion for the mass m are

$$m\ddot{x} + c_1\dot{x}_1 + k_1x_1 + c_1\Omega y_1 = 0 \qquad (7.93)$$

$$m\ddot{y} + c_1\dot{y}_1 + k_1y_1 - c_1\Omega x_1 = 0 \qquad (7.94)$$

As the rotating ring representing the bearing journals is fictitious and massless, the force balances give

$$k_1x_1 + c_1\dot{x}_1 + c_1\Omega y_1 = k_3x_2 + c_3\dot{x}_2 \qquad (7.95)$$

$$k_1y_1 + c_1\dot{y}_1 - c_1\Omega x_1 = k_2y_2 + c_2\dot{y}_2 \qquad (7.96)$$

where

$$x_1 + x_2 = x \quad \text{and} \quad y_1 + y_2 = y$$

On substituting for x_2 and y_2, equations (7.95) and (7.96) become

$$[k_1 + k_3 + (c_1 + c_3)\mathrm{D}]x_1 + c_1\Omega y_1 = (k_3 + c_3\mathrm{D})x \qquad (7.97)$$

and

$$[k_1 + k_2 + (c_1 + c_2)\mathrm{D}]y_1 - c_1\Omega x_1 = (k_2 + c_2\mathrm{D})y \qquad (7.98)$$

where

$$\mathrm{D} \equiv \frac{\mathrm{d}}{\mathrm{d}t}$$

Equations (7.97) and (7.98) may be solved for x_1 and y_1. In the case of light damping, the damping forces are small compared with the elastic force and when small terms of second and higher orders are neglected the expressions become

$$x_1 = \frac{k_3x}{k_1 + k_3} + \frac{(k_1c_3 - k_3c_1)\dot{x}}{(k_1 + k_3)^2} - \frac{c_1k_2\Omega y}{(k_1 + k_2)(k_1 + k_3)} \qquad (7.99)$$

$$y_1 = \frac{k_2y}{k_1 + k_2} + \frac{(k_1c_2 - k_2c_1)\dot{y}}{(k_1 + k_2)^2} + \frac{c_1k_3\Omega x}{(k_1 + k_2)(k_1 + k_3)} \qquad (7.100)$$

Substituting these values into equations (7.93) and (7.94) gives the equations of motion

$$m\ddot{x} + \frac{(k_1{}^2c_3 + k_3{}^2c_1)\dot{x}}{(k_1 + k_3)^2} + \frac{k_1k_3x}{k_1 + k_3} + \frac{c_1k_2k_3\Omega y}{(k_1 + k_2)(k_1 + k_3)} = 0 \qquad (7.101)$$

$$m\ddot{y} + \frac{(k_1{}^2c_2 + k_2{}^2c_1)}{(k_1 + k_2)^2}\dot{y} + \frac{k_1k_2}{k_1 + k_2}y - \frac{c_1k_2k_3\Omega x}{(k_1 + k_2)(k_1 + k_3)} = 0 \qquad (7.102)$$

which for convenience may be put into the form

$$\ddot{x} + (\sigma_a + \rho_a)\dot{x} + \omega_a{}^2x + (\rho_a\rho_b)^{1/2}\Omega y = 0 \qquad (7.103)$$

$$\ddot{y} + (\sigma_b + \rho_b)\dot{y} + \omega_b{}^2y - (\rho_a\rho_b)^{1/2}\Omega x = 0 \qquad (7.104)$$

where σ_a and σ_b are the stationary damping coefficients, ρ_a and ρ_b are the rotary damping coefficients, and ω_a^2 and ω_b^2 are the stiffness/mass coefficients.

We may now examine separately the effects of stationary and rotary damping. When there is no damping ($\sigma_a = \sigma_b = \rho_a = \rho_b = 0$) the equations of motion are uncoupled; free simple harmonic oscillations may take place in each principal direction; the natural angular frequencies are ω_a in the x-direction and ω_b in the y-direction; the critical speeds are given by $\Omega = \omega_a$ and $\Omega = \omega_b$ and the motion is always stable.

With stationary damping alone ($\rho_a = \rho_b = 0$) the equations are again uncoupled. Thus the free motion consists of damped harmonic oscillations in each principal direction and the motion is always stable. Critical speeds will occur at

$$\Omega = (\omega_a^2 - 0{\cdot}25\sigma_a^2)^{1/2} \quad \text{and} \quad \Omega = (\omega_b^2 - 0{\cdot}25\sigma_b^2)^{1/2}$$

and as damping is normally small these will be practically coincident with those for the undamped case.

When only rotary damping is present ($\sigma_a = \sigma_b = 0$) the equations are coupled. Eliminating x from equations (7.103) and (7.104) and taking the solution for y in the form $y = A \exp(st)$ gives the characteristic equation

$$s^4 + (\rho_a + \rho_b)s^3 + (\rho_a\rho_b + \omega_a^2 + \omega_b^2)s^2 + (\rho_a\omega_b^2 + \rho_b\omega_a^2)s + \rho_a\rho_b\Omega^2 + \omega_a^2\omega_b^2 = 0 \qquad (7.105)$$

By the Routh-Hurwitz stability criterion, for all roots of this equation to have negative real parts

$$\begin{vmatrix} \rho_a + \rho_b & 1 & 0 \\ \rho_a\omega_b^2 + \rho_b\omega_a^2 & \rho_a\rho_b + \omega_a^2 + \omega_b^2 & \rho_a + \rho_b \\ 0 & \rho_a\rho_b\Omega^2 + \omega_a^2\omega_b^2 & \rho_a\omega_b^2 + \rho_b\omega_a^2 \end{vmatrix} > 0 \qquad (7.106)$$

That is

$$(\rho_a + \rho_b)(\rho_a\omega_b^2 + \rho_b\omega_a^2) + (\omega_a^2 - \omega_b^2)^2 > (\rho_a + \rho_b)^2\Omega^2 \qquad (7.107)$$

Therefore the value of Ω at the transition between stable and unstable motion of the shaft increases rapidly as the difference between the undamped system natural frequencies in the two principal transverse directions increases. Asymmetry of the bearing stiffness increases the transition speed.

We have limited the discussion to the case in which the asymmetrical bearings supporting the shaft are identical and their corresponding principal stiffness axes are parallel. Pedersen (1969) extended this to include the case in which the principal stiffness axes of the support bearings are independently oriented. His predictions were very like those we have examined.

Exercises

1. Show, for a symmetrical shaft running in unsymmetrical bearings, that, when both rotary and stationary damping are present, the stationary damping causes an increase in the transition speed between stable and unstable running of the shaft.

2. Show, for a symmetrical shaft, carrying an unbalanced concentrated mass, and supported in unsymmetrical bearings, that the forced whirl is a forward whirl when the shaft speed is lower than its first critical speed, a reverse whirl between the critical speeds and a forward whirl when it is above the second critical speed. You may neglect damping.

   Also show that any point on the elastic axis of the shaft describes an elliptical path. The above theoretical findings were confirmed experimentally by Downham (1957).
3. For the system described in question 2 show that the presence of stationary damping decreases the speed range over which the reverse whirl occurs.
4. Set up, in generalised matrix form, the equations of motion for a massless elastic shaft supported in asymmetrical bearings at its ends. The bearings impose only linear restraints on the shaft, but their stiffness characteristics are not identical, neither is the orientation of their principal axes. The shaft has fixed to it n concentrated masses (m_i), the mass centre of each being a distance r_i from the elastic axis of the shaft. Relative to the radius vector of the out of balance of the first mass the other radius vectors are at angles ψ_i.

   How would you determine the natural frequencies, the normal modes and the form of the elastic axis of the shaft at a given rotational speed?

   Express in matrix form the procedure for balancing the first three modes, n being greater than 3.

7.8.2 Symmetrical shaft, hydrodynamic journal bearings

7.8.2.1 *Bearing characteristics*

Often shafts are supported by oil-lubricated journal bearings, the load being reacted by hydrodynamically-generated pressures in the oil film. If the journal of this type of bearing is rotated at a constant angular velocity, Ω, about its own centre, the centre being fixed at a distance $\boldsymbol{r}$ from the bush centre, it will experience a constant force $\boldsymbol{F}$ at an angle ψ to the displacement $\boldsymbol{r}$, as indicated in Fig. 7.11(a).

The magnitudes of the force F and the angle ψ depend upon the value of r/c, where c is the radial clearance of the bearing. The ratio r/c is called the eccentricity ratio, ϵ. Figure 7.11(b) illustrates a typical variation in the magnitudes of F and ψ with ϵ, for a bearing of circular profile when the journal centre, J, is stationary. Thus the force vector $\boldsymbol{F}$ is a non-linear function of the position vector $\boldsymbol{r}$. It is also non-linearly dependent upon the velocity of the journal centre $\mathrm{d}\boldsymbol{r}/\mathrm{d}t$.

In theoretical investigations into the influence of hydrodynamic journal bearings on small vibrations of rotating systems, linear approximations to the above non-linear functions have been employed, see for example Holmes (1960) or Morrison (1962). These have taken the following form, for a horizontal shaft,

$$\begin{bmatrix} F_x \\ F_y \end{bmatrix} = \begin{bmatrix} F_{0x} \\ F_{0y} \end{bmatrix} + \begin{bmatrix} a_{xx} & a_{xy} \\ a_{yx} & a_{yy} \end{bmatrix} \begin{bmatrix} x \\ y \end{bmatrix} + \begin{bmatrix} b_{xx} & b_{xy} \\ b_{yx} & b_{yy} \end{bmatrix} \begin{bmatrix} \dot{x} \\ \dot{y} \end{bmatrix} \tag{7.108}$$

where	F_x, F_y	are the x and y components respectively of $\boldsymbol{F}$
	F_{0x}, F_{0y}	are the values of F_x and F_y when the journal centre is stationary at its steady running position
	x, y	are small displacements from the steady running position
	a_{ij}	are known as the displacement coefficients
and	b_{ij}	are known as the velocity coefficients.

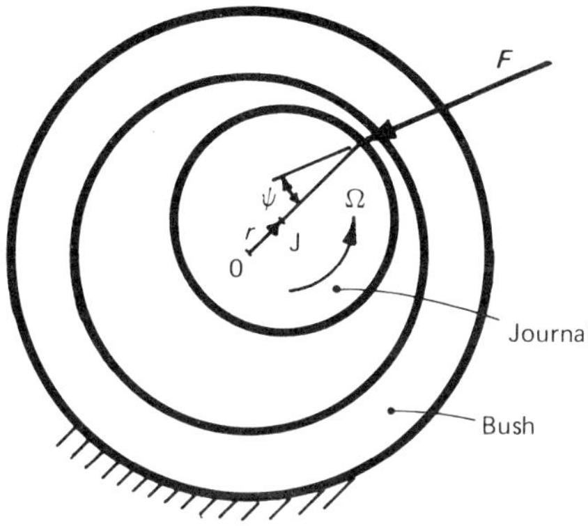

(a)

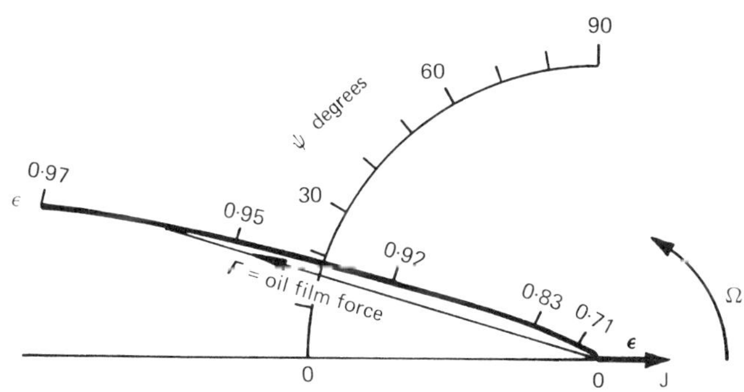

(b)

Fig. 7.11(*a*); (*b*) ($\boldsymbol{F}$ = oil film force in arbitrary units; OJ = displacement of journal centre from bush centre; ϵ = OJ/radial clearance; Ω = angular velocity of journal)

The values of the displacement and velocity coefficients depend upon many factors, including the shape and size of the bush and journal surfaces; the rotational speed of the shaft; the oil properties, particularly its viscosity; and the magnitude and direction of the steady load.

7.8.2.2 *System behaviour*

The small amplitude motion, about its steady running position, of a symmetrical shaft of bending stiffness k, when supported in two hydrodynamic journal bearings and carrying a mass M at its mid-span, is governed by the following equations of motion:

$$M\ddot{x} + k(x - x_1) = M\Omega^2 h \cos \Omega t \tag{7.109}$$

$$M\ddot{y} + k(y - y_1) = M\Omega^2 h \sin \Omega t \tag{7.110}$$

where x, y are the coordinates of the shaft axis at mid-span relative to its steady running position; x_1, y_1 are the coordinates of the shaft axis at the journal, relative to its steady running position and h is the radius of the mass centre from the shaft axis at mid-span. A force balance on the journals gives

$$k(x - x_1) = a_{xx}x + a_{xy}y + b_{xx}\dot{x} + b_{xy}\dot{y} \tag{7.111}$$

and

$$k(y - y_1) = a_{yx}x + a_{yy}y + b_{yx}\dot{x} + b_{yy}\dot{y} \tag{7.112}$$

where the a_{ij} and b_{ij} relate to the combined effect of two bearings.

As the above are linear differential equations the total motion is the sum of the forced motion and the free motion. In general, even for a perfectly circular bearing the oil film characteristics are unsymmetrical, so we should expect two natural frequencies for each shaft speed and for each the whirl frequency coincides with the natural frequency. Consequently, we should expect two resonance type critical speeds in which the whirl velocity coincides with the shaft velocity, this condition is referred to as a synchronous whirl. However, there are two complicating factors which prevent the system behaving completely in this manner: (*a*) for a particular system, under a particular steady load there is a rotational speed above which the free motion is unstable, because the system damping becomes negative; (*b*) when the amplitude becomes large, as at resonance, the linearised equations (7.108) do not hold. The unstable motion arising from the oil film forces was first reported by Newkirk and Taylor (1925). They referred to the phenomenon as 'oil whip', and reported that the whipping occurred over a wide range of shaft speed, that the vibration or whirling frequency bore little, if any, relation to the shaft speed and that the phenomenon was practically independent of balance. More recent investigations, Pinkus (1956), Tondl (1965), Pope and Healy (1966), indicate that the ratio of free whirl velocity to shaft velocity takes values which decrease from about 0·5 when the shaft velocity is much less than twice the lowest natural angular frequency of the rotor, the whirl velocity taking the value of the lowest natural angular frequency of the rotor when the shaft velocity is greater than twice that value. That is, for a rigid rotor the free whirl velocity is about half the shaft velocity. This condition has been called 'half-speed whirl', and there are hydrodynamic reasons for its being nearly half-speed but they are beyond the scope of this discussion, see D. M. Smith (1969) for an excellent book on journal bearings in turbo meachinery, which relates a lifetime of practical experience in this field.

A number of workers, Holmes (1960 and 1963), Morrison (1962), Glienecke (1966), have examined the linearised equations (7.109) to (7.112) for stability by criteria such as those given in section 7.5. Their theoretical results predict the existence of negative damping, but experimental confirmation of the threshold speed for unstable motion does not appear to have been accomplished yet.

Empirically it has been found that the threshold speed, at which unstable free motion may arise, can be increased for a particular rotor by subjecting it to a unidirectional constant force, such as a gravity force, or by modifying the bearing profile. Non-circular bearings, for example elliptical and three-lobed bearings, are often used for this purpose in medium speed applications and tilting pad bearings, which are inherently very stable are used for high speed applications.

7.8.3 Unsymmetrical shaft, symmetrical bearings

Bearing flexibility introduces no new phenomena for the above combination. Examination of its influence will therefore be left to the reader in the following exercises.

Exercises

1. Show that the motion of a rotating asymmetrical shaft, running in bearings with symmetrical stiffness and damping, is unstable in the speed range between the critical speeds.
2. Show that the motion of a rotating asymmetrical shaft, running in symmetrical bearings, is stable up to the lower critical speed and is unstable at all speeds above, when rotary damping is present and stationary damping is absent.
3. Find the critical speeds of a vertical shaft given $k_{uu} = 100$ kN/m, $k_{vv} = 150$ kN/m, $k_1 = k_2 = 200$ kN/m, and $M = 2$ kg. Find expressions for the amplitude of the mass M when $\Omega = 600$ rad/s and there is an out-of-balance (mh) in the u direction of 0·3 kg mm and in the v direction of 0·4 kg mm.
4. For the system described in question 3 show that the amplitude grows exponentially and the whirl frequency is Ω for free vibrations when the shaft is spinning at an angular velocity between the two critical speeds.
5. When the shaft described in question 3 is horizontal, show that it has a sharp resonance at $\Omega = 96{\cdot}7$ rad/s. Find the motion of the mass under these circumstances.

References

ARIARATNAM, S. T.
'The vibration of unsymmetrical rotating shafts', *J. Appl. Mech.*, **32**, 1965, 157–62.

BISHOP, R. E. D. and GLADWELL, G. M. L.
'The vibration and balancing of an unbalanced flexible rotor', *J. Mech. Eng. Sci.*, **1**, No. 1, 1959, 66–77.

DOWNHAM, E.
'Theory of shaft whirling', *Engr.*, **204**, 1957, 518–22, 552–5.

GLIENECKE, J.
'Experimental investigation of the stiffness and damping coefficients of turbine bearings and their application to instability prediction', *Proc. Instn. Mech. Engrs.*, **181**, part 3B, 1966, 116–29.

HOLMES, R.
'The vibration of a rigid shaft on short sleeve bearings', *J. Mech. Eng. Sci.*, **2**, No. 4, 1960, 337–41.

HOLMES, R.
'Oil whirl characteristics of a rigid rotor in 360° journal bearings', *Proc. Instn. Mech. Engrs.*, **177**, 1963, 291–307.

MORRISON, D.
'Influence of plain journal bearings on the whirling action of an elastic rotor', *Proc. Instn. Mech. Engrs.*, **176**, 1962, 542–53.

NEWKIRK, B. L. and TAYLOR, H. D.
'Shaft whipping due to oil action in journal bearings', *General Electric Rev.*, **28** 1925, 559–68.

PEDERSEN, P. T.
'Stability of rotors with unsymmetrical flexible bearings', *Proc. 12th International Congress of Applied Mechanics*, Springer-Verlag, 1969.

PINKUS, O
'Experimental investigation of resonant whip', *Trans. Amer. Soc. Mech. Engrs.*, **78**, 1956, 975–83.

POPE, A. W. and HEALY, S. P.
'Anti-vibration journal bearings', *Proc. Instn. Mech. Engrs.*, **181**, part 3B, 1966, 98–115.

SMITH, D. M.
'The motion of a rotor carried by a flexible shaft in flexible bearings', *Proc. Roy. Soc., Series A*, **142**, 1933, 92–118.

SMITH, D. M.
Journal bearings in turbomachinery, Chapman & Hall, 1969, p. 81.

TONDL, A.
Some problems of rotor dynamics, Chapman & Hall, London, 1965.

Chapter 8

Gyrodynamics

Only systems having negligible moments of inertia were considered in chapter 7 so that the changes in angular velocities were unimportant. A large and important class of practical systems was thereby studied. Another important class, the systems being comprised of at least one rigid massive rotor, will now be studied. Here the couples which cause angular accelerations must be considered. Gyroscopic instruments are obvious examples, high speed turbine rotors are also in this class.

We shall commence the study with relatively simple systems, firstly an unrestrained symmetrical rigid rotor and then the same rotor elastically restrained. Small amplitude oscillations, only, will be considered. Later sections will present the more general case of an unsymmetrical rotor.

8.1 Symmetrical rigid rotors

8.1.1 The equations of motion, a rotor supported at its centre of mass

A general study of the dynamics of a rigid body was commenced in chapter 2, section 2.8.2. There the terms inertia tensor and principal moments of inertia were defined.

Any rigid body has three mutually perpendicular axes ($0X_1$, $0X_2$, $0X_3$; 0 being its centre of mass) such that its rotational kinetic energy may be expressed as

$$2T = I_1\omega_1^2 + I_2\omega_2^2 + I_3\omega_3^2 \tag{8.1}$$

where the principal moment of inertia

$$I_i = \int_v (x_j^2 + x_k^2)\,\mathrm{d}m$$

and ω_i is the angular velocity of the body about $0X_i$, i, j, and k taking the values 1, 2, and 3.

In section 2.8.2. we restricted the motion to small amplitude oscillations about the principal axes; now we wish to consider a more general motion. If two of the principal moments of inertia are equal and the body spins about the third principal axis of inertia, the rigid body is called a symmetrical rotor. A right circular cylinder spinning about its longitudinal axis is a symmetrical rotor. The motion we shall consider is constant velocity spin, Ω, about the third principal axis together with small amplitude oscillations about the other two principal axes.

Obviously, in the plane containing them, the axes associated with the two equal principal moments of inertia may be positioned arbitrarily. For this reason, small

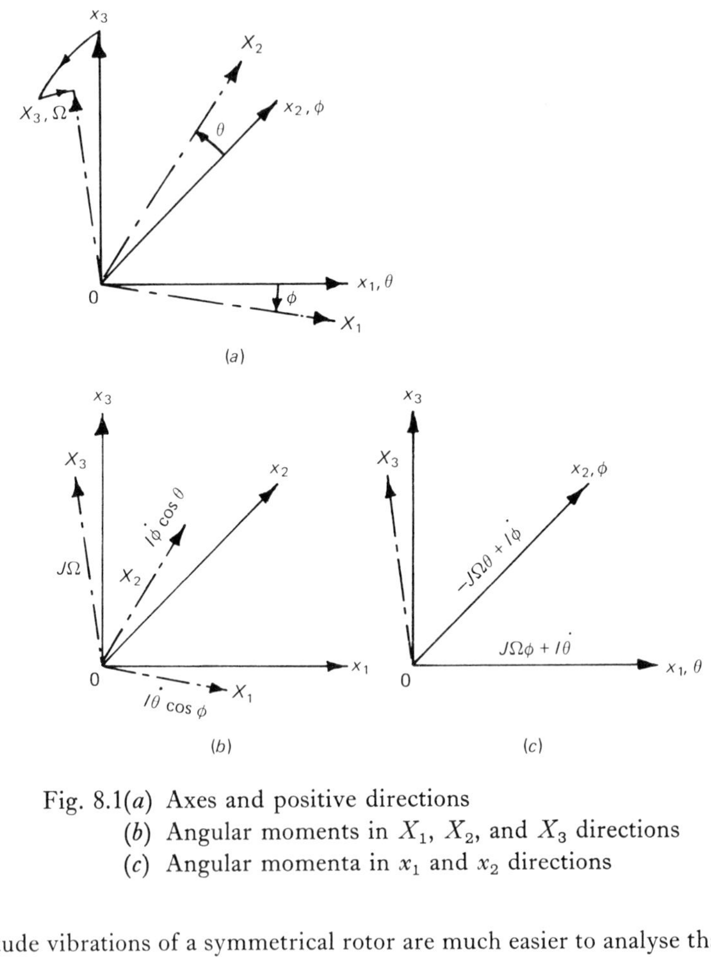

Fig. 8.1(*a*) Axes and positive directions
(*b*) Angular moments in X_1, X_2, and X_3 directions
(*c*) Angular momenta in x_1 and x_2 directions

amplitude vibrations of a symmetrical rotor are much easier to analyse than those of an asymmetrical rotor.

From Newton's laws it may easily be deduced (see section 8.2.3) that a torque $\boldsymbol{G}_G$ applied to a rigid body about its centre of mass causes an angular acceleration of the body, the rate of change of angular momentum about its centre of mass being equal to $\boldsymbol{G}_G$. Let the symmetrical rotor have principal moments of inertia I, I, and J about $0X_1$, $0X_2$, and $0X_3$ respectively; $0X_3$ is the spin axis. Consider a stationary set of axes $0x_1$, $0x_2$, and $0x_3$, $0x_3$ coinciding with $0X_3$ in the quiescent state, see Fig. 8.1(*a*). For small movements about $0x_1(\theta)$ and $0x_2(\phi)$ whilst spinning with angular velocity Ω about $0X_3$, the angular momenta about $0x_1(h_1)$, $0x_2(h_2)$, and $0X_3(H_3)$ are given by

$$h_1 = I\dot{\theta} + J\Omega\phi \tag{8.2}$$

$$h_2 = I\dot{\phi} - J\Omega\theta \tag{8.3}$$

$$H_3 = J\Omega \tag{8.4}$$

when second and higher order small quantities are neglected. As mentioned earlier, $0X_1$ and $0X_2$ may be positioned arbitrarily due to symmetry, here we have held them in the $x_1 0 x_3$ and $x_2 0 x_3$ planes respectively.

The instantaneous couples G_θ and G_ϕ about the $0x_1$ and $0x_2$ axes respectively are

$$G_\theta = \frac{\mathrm{d}h_1}{\mathrm{d}t} = I\ddot{\theta} + J\Omega\dot{\phi} \tag{8.5}$$

$$G_\phi = \frac{\mathrm{d}h_2}{\mathrm{d}t} = I\ddot{\phi} - J\Omega\dot{\theta} \tag{8.6}$$

Suppose now that the rotor is firmly supported at its centre of mass on a frictionless spherical pivot and that, except for being forced to rotate with a constant angular velocity Ω about its principal axis $0X_3$, it is unconstrained. The equations of motion then become

$$G_\theta = I\ddot{\theta} + J\Omega\dot{\phi} = 0 \tag{8.7}$$

$$G_\phi = I\ddot{\phi} - J\Omega\dot{\theta} = 0 \tag{8.8}$$

If we let $\psi = \theta + i\phi$ these combine to give

$$I\ddot{\psi} - iJ\Omega\dot{\psi} = 0 \tag{8.9}$$

the solution of which is easily shown to be

$$\psi = A + B\exp\left(i\frac{J\Omega}{I}t\right) \tag{8.10}$$

Therefore, the motion may be a static displacement and/or a forward whirl of angular velocity $J\Omega/I$.

In gyro-compasses $J\Omega$ is much greater than I; a very high natural frequency, synonomous with a very stiff system, results. Such a system is not easily deflected from its set position by random impulses.

8.1.2 Rotor with symmetrical elastic restraints, linear, and angular motions uncoupled

Many forms of elastic restraint have been devised for rotors of gyroscopic instruments, particularly to enable quantities such as angular velocity to be measured.

For the time being, however, we shall restrict our study to the case in which the elastic restraint is symmetrical and the elastic forces due to angular displacements are not coupled with those due to linear displacements. A light symmetrical elastic shaft rotating in identical symmetrical bearings and carrying a centrally mounted rotor is such a case, we shall examine it. Assuming the rotor to be perfectly balanced the equations of motion may be written as

$$G_\theta = -S\theta = I\ddot{\theta} + J\Omega\dot{\phi} \tag{8.11}$$

$$G_\phi = -S\phi = I\ddot{\phi} - J\Omega\dot{\theta} \tag{8.12}$$

$$F_1 = -kx_1 = m\ddot{x}_1 \tag{8.13}$$

$$F_2 = -kx_2 = m\ddot{x}_2 \tag{8.14}$$

where S is the stiffness of the shaft to angular deflections, k its stiffness to linear deflections, and m is the mass of the rotor.

Obviously, for this case, the linear and angular motions are not coupled and as the linear motions have been studied in chapter 7 they will not be reconsidered here.

8.1.2.1 *Angular motion only*

Equations (8.11) and (8.12) may be expressed in terms of the complex quantity ψ $(= \theta + i\phi)$ to give

$$I\ddot{\psi} - iJ\Omega\dot{\psi} + S\psi = 0 \tag{8.15}$$

the solution of which is

$$\psi = A\,e^{i\omega_1 t} + B\,e^{i\omega_2 t}$$

where

$$\omega_1 = \frac{J\Omega}{2I}\left\{1 + \sqrt{\left(1 + \frac{4IS}{J^2\Omega^2}\right)}\right\} \tag{8.16}$$

$$\omega_2 = \frac{J\Omega}{2I}\left\{1 - \sqrt{\left(1 + \frac{4IS}{J^2\Omega^2}\right)}\right\}$$

Therefore, relative to a stationary coordinate system the free vibratory motion of the system, when the rotor spins at high speed, is comprised of a high frequency forward whirl and a low frequency reverse whirl; the amplitudes A and B depending upon the initial conditions. Note that the natural frequencies are functions of the rotational speed of the shaft. Figure 8.2 illustrates the variation of ω_1 and ω_2 with Ω when $J = 0{\cdot}3$ kg m^2, $I = 0{\cdot}2$ kg m^2, and $S = 20\,000$ N m/rad.

As the line $\omega = \Omega$ does not cut either the ω_1 versus Ω or the ω_2 versus Ω curve one would not expect a resonance or whirling critical speed due to out of balance type excitation for this system, that is when $I < J$. However, many experimenters have found a critical speed with a reverse whirl corresponding to the speed at which the ω_2 versus Ω curve crosses the $\omega = -\Omega$ line.

Stodola (1927) suggested that the unpredicted critical speed with a reverse whirl was due to loosening of the rotor on the shaft; Schulte and Riester (1954) carefully eliminated this possibility and still found the unpredicted critical speed. The rotor proportions used by Schulte and Riestei were such that the forward whirl critical speed existed for each, that is $I > J$: out of nine cases tested the unpredicted reverse whirl critical speed was detected in six, but they observed that the forward whirl was much more vigorous than the reverse whirl. This observation leads one to suspect that the reverse whirl found experimentally is associated with slight asymmetry in the bearing stiffness which, as discussed in section 8.1.3, gives rise to a weak critical speed with a reverse whirl.

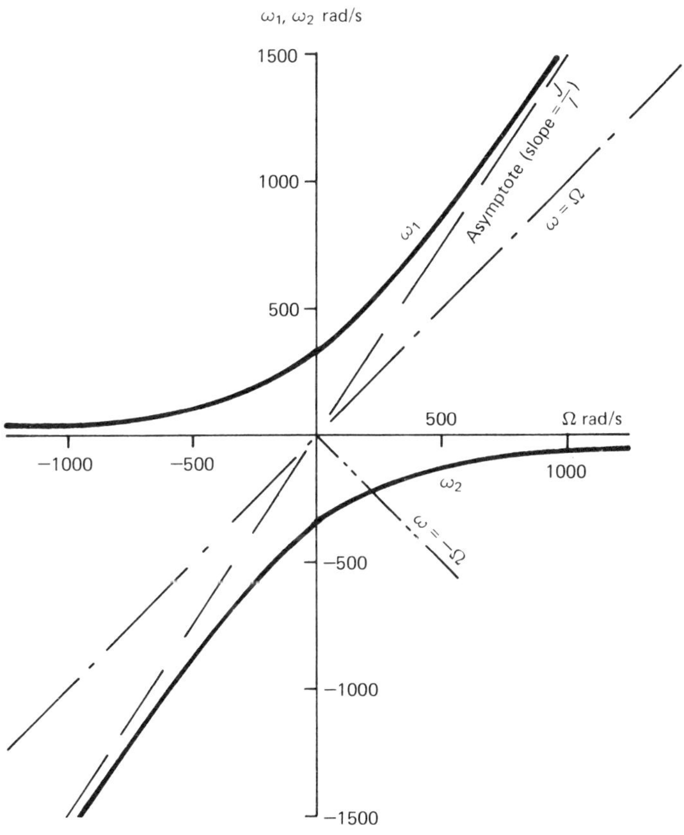

Fig. 8.2 Variation of natural angular frequencies ω_i with rotor angular velocity Ω for a simple gyroscopic system consisting of a symmetrical rotor with symmetrical elastic constraints

The case in which $I > J$, that is rotors long compared with their diameter, is given as an exercise below.

Exercises

1. For the system examined above, draw the θ and ϕ amplitudes as functions of exciting frequency if the system spins at an angular velocity Ω of 500 rad/s and

is excited by a constant amplitude torque on the θ axis. Take the torque amplitude as 100 N m and frequency range as -1500 rad/s to 1500 rad/s.

2. Plot curves showing the variation of ω_1 and ω_2 with Ω when $J = 0{\cdot}3$ kg m^2, $I = 0{\cdot}4$ kg m^2, and $S = 20\,000$ N m/rad. Hence or otherwise show that there will be a critical speed with a forward whirl velocity of approximately 460 rad/s when the system is subjected to out-of-balance type excitation.
[**Note:** *Experimentalists also find a reverse whirl 'critical' speed corresponding to* $\omega_2 = -\Omega$, *as in the case studied in the text.*]
3. Arnold (1947) described a gyrostatic vibration absorber to improve the surface finish produced by metal-planing machines. The concept is illustrated in Fig. 8.2(*a*), in which the gyroscopic action of the rotor I modifies the torsional oscillations of the disc J.

About its spin axis, the rotor has a principal moment of inertia I and constant angular velocity Ω. It rotates about a spindle which is supported in frame A, which in turn is supported by bearings CC in frame B. About CC, the rotor and frame A have a combined moment of inertia of K, and the rotor spindle axis intersects the axis CC at right angles. The frame A is unbalanced. the unbalance is equivalent to a mass M at a distance d from CC and below it. The frame B is rigidly attached to the disc, and the moment of inertia of the disc and associated parts when all axes are mutually perpendicular is J. The disc is attached to a rigid support by means of a massless elastic shaft of torsional stiffness S.

If a disturbing torque $T \sin pt$ is applied to the disc, rotating it through an angle θ, show that the equation of motion for small oscillations about the quiescent position is

$$JK\frac{\mathrm{d}^4\theta}{\mathrm{d}t^4} + (MgdJ + SK + I^2\Omega^2)\frac{\mathrm{d}^2\theta}{\mathrm{d}t^2} + MgdS\theta = (Mgd - p^2K)T \sin pt$$

If $J = 25$ kg m^2, $I = 0{\cdot}5$ kg m^2, $K = 0{\cdot}1$ kg m^2, $M = 2$ kg, $d = 0{\cdot}2$ m, $S = 100\,000$ N m/rad and $\Omega = 200\pi$ rad/s, show that the resonant angular frequencies are given by $p^2 = 43\,500$ rad^2/s^2 or $p^2 = 3{\cdot}6$ rad^2/s^2, whilst with the rotor not rotating, that is $\Omega = 0$, the resonant angular frequency is given by $p^2 = 4000$ rad^2/s^2.
4. A circular vertical shaft, of total length L, flexural rigidity EI, cross-sectional area A, and mass density ρ, rotates at a constant angular velocity ω, with its upper end restrained in a long bearing and its lower end in a short bearing. To the shaft is attached, at a distance a from the upper bearing, a rigid symmetrical rotor of mass M and principal moments of inertia J, K, and K; J being about the spin axis.

Write down the boundary conditions relating to the shaft which must be satisfied in setting up the frequency equation.
5. The system shown in Fig. 8.2(*b*) is a gyroscopic angular accelerometer. The displacement of the vertical gimbal, provided that it is small, is proportional to

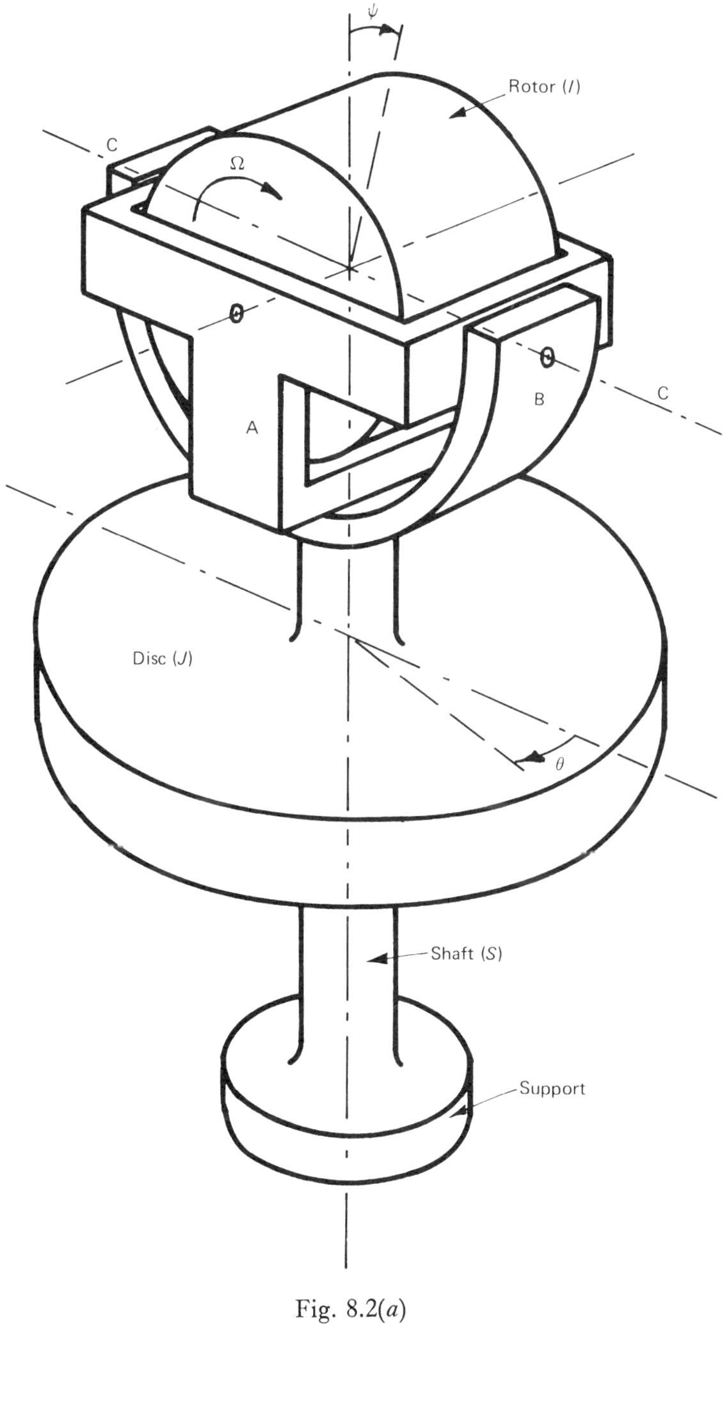

Fig. 8.2(*a*)

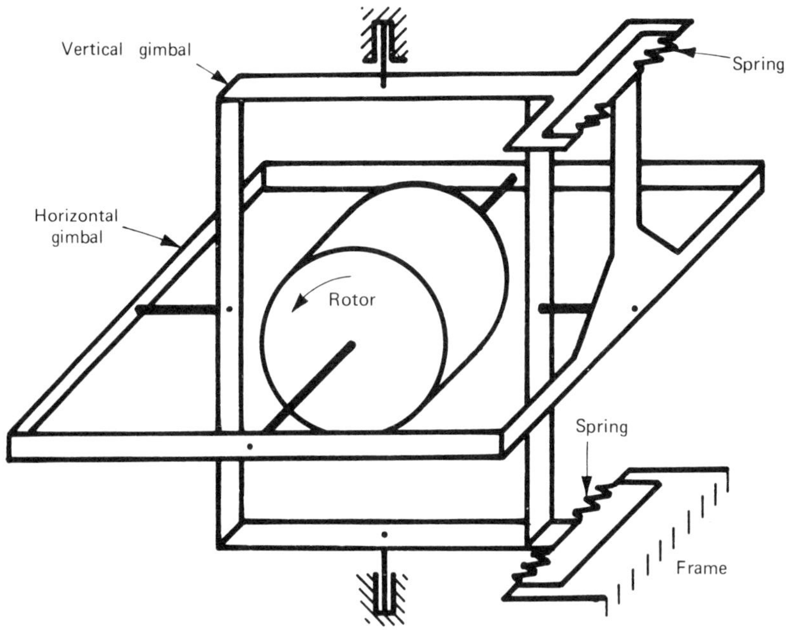

Fig. 8.2(*b*)

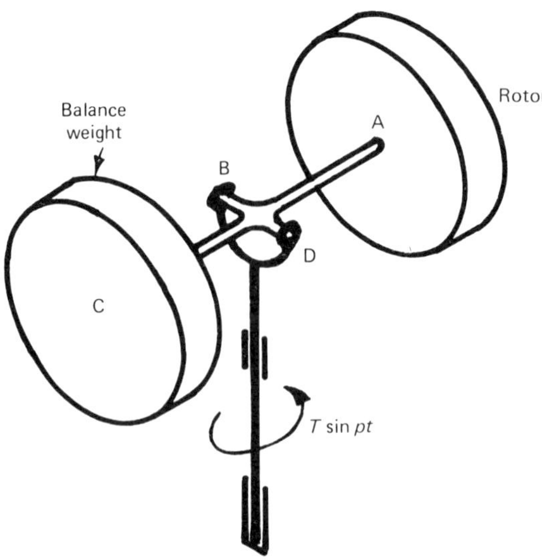

Fig. 8.2(*c*)

the angular acceleration about the vertical axis. Prove this and derive the frequency equation for the system. You may neglect the mass of the gimbals.

6. Derive the equations of motion for forced vibration of the gyroscopic system shown in Fig. 8.2(*c*).

   The rigid member ABCD is supported in bearings at B and D; it carries a perfectly balanced rotor at A which is free to rotate about the axis AC. The weight of the rotor is balanced, about BD, by a body identical to the rotor which is rigidly attached at C. The bearings at B and D are in the forked end of a vertical spindle which is free to rotate in bearings about its own axis.

   The moment of inertia of the rotor about the axis AC is J and it spins at a constant angular velocity Ω. The moment of inertia of ABCD and attachments about the axis BD is I, and that of the whole system about the axis of the vertical spindle is K. It may be assumed that the friction due to the bearings is proportional to and in phase with the angular velocity about the axis under consideration. An exciting torque $T \sin pt$ is applied to the vertical spindle.

   At what value of the exciting torque frequency would you expect the amplitude of vibration of the rotor to be a maximum? Would this be equal to the frequency of free vibration of the system?

8.1.3 Rotor with unsymmetrical elastic bearings, angular motion only

Practically, it is unlikely that elastic restraint on a rotor due to the bearings will be symmetrical, unless great care has been taken in the design and manufacture of the whole system, especially the bearings. It is to elastic asymmetry arising from the bearings that we now give consideration. Equations (8.11) and (8.12) are modified to allow the rotational stiffness S_1 on the θ axis to differ from S_2 on the ϕ axis. The equations of motion become

$$I\ddot{\theta} + S_1\theta - J\Omega\dot{\phi} = 0 \tag{8.17}$$

$$J\Omega\dot{\theta} + I\ddot{\phi} + S_2\phi = 0 \tag{8.18}$$

As S_1 differs from S_2 we cannot combine the above equations as we did previously. However, we may eliminate ϕ by the usual procedure to give the following fourth-order equation for θ

$$\{(I\mathrm{D}^2 + S_1)(I\mathrm{D}^2 + S_2) + J^2\Omega^2\mathrm{D}^2\}\theta = 0 \tag{8.19}$$

Assuming a solution of the form $\theta = \Theta\,\mathrm{e}^{i\lambda t}$ we find

$$\lambda^4 - (\omega_1{}^2 + \omega_2{}^2 + p^2)\lambda^2 + \omega_1{}^2\omega_2{}^2 = 0 \tag{8.20}$$

where $\omega_1{}^2 = S_1/I$

$\omega_2{}^2 = S_2/I$

and $p^2 = J^2\Omega^2/I^2$

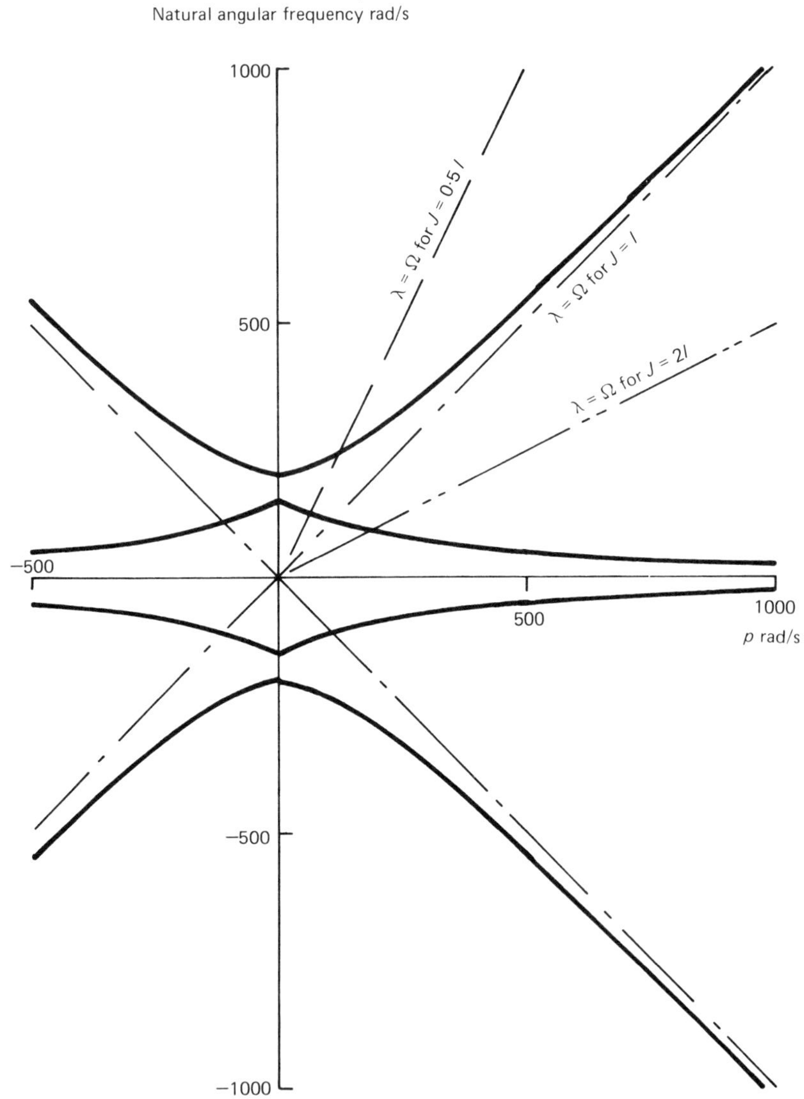

Fig. 8.3 The variation of natural angular frequencies with $p(=J\Omega/I)$ for a symmetrical rotor supported in unsymmetrical elastic bearings ($\omega_1 = 150$ rad/s, $\omega_2 = 200$ rad/s)

For particular values of ω_1, ω_2, and p, the four values of λ may be found easily. A plot of λ as a function of p for $\omega_1 = 150$ rad/s and $\omega_2 = 200$ rad/s is given in Fig. 8.3. From this it may be seen that, when the bearings are unsymmetrical, the four roots of equation (8.20), corresponding to two natural frequencies, exist at all rotational speeds. The higher of the frequencies corresponds with the higher irrotational natural frequency (at zero rotational speed) and increases with rotational speed, becoming asymptotic to the line $\lambda = p$. The lower of the frequencies decreases from the lower irrotational natural frequency, becoming asymptotic with the axis $\lambda = 0$.

When forced vibrations due to out-of-balance are considered it is found that short rotors ($J > I$) have one critical speed, where the line $\lambda = \Omega$ cuts the lower positive branch of the λ versus p curve, and that long rotors ($J < I$) have two critical speeds. In the study of a rotor supported in symmetrical bearings we did not find the positive branch which approaches zero asymptotically or the negative branch which approaches $\lambda = -p$ asymptotically. For perfectly symmetrical supports it is not possible to excite these angular velocities. Therefore, we would expect the critical speed associated with the lower positive branch of the λ versus p curve to be sharply resonant and slow to build-up.

Exercises

1. For the system investigated in the text, show that at any rotational velocity Ω, greater than zero, the normal modes involve precession of the rotor axis in a cone of elliptical section. Show that the precession is forward for the higher frequency and backward for the lower frequency.
2. For the system investigated in the text, show that in the case of a long rotor ($J < I$) the build-up of amplitude should be much slower at the lower critical speed than at the higher one. Also show that the lower critical speed is more sharply resonant than the higher one. Which directions would you expect the resulting whirls to take, at the critical speeds?
3. By assuming symmetrical bearings as the limiting case of asymmetrical bearings, explain why it is not possible to excite a critical speed when the rotor is short ($J > I$).

8.1.4 Rotor on unsymmetrical shaft, angular motion only

When the stiffnesses about the two principal axes of the shaft cross-section differ, long rotors exhibit unstable behaviour within a speed range, the lower end of which is greater than the angular frequency of the lower irrotational natural frequency. To show that this is expected from analysis, we shall examine a balanced system, in rotating coordinates.

Let ξ and ζ be angular coordinates of the rotor relative to the principal axes of the shaft. Then from Fig. 8.4 it may be seen that

$$\theta = \xi \cos \Omega t - \zeta \sin \Omega t \qquad [8.21(a)]$$

$$\phi = \xi \sin \Omega t + \zeta \cos \Omega t$$

Also

$$G_\xi = G_\theta \cos \Omega t + G_\phi \sin \Omega t$$

$$G_\zeta = -G_\theta \sin \Omega t + G_\phi \cos \Omega t \qquad [8.21(b)]$$

From equations (8.7) and (8.8) we have

$$\begin{bmatrix} G_\theta \\ G_\phi \end{bmatrix} = \begin{bmatrix} I\mathrm{D}^2 & J\Omega\mathrm{D} \\ -J\Omega\mathrm{D} & I\mathrm{D}^2 \end{bmatrix} \begin{bmatrix} \theta \\ \phi \end{bmatrix}$$

Therefore,

$$\begin{bmatrix} G_\xi \\ G_\zeta \end{bmatrix} = \begin{bmatrix} \cos\Omega t & \sin\Omega t \\ -\sin\Omega t & \cos\Omega t \end{bmatrix} \begin{bmatrix} I\mathrm{D}^2 & J\Omega\mathrm{D} \\ -J\Omega\mathrm{D} & I\mathrm{D}^2 \end{bmatrix} \begin{bmatrix} \cos\Omega t & -\sin\Omega t \\ \sin\Omega t & \cos\Omega t \end{bmatrix} \begin{bmatrix} \xi \\ \zeta \end{bmatrix} \quad (8.22)$$

where $\mathrm{D} \equiv \mathrm{d}/\mathrm{d}t$ and the matrix operator containing it operates on all following matrices, to give

$$\begin{bmatrix} G_\xi \\ G_\zeta \end{bmatrix} = \begin{bmatrix} I(\mathrm{D}^2 - \Omega^2) + J\Omega^2 & -2I\Omega\mathrm{D} + J\Omega\mathrm{D} \\ 2I\Omega\mathrm{D} - J\Omega\mathrm{D} & I(\mathrm{D}^2 - \Omega^2) + J\Omega^2 \end{bmatrix} \begin{bmatrix} \xi \\ \zeta \end{bmatrix} \quad (8.23)$$

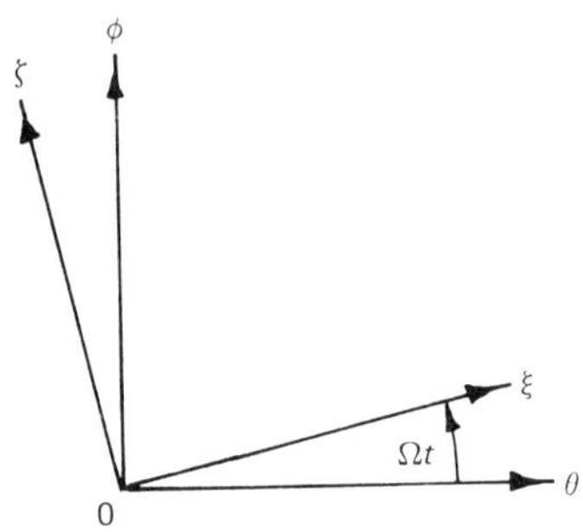

Fig. 8.4 Definition of rotating coordinates ξ and ζ

Let S_ξ and S_ζ be the stiffnesses to angular displacements about the ξ and ζ axes respectively, these being principal axes. Then the equation of motion becomes

$$\begin{bmatrix} I(\mathrm{D}^2 - \Omega^2) + J\Omega^2 + S_\xi & -2I\Omega\mathrm{D} + J\Omega\mathrm{D} \\ 2I\Omega\mathrm{D} - J\Omega\mathrm{D} & I(\mathrm{D}^2 - \Omega^2) + J\Omega^2 + S_\zeta \end{bmatrix} \begin{bmatrix} \xi \\ \zeta \end{bmatrix} = 0 \quad (8.24)$$

Upon substituting $\xi = A\,\mathrm{e}^{\gamma t}$ and $\zeta = B\,\mathrm{e}^{\gamma t}$ the characteristic equation is found to be

$$\gamma^4 + \left\{\omega_\xi^2 + \omega_\zeta^2 + \Omega^2\left[1 + \left(\frac{J}{I} - 1\right)^2\right]\right\}\gamma^2 + \Omega^4\left(\frac{J}{I} - 1\right)^2 + \Omega^2(\omega_\xi^2 + \omega_\zeta^2)\left(\frac{J}{I} - 1\right) + \omega_\xi^2\omega_\zeta^2 = 0 \quad (8.25)$$

where $\omega_\xi^2 = S_\xi/I$

and $\omega_\zeta^2 = S_\zeta/I$

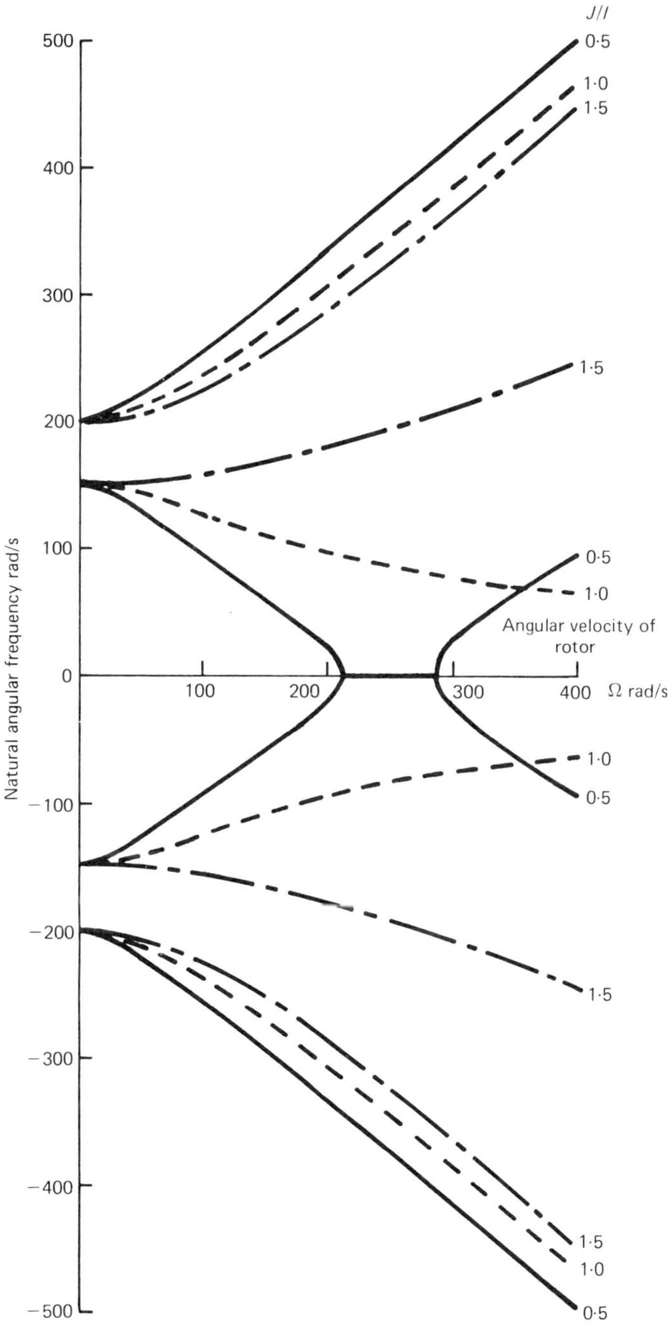

Fig. 8.5 The variation of natural angular frequencies with rotor angular velocity Ω, referred to a coordinate system which rotates with the rotor, for a symmetrical rotor on an unsymmetrical shaft

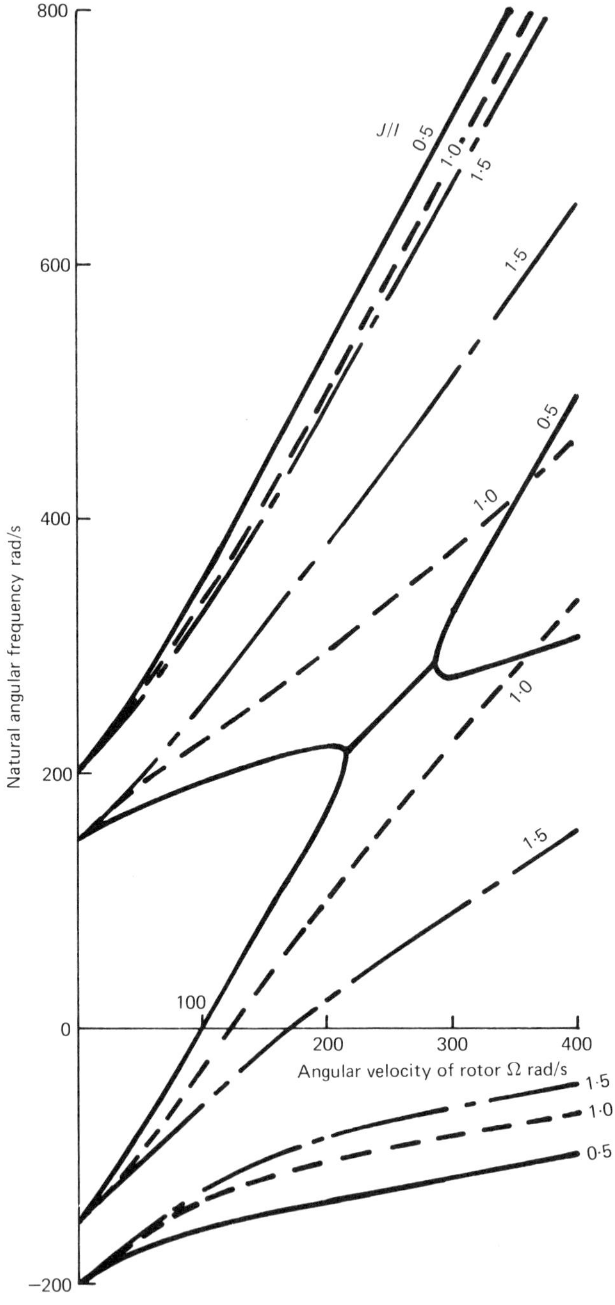

Fig. 8.6 The variation of natural angular frequencies with rotor angular velocity, referred to a stationary coordinate system, for a symmetrical rotor on an unsymmetrical shaft

The sum of the last three terms is zero when

$$\Omega^2\left(\frac{J}{I} - 1\right) = -\omega_\xi{}^2 \quad \text{or} \quad -\omega_\zeta{}^2 \tag{8.26}$$

These equalities can only hold when J is less than I. Equation (8.25) will have a real root between shaft rotational speeds of

$$\Omega_1 = \frac{\omega_\xi}{(1 - J/I)^{1/2}} \quad \text{and} \quad \Omega_2 = \frac{\omega_\zeta}{(1 - J/I)^{1/2}} \tag{8.27}$$

That is for $J < I$, the motion of the system is unstable between shaft rotational speeds of Ω_1 and Ω_2, the amplitude growing exponentially.

When $J \geqslant I$ the system has no critical speeds.

Three cases are illustrated in Fig. 8.5 and 8.6. Figure 8.5 shows, for a range of rotor angular velocity Ω, the variations in the natural angular frequencies relative to the coordinate system which rotates with the rotor, the magnitudes of the natural angular frequencies being given by the imaginary parts of the roots of equation (8.25). Figure 8.6 shows the variations in the natural angular frequencies relative to a stationary coordinate system over the same range of rotor angular velocity Ω. From these it is obvious that if a constant couple, fixed in space, acts upon the rotor it will cause a resonance similar to the secondary resonance found in section 7.6.2.

8.1.5 Rotor supported on an overhung shaft

Rotors in turbo-machinery are in many instances supported in such a way that their linear and angular motions are coupled by the elastic characteristics of the shaft. We shall study one such system, a rotor rigidly attached to an overhung shaft; the bearings supporting the shaft are to be assumed unsymmetrical. The elastic restraints imposed by the shaft are typified by the following relationship:

$$\begin{bmatrix} X \\ Y \\ G_\theta \\ G_\phi \end{bmatrix} = \begin{bmatrix} S_{11} & 0 & S_{13} & 0 \\ 0 & S_{22} & 0 & S_{24} \\ S_{31} & 0 & S_{33} & 0 \\ 0 & S_{42} & 0 & S_{44} \end{bmatrix} \begin{bmatrix} x \\ y \\ \theta \\ \phi \end{bmatrix} \tag{8.28}$$

Therefore the equations of motion are

$$\begin{bmatrix} m\mathrm{D}^2 & 0 & 0 & 0 \\ 0 & m\mathrm{D}^2 & 0 & 0 \\ 0 & 0 & I\mathrm{D}^2 & J\Omega\mathrm{D} \\ 0 & 0 & -J\Omega\mathrm{D} & I\mathrm{D}^2 \end{bmatrix} \begin{bmatrix} x \\ y \\ \theta \\ \phi \end{bmatrix} = - \begin{bmatrix} X \\ Y \\ G_\theta \\ G_\phi \end{bmatrix} \tag{8.29}$$

Assuming that the system may execute undamped natural oscillations, the frequency equation is given by

$$\begin{vmatrix} S_{11} - m\lambda^2 & 0 & S_{13} & 0 \\ 0 & S_{22} - m\lambda^2 & 0 & S_{24} \\ S_{31} & 0 & S_{33} - I\lambda^2 & J\Omega\lambda i \\ 0 & S_{42} & -J\Omega\lambda i & S_{44} - I\lambda^2 \end{vmatrix} = 0 \qquad (8.30)$$

Inspection of the frequency equation leads us to expect four natural frequencies; two in which the motion is mainly linear, and two in which it is mainly angular. The frequencies associated with the mainly linear motion increase from their irrotational values to those corresponding to a direction-fixed, position-free restraint as the rotational speed becomes very large. The frequencies associated with the mainly angular motions behave in much the same way as found in Fig. 8.3. The variations of λ for a system with the following physical characteristics are shown in Fig. 8.7.

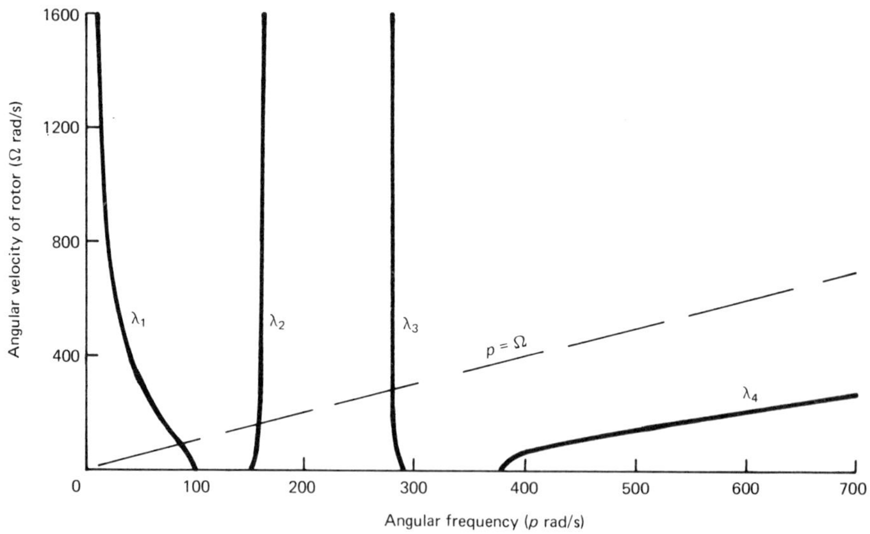

Fig. 8.7 The variation of the natural angular frequencies λ_i of a rotor supported on an overhung shaft, with rotor angular velocity Ω

$$\begin{aligned} S_{11} &= 32 \text{ kN/m} & S_{13} &= S_{31} = -2 \text{ kN m/m (or kN/rad)} \\ S_{22} &= 11 \text{ kN/m} & S_{24} &= S_{42} = -1 \text{ kN m/m (or kN/rad)} \\ S_{33} &= 230 \text{ N m/rad} & S_{44} &= 185 \text{ N m/rad} \\ m &= 0{\cdot}5 \text{ kg} & I &= 0{\cdot}0027 \text{ kg m}^2 \\ J &= 0{\cdot}0058 \text{ kg m}^2 \end{aligned}$$

A system with these physical parameters was investigated experimentally by Downham (1957). In addition to the critical speeds found in Fig. 8.7, that is where the forcing frequency Ω is equal to a natural frequency λ, Downham found a fourth critical speed at a value of $\lambda = 220$ rad/s, approximately, for which he could not give a physical explanation. He also found an extra group of natural frequencies which appeared to be associated with the additional critical speed and again gave no physical explanation. These observed phenomena may be associated with slight asymmetry of the rotor, which Smith (1933) found to lead to 'quasi-harmonic' motion.

8.2 Unsymmetrical rotors

The rotors of the systems analysed in the previous sections were symmetrical with respect to their spin axes. This simplified the analysis. We may remove this restriction by studying in greater depth the motion of a rigid body which spins about one of its principal axes of inertia, the three principal moments of inertia being unequal. We shall carry out this study firstly by an energy method using Lagrange's equations, and secondly by the Newtonian approach leading to Euler's equations.

The first task in the energy approach is the choice of a suitable set of generalised coordinates.

8.2.1 Generalised coordinates for a rigid body

Let three mutually perpendicular axes, fixed relative to the rigid body, coincide with the triad of unit vectors $\boldsymbol{i}, \boldsymbol{j}, \boldsymbol{k}$ with origin at 'a'. Let 'a' be in position $\boldsymbol{r}$ from the origin of a stationary triad $\boldsymbol{I}, \boldsymbol{J}, \boldsymbol{K}$. The position and orientation of the triad $\boldsymbol{i}, \boldsymbol{j}, \boldsymbol{k}$ may be related to the triad $\boldsymbol{I}, \boldsymbol{J}, \boldsymbol{K}$ by six independent quantities, for example $\boldsymbol{r}\cdot\boldsymbol{I}$, $\boldsymbol{r}\cdot\boldsymbol{J}$, $\boldsymbol{r}\cdot\boldsymbol{K}$, $\boldsymbol{i}\cdot\boldsymbol{I}$, $\boldsymbol{i}\cdot\boldsymbol{J}$, and $\boldsymbol{j}\cdot\boldsymbol{K}$. That is, a rigid body has six degrees of freedom, three linear and three angular, and hence requires six generalised coordinates to specify its position completely. In many systems it is convenient to take these as the three cartesian coordinates of the mass centre of the rigid body and three angles which locate the orientation of the triad $\boldsymbol{i}, \boldsymbol{j}, \boldsymbol{k}$, often taken as coinciding with the principal axes of inertia of the body.

There are a number of sets of angles which may be used to represent the orientation of $\boldsymbol{i}, \boldsymbol{j}, \boldsymbol{k}$. A convenient set for many purposes was introduced by Euler and are known as the Eulerian angles. To define these we shall set up another triad $\boldsymbol{I}', \boldsymbol{J}', \boldsymbol{K}'$ with origin at 'a' and parallel respectively to $\boldsymbol{I}, \boldsymbol{J}, \boldsymbol{K}$. Then the angles ϕ, θ, ψ, see Fig. 8.8 form a set of Eulerian angles. ϕ is a rotation about $\boldsymbol{K}'$, that is, it is the angle between the plane containing $\boldsymbol{I}'$ and $\boldsymbol{K}'$ and the plane containing $\boldsymbol{K}'$ and $\boldsymbol{k}$. From an initial state of $\boldsymbol{i}, \boldsymbol{j}, \boldsymbol{k}$, being coincident with $\boldsymbol{I}', \boldsymbol{J}', \boldsymbol{K}'$, the rotation ϕ leaves the triad $\boldsymbol{i}, \boldsymbol{j}, \boldsymbol{k}$ in position $\boldsymbol{i}', \boldsymbol{j}', \boldsymbol{K}'$. θ is a rotation about $\boldsymbol{j}'$. It is the angle between $\boldsymbol{K}'$ and $\boldsymbol{k}$ in the plane containing them. When rotations ϕ and θ have been applied the triad $\boldsymbol{i}, \boldsymbol{j}, \boldsymbol{k}$ takes up the position $\boldsymbol{i}'', \boldsymbol{j}', \boldsymbol{k}$. ψ is a rotation about $\boldsymbol{k}$. It is the angle between $\boldsymbol{i}''$ and $\boldsymbol{i}$.

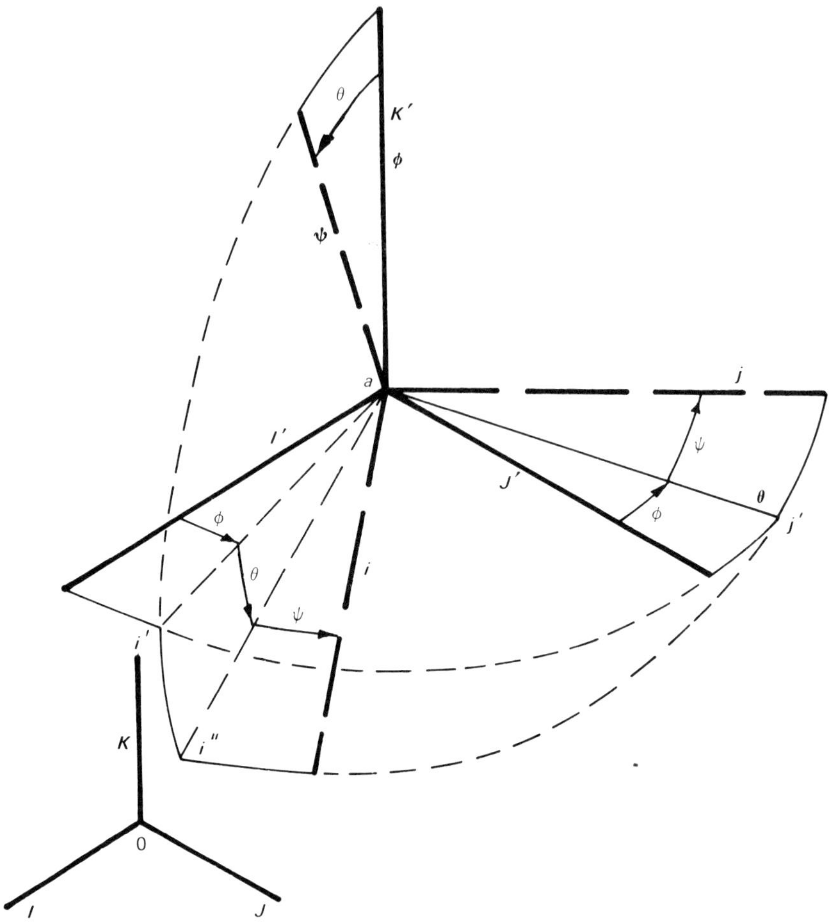

Fig. 8.8 Eulerian angles

Note that the above angles ϕ, θ, ψ may be varied independently and may therefore be used as generalised coordinates. Obviously there are two other sets of Eulerian angles.

Alternatively we may obtain the same result by a set of matrix rotations as follows. The first rotation is about $\boldsymbol{K}'$ and

$$\begin{bmatrix} \delta\gamma_{1b} \\ \delta\gamma_{2b} \\ \delta\gamma_{3b} \end{bmatrix} = \begin{bmatrix} \cos\phi & \sin\phi & 0 \\ -\sin\phi & \cos\phi & 0 \\ 0 & 0 & 1 \end{bmatrix} \begin{bmatrix} \delta\gamma_{1a} \\ \delta\gamma_{2a} \\ \delta\gamma_{3a} \end{bmatrix} \tag{8.31}$$

where $\delta\gamma_{1a}$, $\delta\gamma_{2a}$, and $\delta\gamma_{3a}$ are virtual displacements about the principal axes of inertia I_1, I_2, and I_3 respectively in the initial position, shown in Fig. 8.9(a).

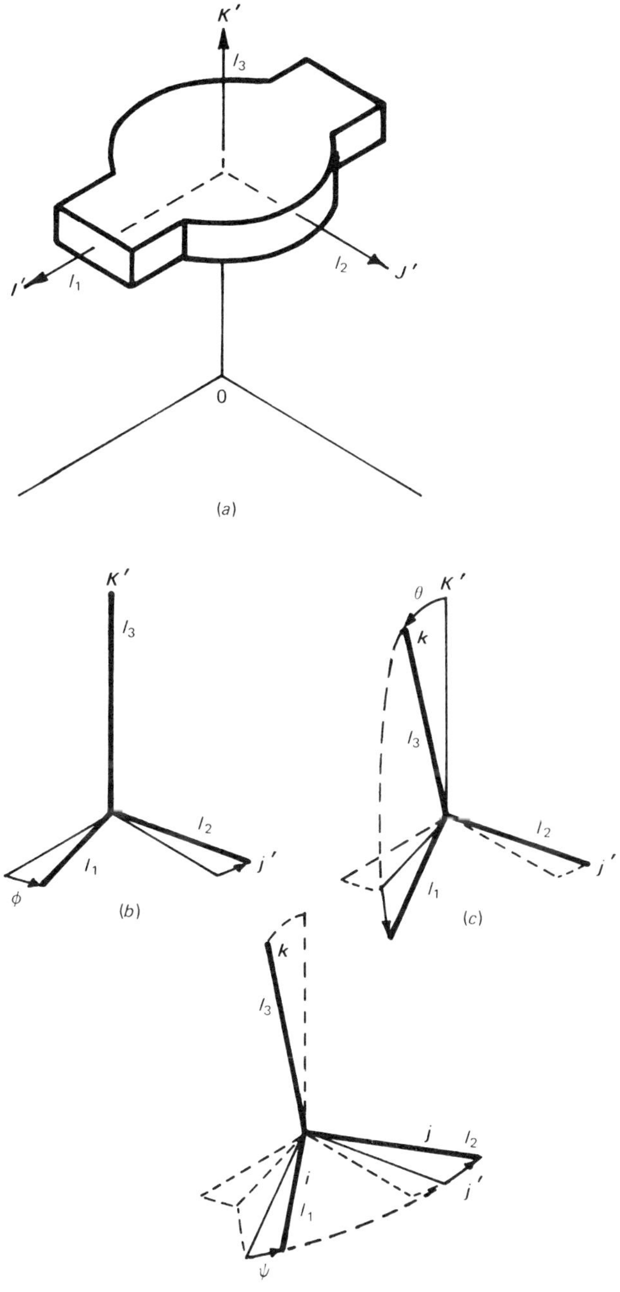

Fig. 8.9(*a*), (*b*), (*c*), (*d*)

$\delta\gamma_{1b}$, $\delta\gamma_{2b}$, $\delta\gamma_{3b}$ are the resulting displacements about the same axis in position (*b*), Fig. 8.9(*b*). The angular velocities are thus related

by
$$\begin{bmatrix} \omega_{1b} \\ \omega_{2b} \\ \omega_{3b} \end{bmatrix} = \begin{bmatrix} \cos\phi & \sin\phi & 0 \\ -\sin\phi & \cos\phi & 0 \\ 0 & 0 & 1 \end{bmatrix} \begin{bmatrix} \omega_{1a} \\ \omega_{2a} \\ \omega_{3a} \end{bmatrix} \tag{8.32}$$

but we have $\omega_{1a} = \omega_{2a} = 0$, $\omega_{3a} = \dot{\phi}$, therefore

$$\begin{bmatrix} \omega_{1b} \\ \omega_{2b} \\ \omega_{3b} \end{bmatrix} = \begin{bmatrix} 0 \\ 0 \\ \dot{\phi} \end{bmatrix} \tag{8.33}$$

The second rotation is about $\boldsymbol{j}'$ the axis coinciding with the principal axis of inertia I_2 and

$$\begin{bmatrix} \delta\gamma_{1c} \\ \delta\gamma_{2c} \\ \delta\gamma_{3c} \end{bmatrix} = \begin{bmatrix} \cos\theta & 0 & -\sin\theta \\ 0 & 1 & 0 \\ \sin\theta & 0 & \cos\theta \end{bmatrix} \begin{bmatrix} \delta\gamma_{1b} \\ \delta\gamma_{2b} \\ \delta\gamma_{3b} \end{bmatrix} \tag{8.34}$$

The angular velocities are thus related by

$$\begin{bmatrix} \omega_{1c} \\ \omega_{2c} \\ \omega_{3c} \end{bmatrix} = \begin{bmatrix} \cos\theta & 0 & -\sin\theta \\ 0 & 1 & 0 \\ \sin\theta & 0 & \cos\theta \end{bmatrix} \begin{bmatrix} \omega_{1b} \\ \omega_{2b} \\ \omega_{3b} \end{bmatrix} \tag{8.35}$$

$$\begin{bmatrix} \omega_{1b} \\ \omega_{2b} \\ \omega_{3b} \end{bmatrix} = \begin{bmatrix} 0 \\ \dot{\theta} \\ \dot{\phi} \end{bmatrix} \tag{8.36}$$

giving
$$\begin{bmatrix} \omega_{1c} \\ \omega_{2c} \\ \omega_{3c} \end{bmatrix} = \begin{bmatrix} -\dot{\phi}\sin\theta \\ \dot{\theta} \\ \dot{\phi}\cos\theta \end{bmatrix} \tag{8.37}$$

The third rotation is about $\boldsymbol{k}$ the axis coinciding with the latest position of the principal axis of inertia I_3, and

$$\begin{bmatrix} \delta\gamma_{1d} \\ \delta\gamma_{2d} \\ \delta\gamma_{3d} \end{bmatrix} = \begin{bmatrix} \cos\psi & \sin\psi & 0 \\ -\sin\psi & \cos\psi & 0 \\ 0 & 0 & 1 \end{bmatrix} \begin{bmatrix} \delta\gamma_{1c} \\ \delta\gamma_{2c} \\ \delta\gamma_{3c} \end{bmatrix} \tag{8.38}$$

giving
$$\begin{bmatrix} \omega_{1d} \\ \omega_{2d} \\ \omega_{3d} \end{bmatrix} = \begin{bmatrix} \cos\psi & \sin\psi & 0 \\ -\sin\psi & \cos\psi & 0 \\ 0 & 0 & 1 \end{bmatrix} \begin{bmatrix} -\dot{\phi}\sin\theta \\ \dot{\theta} \\ \dot{\phi}\cos\theta + \dot{\psi} \end{bmatrix} \tag{8.39}$$

$$= \begin{bmatrix} -\sin\theta\cos\psi & \sin\psi & 0 \\ \sin\theta\sin\psi & \cos\psi & 0 \\ \cos\theta & 0 & 1 \end{bmatrix} \begin{bmatrix} \dot{\phi} \\ \dot{\theta} \\ \dot{\psi} \end{bmatrix} \tag{8.40}$$

8.2.2 The kinetic energy of a rigid body

Before writing down the kinetic energy it is necessary to be able to express the rotations $\delta\gamma_i$, $\delta\gamma_j$, $\delta\gamma_k$ in the $\boldsymbol{i}, \boldsymbol{j}, \boldsymbol{k}$ directions respectively in terms of the virtual displacements $\delta\phi$, $\delta\theta$, $\delta\psi$. We have

$$\delta\gamma_i\boldsymbol{i} + \delta\gamma_j\boldsymbol{j} + \delta\gamma_k\boldsymbol{k} = \delta\phi\boldsymbol{K}' + \delta\theta\boldsymbol{j}' + \delta\psi\boldsymbol{k}$$

giving, in matrix notation

$$\begin{bmatrix} \delta\gamma_i \\ \delta\gamma_j \\ \delta\gamma_k \end{bmatrix} = \begin{bmatrix} -\sin\theta\cos\psi & \sin\psi & 0 \\ \sin\theta\sin\psi & \cos\psi & 0 \\ \cos\theta & 0 & 1 \end{bmatrix} \begin{bmatrix} \delta\phi \\ \delta\theta \\ \delta\psi \end{bmatrix} \tag{8.41}$$

Obviously the velocities are related by

$$\begin{bmatrix} \dot{\gamma}_i \\ \dot{\gamma}_j \\ \dot{\gamma}_k \end{bmatrix} = \begin{bmatrix} -\sin\theta\cos\psi & \sin\psi & 0 \\ \sin\theta\sin\psi & \cos\psi & 0 \\ \cos\theta & 0 & 1 \end{bmatrix} \begin{bmatrix} \dot{\phi} \\ \dot{\theta} \\ \dot{\psi} \end{bmatrix} \tag{8.42}$$

which is the same as we obtained previously, see equation (8.40). If the rigid body has mass m and principal moments of inertia I_1, I_2, and I_3 and its principal axes of inertia u_1, u_2, u_3 coincide with the triad $\boldsymbol{i}, \boldsymbol{j}, \boldsymbol{k}$, then its kinetic energy T is given by

$$2T = [\dot{x}_1 \quad \dot{x}_2 \quad \dot{x}_3] \begin{bmatrix} m & 0 & 0 \\ 0 & m & 0 \\ 0 & 0 & m \end{bmatrix} \begin{bmatrix} \dot{x}_1 \\ \dot{x}_2 \\ \dot{x}_3 \end{bmatrix} + [\dot{\gamma}_1 \quad \dot{\gamma}_2 \quad \dot{\gamma}_3] \begin{bmatrix} I_1 & 0 & 0 \\ 0 & I_2 & 0 \\ 0 & 0 & I_3 \end{bmatrix} \begin{bmatrix} \dot{\gamma}_1 \\ \dot{\gamma}_2 \\ \dot{\gamma}_3 \end{bmatrix} \tag{8.43}$$

where $\dot{x}_l$ are the linear velocities of its centre of mass in three mutually perpendicular directions and $\dot{\gamma}_l$ are its angular velocities about its principal axes of inertia.

The linear motion has been examined in chapter 2 so we shall concentrate now on the equations of motion relating to angular motion. That is, the equation of motion for a rigid body free to rotate on a fixed support at its centre of mass.

8.2.2.1 Angular motion about the centre of mass

From equations (8.42) and (8.43) we find

$$2T = [\dot{\phi} \quad \dot{\theta} \quad \dot{\psi}] \begin{bmatrix} -\sin\theta\cos\psi & \sin\theta\sin\psi & \cos\theta \\ \sin\psi & \cos\psi & 0 \\ 0 & 0 & 1 \end{bmatrix} \begin{bmatrix} I_1 & 0 & 0 \\ 0 & I_2 & 0 \\ 0 & 0 & I_3 \end{bmatrix}$$

$$\begin{bmatrix} -\sin\theta\cos\psi & \sin\psi & 0 \\ \sin\theta\sin\psi & \cos\psi & 0 \\ \cos\theta & 0 & 1 \end{bmatrix} \begin{bmatrix} \dot{\phi} \\ \dot{\theta} \\ \dot{\psi} \end{bmatrix} \tag{8.44}$$

given the generalised forces Φ, Θ, Ψ; Lagrange's equations of motion then apply.

Rather than find the general equations of motion for a rigid body in any orientation, we shall restrict our examination to the case of a rigid body spinning about a principal axis, say $\boldsymbol{k}$ in Fig. 8.9. We shall consider the case in which the spin axis executes small motions α and β, where α is given by $\phi = \pi/2 - \alpha$ and β by $\theta = \pi/2 + \beta$.

Substituting these values of ϕ and θ, equations (8.44) becomes

$$2T = [\dot{\alpha} \quad \dot{\beta} \quad \dot{\psi}] \begin{bmatrix} \cos\psi & -\sin\psi & \beta \\ \sin\psi & \cos\psi & 0 \\ 0 & 0 & 1 \end{bmatrix} \begin{bmatrix} I_1 & 0 & 0 \\ 0 & I_2 & 0 \\ 0 & 0 & I_3 \end{bmatrix} \begin{bmatrix} \cos\psi & \sin\psi & 0 \\ -\sin\psi & \cos\psi & 0 \\ \beta & 0 & 1 \end{bmatrix} \begin{bmatrix} \dot{\alpha} \\ \dot{\beta} \\ \dot{\psi} \end{bmatrix}$$

$$= I_1(\dot{\alpha}\cos\psi + \dot{\beta}\sin\psi)^2 + I_2(\dot{\alpha}\sin\psi - \dot{\beta}\cos\psi)^2 + I_3(\dot{\psi} + \dot{\alpha}\beta)^2 \tag{8.45}$$

The term $I_3\beta^2\dot{\alpha}^2$, being a small term of higher order than the others, will be neglected. Remembering that we wish to maintain the spin velocity constant, we shall take $\ddot{\psi} = 0$. Substituting for T in Lagrange's equations of motion gives expressions for the generalised forces G_α and G_β

$$G_\alpha = \frac{\mathrm{d}}{\mathrm{d}t}\left(\frac{\partial T}{\partial \dot{\alpha}}\right) - \frac{\partial T}{\partial \alpha}$$

$$= \left(\frac{I_1 + I_2}{2}\right)\ddot{\alpha} + \left(\frac{I_1 - I_2}{2}\right)\frac{\mathrm{d}}{\mathrm{d}t}(\dot{\alpha}\cos 2\psi + \dot{\beta}\sin 2\psi) + I_3\dot{\beta}\dot{\psi} \tag{8.46}$$

$$G_\beta = \frac{\mathrm{d}}{\mathrm{d}t}\left(\frac{\partial T}{\partial \dot{\beta}}\right) - \frac{\partial T}{\partial \beta}$$

$$= \left(\frac{I_1 + I_2}{2}\right)\ddot{\beta} - \left(\frac{I_1 - I_2}{2}\right)\frac{\mathrm{d}}{\mathrm{d}t}(-\dot{\alpha}\sin 2\psi + \dot{\beta}\cos 2\psi) - I_3\dot{\alpha}\dot{\psi} \tag{8.47}$$

Note that when the rotor is symmetrical these have the same form as the equations (8.7) and (8.8) derived in section 8.1.1.

Thus in a stationary coordinate system we find the equations of motion to be linear with periodic coefficients and again it is more convenient to solve for the motion in a uniformly rotating coordinate system, ξ, ζ, where from Fig. 8.10

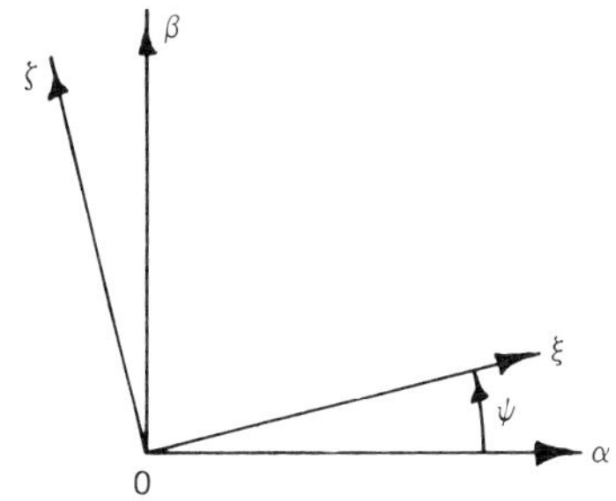

Fig. 8.10 Definition of coordinates ξ and ζ

$$\begin{bmatrix}\alpha\\ \beta\end{bmatrix} = \begin{bmatrix}\cos\psi & -\sin\psi\\ \sin\psi & \cos\psi\end{bmatrix}\begin{bmatrix}\xi\\ \zeta\end{bmatrix} \tag{8.48}$$

giving

$$\begin{bmatrix}\dot\alpha\\ \dot\beta\end{bmatrix} = \begin{bmatrix}-\dot\psi\sin\psi & -\dot\psi\cos\psi & \cos\psi & -\sin\psi\\ \dot\psi\cos\psi & -\dot\psi\sin\psi & \sin\psi & \cos\psi\end{bmatrix}\begin{bmatrix}\xi\\ \zeta\\ \dot\xi\\ \dot\zeta\end{bmatrix} \tag{8.49}$$

Substituting for α, β, $\dot\alpha$, and $\dot\beta$ in the expression for kinetic energy gives

$$\begin{aligned} 2T &= I_1(\dot\xi - \dot\psi\zeta)^2 + I_2(\dot\zeta + \dot\psi\xi)^2 \\ &\quad + I_3\{\dot\psi + (\xi\sin\psi + \zeta\cos\psi) \\ &\quad \times [(\dot\xi - \dot\psi\zeta)\cos\psi - (\dot\zeta + \dot\psi\xi)\sin\psi]\}^2 \end{aligned} \tag{8.50}$$

Again, as in the derivation of equations (8.46) and (8.47), we shall neglect the small high order terms such as

$$I_3\xi\dot\xi\cos\psi\sin\psi \quad \text{and} \quad I_3\dot\psi\xi^2\sin^2\psi$$

Hence

$$G_\xi = I_1\ddot\xi - (I_1 + I_2 - I_3)\dot\psi\dot\zeta - (I_2 - I_3)\dot\psi^2\xi \tag{8.51}$$

$$G_\zeta = I_2\ddot\zeta + (I_1 + I_2 - I_3)\dot\psi\dot\xi - (I_1 - I_3)\dot\psi^2\zeta \tag{8.52}$$

When the rotor is symmetrical these reduce to the form derived in section 8.1.4.

8.2.3 Euler's equations of motion for a rigid body

We shall now set up the equations of motion by the alternative, Newtonian, approach. Initially we must develop an expression for the momentum of the body.

A rigid body may be regarded as a system of particles which move under particular internal constraints. The momentum ($\boldsymbol{p}$) of a single particle of constant mass (m) is given by

$$\boldsymbol{p} = m\dot{\boldsymbol{r}} \tag{8.53}$$

where $\boldsymbol{r}$ is its position in a stationary coordinate system. By Newton's second law of motion, the time rate of change of momentum of a particle is equal to the force, $\boldsymbol{F}$, necessary to cause the change in motion. That is

$$\boldsymbol{F} = \dot{\boldsymbol{p}} = m\ddot{\boldsymbol{r}} \tag{8.54}$$

For a rigid body

$$\boldsymbol{F} = \int_v \frac{\mathrm{d}}{\mathrm{d}t}(\dot{\boldsymbol{r}}\,\mathrm{d}m) = \frac{\mathrm{d}}{\mathrm{d}t}(M\dot{\boldsymbol{r}}_G) \tag{8.55}$$

where $\boldsymbol{r}_G$ is the vector position of the centre of mass and M is the total mass of the body.

Relating again to a particle, consider the moment of the force, equation (8.54), about any point s moving in space, the instantaneous position of which is given by $\boldsymbol{s}$ in a stationary coordinate system. Let the moment about s be $\boldsymbol{G}_s$, and let the vector from s to the particle be $\boldsymbol{t}$, then

$$\boldsymbol{G}_s = \boldsymbol{t} \times \boldsymbol{F} = \boldsymbol{t} \times \frac{\mathrm{d}}{\mathrm{d}t}(m\dot{\boldsymbol{r}}) \tag{8.56}$$

Equation (8.56) may be manipulated into two forms

$$\begin{aligned}(i)\quad \boldsymbol{G}_s = \boldsymbol{t} \times \boldsymbol{F} &= \frac{\mathrm{d}}{\mathrm{d}t}(\boldsymbol{t} \times m\dot{\boldsymbol{r}}) + \dot{\boldsymbol{s}} \times m\dot{\boldsymbol{r}} \\ &= \frac{\mathrm{d}}{\mathrm{d}t}(\boldsymbol{H}_s) + \dot{\boldsymbol{s}} \times m\dot{\boldsymbol{r}}\end{aligned} \tag{8.57}$$

$$\begin{aligned}(ii)\quad \boldsymbol{G}_s = \boldsymbol{t} \times \boldsymbol{F} &= \frac{\mathrm{d}}{\mathrm{d}t}(\boldsymbol{t} \times m\dot{\boldsymbol{t}}) + \boldsymbol{t} \times \frac{\mathrm{d}}{\mathrm{d}t}(m\dot{\boldsymbol{s}}) \\ &= \frac{\mathrm{d}}{\mathrm{d}t}(\boldsymbol{h}_s) + \boldsymbol{t} \times \frac{\mathrm{d}}{\mathrm{d}t}(m\dot{\boldsymbol{s}})\end{aligned} \tag{8.58}$$

where $\boldsymbol{H}_s$ is known as the total angular momentum of the particle with respect to s and $\boldsymbol{h}_s$ is the relative angular momentum of the particle with respect to s.

For a rigid body equations (8.57) and (8.58) become, when integrated over the volume of the rigid body,

$$\boldsymbol{G}_s = \frac{\mathrm{d}}{\mathrm{d}t}(\boldsymbol{H}_s) + \dot{\boldsymbol{s}} \times M\dot{\boldsymbol{r}}_G = \frac{\mathrm{d}}{\mathrm{d}t}(\boldsymbol{h}_s) + \boldsymbol{t}_G \times \frac{\mathrm{d}}{\mathrm{d}t}(M\dot{\boldsymbol{s}}) \tag{8.59}$$

where the subscript G relates the quantities to the centre of mass of the body. Therefore, if s is a stationary point

$$\boldsymbol{G}_s = \frac{\mathrm{d}}{\mathrm{d}t}(\boldsymbol{H}_s) \tag{8.60}$$

hence torque is equal to the rate of change of angular momentum about any fixed point, and if s coincides with the centre of mass G of the body.

$$\boldsymbol{G}_G = \frac{\mathrm{d}}{\mathrm{d}t}(\boldsymbol{h}_G) = \frac{\mathrm{d}}{\mathrm{d}t}(\boldsymbol{H}_G) \tag{8.61}$$

hence torque about the centre of mass of a body is equal to the rate of change of angular momentum with respect to the centre of mass.

For a rigid body

$$\boldsymbol{h}_G = \int_v \boldsymbol{\rho} \times \boldsymbol{\omega} \times \boldsymbol{\rho}\, \mathrm{d}m \qquad [8.62(a)]$$

where $\boldsymbol{\rho}$ is the position vector of the elemental mass relative to the mass centre and ω is the angular velocity about it. The vector triple product may be expressed in the form

$$\boldsymbol{h}_G = \int_v \{(\boldsymbol{\rho}\cdot\boldsymbol{\rho})\boldsymbol{\omega} - (\boldsymbol{\rho}\cdot\boldsymbol{\omega})\boldsymbol{\rho}\}\, \mathrm{d}m \qquad [8.62(b)]$$

In a stationary rectangular coordinate system with origin at G, the centre of mass, let

$$\boldsymbol{\rho} = x\boldsymbol{i} + y\boldsymbol{j} + z\boldsymbol{k} \qquad [8.62(c)]$$

and

$$\boldsymbol{\omega} = \omega_x\boldsymbol{i} + \omega_y\boldsymbol{j} + \omega_z\boldsymbol{k} \qquad [8.62(d)]$$

the equation [8.62(*b*)] becomes

$$\boldsymbol{h}_G = [\boldsymbol{i} \;\; \boldsymbol{j} \;\; \boldsymbol{k}] \begin{bmatrix} A & -F & E \\ -F & B & -D \\ -E & -D & C \end{bmatrix} \begin{bmatrix} \omega_x \\ \omega_y \\ \omega_z \end{bmatrix} \qquad [8.62(e)]$$

where A, B, and C are the moments of inertia and F, E, D are the products of inertia, see section 2.8.2.

When the triad $\boldsymbol{i}, \boldsymbol{j}, \boldsymbol{k}$ coincides with the principal axes of inertia of the body, the products of inertia are each zero, by definition, and the moments of inertia are its principal moments of inertia I_1, I_2, and I_3.

In a fixed coordinate system the moments and products of inertia are continuously changing and their time rate of change enters the expression for the applied couple. However, they are constant in a coordinate system which rotates with the body and for convenience we shall work in such a system.

If we let a triad of unit vectors $\boldsymbol{r}$, $\boldsymbol{s}$, $\boldsymbol{t}$ coincide with and rotate with the principal axes of inertia of a body, I_1, I_2, and I_3 respectively, the angular momentum about its centre of mass is given by

$$\boldsymbol{h} = I_1\omega_1\boldsymbol{r} + I_2\omega_2\boldsymbol{s} + I_3\omega_3\boldsymbol{t} \tag{8.63}$$

Obviously the angular velocity, $\boldsymbol{\omega}$, of the triad of unit vectors $\boldsymbol{r}$, $\boldsymbol{s}$, $\boldsymbol{t}$ relative to a stationary coordinate system $\boldsymbol{i}, \boldsymbol{j}, \boldsymbol{k}$ must be

$$\boldsymbol{\omega} = \omega_1 \boldsymbol{r} + \omega_2 \boldsymbol{s} + \omega_3 \boldsymbol{t} \tag{8.64}$$

Hence,

$$\boldsymbol{G} = \frac{\mathrm{d}\boldsymbol{h}}{\mathrm{d}t} = I_1\dot{\omega}_1\boldsymbol{r} + I_1\omega_1\frac{\mathrm{d}\boldsymbol{r}}{\mathrm{d}t} + I_2\dot{\omega}_2\boldsymbol{s} + I_2\omega_2\frac{\mathrm{d}\boldsymbol{s}}{\mathrm{d}t} + I_3\dot{\omega}_3\boldsymbol{t} + I_3\omega_3\frac{\mathrm{d}\boldsymbol{t}}{\mathrm{d}t} \tag{8.65}$$

Now

$$\frac{\mathrm{d}\boldsymbol{r}}{\mathrm{d}t} = \boldsymbol{\omega} \times \boldsymbol{r} = -\omega_2\boldsymbol{t} + \omega_3\boldsymbol{s} \tag{8.66}$$

with similar expressions for $\mathrm{d}\boldsymbol{s}/\mathrm{d}t$ and $\mathrm{d}\boldsymbol{t}/\mathrm{d}t$, therefore,

$$\begin{aligned}\boldsymbol{G} = {} & \{I_1\dot{\omega}_1 - (I_2 - I_3)\omega_2\omega_3\}\boldsymbol{r} \\ & + \{I_2\dot{\omega}_2 - (I_3 - I_1)\omega_3\omega_1\}\boldsymbol{s} \\ & + \{I_3\dot{\omega}_3 - (I_1 - I_2)\omega_1\omega_2\}\boldsymbol{t}\end{aligned} \tag{8.67}$$

The scalar equations

$$G_1 = I_1\dot{\omega}_1 - (I_2 - I_3)\omega_2\omega_3 \tag{8.68}$$

$$G_2 = I_2\dot{\omega}_2 - (I_3 - I_1)\omega_3\omega_1 \tag{8.69}$$

$$G_3 = I_3\dot{\omega}_3 - (I_1 - I_2)\omega_1\omega_2 \tag{8.70}$$

are known as the Euler dynamical equations.

In terms of the generalised coordinates adopted in section 8.2.2.1.

$$\begin{bmatrix}\omega_1\\ \omega_2\\ \omega_3\end{bmatrix} = \begin{bmatrix}\cos\psi & \sin\psi & 0\\ -\sin\psi & \cos\psi & 0\\ \beta & 0 & 1\end{bmatrix}\begin{bmatrix}\dot{\alpha}\\ \dot{\beta}\\ \dot{\psi}\end{bmatrix} \tag{8.71}$$

As G_i are relative to axes which move with the body and α, β, ψ are in a stationary coordinate system we shall transform to the rotating coordinate system ξ, ζ defined in Fig. 8.10. Then

$$\begin{bmatrix}\omega_1\\ \omega_2\\ \omega_3\end{bmatrix} = \begin{bmatrix}\cos\psi & \sin\psi & 0\\ -\sin\psi & \cos\psi & 0\\ \beta & 0 & 1\end{bmatrix}\begin{bmatrix}-\dot{\psi}\sin\psi & -\dot{\psi}\cos\psi & \cos\psi & -\sin\psi & 0\\ \dot{\psi}\cos\psi & -\dot{\psi}\sin\psi & \sin\psi & \cos\psi & 0\\ 0 & 0 & 0 & 0 & 1\end{bmatrix}\begin{bmatrix}\xi\\ \zeta\\ \dot{\xi}\\ \dot{\zeta}\\ \dot{\psi}\end{bmatrix}$$

$$= \begin{bmatrix}-\dot{\psi}\zeta + \dot{\xi}\\ \dot{\psi}\xi + \dot{\zeta}\\ \dot{\psi}\end{bmatrix} \tag{8.72}$$

ignoring second and higher order quantities, and it can be shown that

$$\begin{bmatrix} \dot{\omega}_1 \\ \dot{\omega}_2 \end{bmatrix} = \begin{bmatrix} \ddot{\xi} - \dot{\psi}\dot{\zeta} \\ \ddot{\zeta} + \dot{\psi}\dot{\xi} \end{bmatrix} \tag{8.73}$$

Therefore,

$$G_1 = I_1\ddot{\xi} - (I_1 + I_2 - I_3)\dot{\psi}\dot{\zeta} - (I_2 - I_3)\dot{\psi}^2\xi \tag{8.74}$$

$$G_2 = I_2\ddot{\zeta} + (I_1 + I_2 - I_3)\dot{\psi}\dot{\xi} - (I_1 - I_3)\dot{\psi}^2\zeta \tag{8.75}$$

which are the same as those found by the energy approach, using Lagrange's equations of motion.

8.2.4 Unsymmetrical rotor on symmetrical restraint, angular motion only

When the rotor is restrained by a symmetrical elastic shaft running in symmetrical bearings and the angular and linear motions are not coupled, the equations for angular motion are

$$\{I_1\mathrm{D}^2 - (I_2 - I_3)\Omega^2 + S\}\xi - (I_1 + I_2 - I_3)\Omega\mathrm{D}\zeta = 0 \tag{8.76}$$

$$(I_1 + I_2 - I_3)\Omega\mathrm{D}\xi + \{I_2\mathrm{D}^2 - (I_1 - I_3)\Omega^2 + S\}\zeta = 0 \tag{8.77}$$

where Ω is the spin velocity on the axis of I_3 and S is the stiffness of the shaft. The characteristic equation is, therefore,

$$I_1I_2\lambda^4 + \{(I_1 + I_2)S - \Omega^2[I_1(I_1 - I_3) + I_2(I_2 - I_3) - (I_1 + I_2 - I_3)^2]\}\lambda^2 + \Omega^2\{\Omega^2(I_1 - I_3)(I_2 - I_3) - S(I_1 + I_2 - 2I_3)\} + S^2 = 0 \tag{8.78}$$

If both I_1 and I_2 are less than I_3, the coefficients of λ^4, λ^2, and λ^0 are each positive. This is more easily seen if we let

$$A = \frac{I_1}{I_3} \quad \text{and} \quad B = \frac{I_2}{I_3}$$

then

$$AB\lambda^4 + \{[1 + A(B - 1) + B(A - 1)]\Omega^2 + (A + B)\omega^2\}\lambda^2 + \{\omega^2 - (A - 1)\Omega^2\}\{\omega^2 - (B - 1)\Omega^2\} = 0 \tag{8.79}$$

where $$\omega^2 = S/I_3$$

Obviously for $0{\cdot}5 \leqslant A < 1$ and $0{\cdot}5 \leqslant B < 1$ each coefficient is positive. Also if we represent the equation in the form

$$a\lambda^4 + b\lambda^2 + c = 0 \tag{8.80}$$

it is easy to show that for this case

$$b^2 - 4ac > 0 \tag{8.81}$$

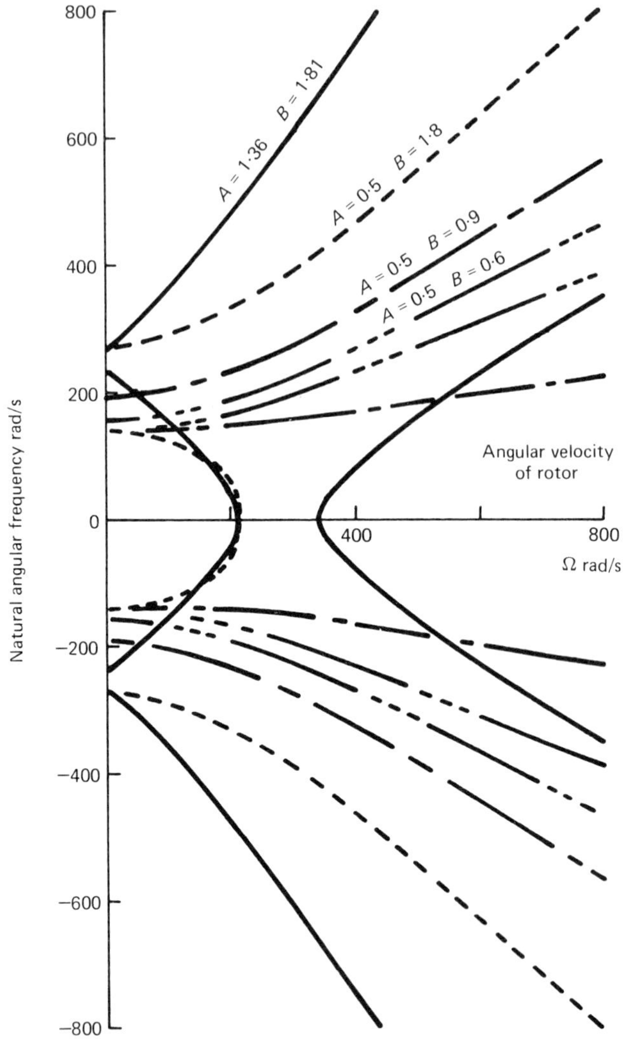

Fig. 8.11 The variation of natural angular frequencies λ_i with rotor angular velocity Ω, relative to a coordinate system which rotates with the rotor, for an unsymmetrical rotor supported on symmetrical elastic constraints ($A = I_1/I_3$, $B = I_2/I_3$, $S/I_3 = 40\,000$ rad^2/s^2)

That is λ^2 is real and negative. Hence for $0{\cdot}5 \leqslant A < 1$ and $0{\cdot}5 \leqslant B < 1$ the system is oscillatory. Figure 8.11 shows plots of the natural angular frequencies, relative to a coordinate system which rotates with the rotor, as functions of the rotor spin velocity Ω for two cases, $A = 0{\cdot}5$, $B = 0{\cdot}6$, $\omega = 200$ rad/s, and $A = 0{\cdot}5$, $B = 0{\cdot}9$, $\omega = 200$ rad/s. The natural frequencies are given by the imaginary components of the roots of equation (8.79). It may be seen that for these cases there are no critical speeds.

When the mass distribution of the rotor is such that $0{\cdot}5 \leqslant A < 1$ and $1 < B$ (or $0{\cdot}5 \leqslant B < 1$ and $1 < A$) the system will be unstable above a value of Ω given by $\omega/(B - 1)^{1/2}$ (or $\omega/(A - 1)^{1/2}$).

That is the system has one critical speed, below it the motion is oscillatory, above it it is exponentially unstable. A case is illustrated in Fig. 8.11 for $A = 0{\cdot}5$, $B = 1{\cdot}8$.

When both $A > 1$ and $B > 1$, the system has two critical speeds given by $\omega/(A - 1)^{1/2}$ and $\omega/(B - 1)^{1/2}$ respectively, between which the motion is exponentially unstable. Outside this speed range the motion is oscillatory.

Exercises

1. A steel shaft 0·01 m diameter carries an unsymmetrical rotor at one end and is free to rotate in a long bearing at its other end. The effective length of the shaft is 0·5 m and the principal moments of inertia of the rotor are $I_1 = 0{\cdot}034$ kg m^2, $I_2 = 0{\cdot}026$ kg m^2, and $I_3 = 0{\cdot}02$ kg m^2; I_3 being about the spin axis. If the mass of the rotor is 2 kg how will the system behave in the speed range 0 to 10 000 rev/min when the mass centre of the rotor is 0·1 mm from the spin axis?
2. Show, for a system in which the linear motion of and angular motion about the centre of mass are uncoupled, that the conclusions reached in the previous section in connections with the angular motion apply also if the shaft has unsymmetrical stiffnesses.

References

ARNOLD, R. N.
'The tuned and damped gyrostatic vibration absorber', *Proc. Inst. Mech. Engrs.*, **157**, no. 25, 1947, 1–19.

DOWNHAM, E.
'Theory of shaft whirling', *Engr.*, **204**, 1957, 588–91, 624–8.

SCHULTE, C. A. and RIESTER, R. A.
'Effect of mass distribution on critical speeds', *Proc. 2nd U.S. National Congress of Applied Mechanics*, 1954 (published by Am. Soc. Mech. Engrs.), 101–110.

SMITH, D. M.
'The motion of a rotor carried by a flexible shaft in flexible bearings', *Proc. Roy. Soc., Series A*, **142**, 1933, 116.

STODOLA, A.
Steam and gas turbines, Vol. 1, McGraw-Hill, 1927 (translated by L. C. Loewenstein), p. 432.

Further reading

ARNOLD, R. N. and MAUNDER, L.
Gyrodynamics and its engineering applications, Academic Press, New York and London, 1961.

Appendix 1

Linear finite difference equations

In analysing mathematically the static or dynamic behaviour of systems composed of interacting sets of identical subsystems, finite difference equations arise. The systems examined in this book involve only linear equations with constant real coefficients. They have the general form

$$a_n U_{r+n} + a_{n-1} U_{r+n-1} + \cdots + a_1 U_{r+1} + a_0 U_r = f(r) \tag{1}$$

the solution of which is comprised of the sum of a particular solution, P, satisfying the complete equation and a set of n complementary solutions $_iU_r$ $(=Q)$ which satisfy the homogeneous equation

$$a_n U_{r+n} + a_{n-1} U_{r+n-1} + \cdots + a_1 U_{r+1} + a_0 U_r = 0 \tag{2}$$

The analogy with differential equations is clear.

For convenience the shifting operator $\mathbf{E}$ may be introduced. It has the following properties:

$$\begin{aligned}
&\text{(i)} & \mathbf{E}^n y_r &= y_{r+n} \\
&\text{(ii)} & (\mathbf{E}^m + \mathbf{E}^n) y_r &= (\mathbf{E}^n + \mathbf{E}^m) y_r = y_{r+m} + y_{r+n} \\
&\text{(iii)} & \mathbf{E}(y_r + z_r) &= y_{r+n} + z_{r+n} \\
&\text{(iv)} & \mathbf{E}^m \mathbf{E}^n c y_r &= c\mathbf{E}^n \mathbf{E}^m y_r = c y_{r+m+n} \\
&\text{(v)} & \mathbf{E}\mathbf{E}^{-1} y_r &= \mathbf{E}^{-1}\mathbf{E} y_r = y_r
\end{aligned} \tag{3}$$

If y_r has the form $A\alpha^r$ $(=A\,\mathrm{e}^{\lambda r})$, a typical solution to equation (2), then

$$\phi(\mathbf{E}) y_r = \phi(\mathbf{E}) A\alpha^r = \phi(\alpha) A\alpha^r \tag{4}$$

Equation (1) may now be written as

$$\Phi(\mathbf{E}) U_r = f(r) \tag{5}$$

and upon substituting a solution of the form

$$U_r = A\alpha^r \tag{6}$$

in the homogeneous equation we obtain

$$(a_n\alpha^n + a_{n-1}\alpha^{n-1} + \cdots + a_1\alpha + a_0) A\alpha^r = 0 \tag{7}$$

The polynomial in α is zero for non-trivial solutions. Let the n roots be the distinct roots p_i ($i = 1$ to k) plus the s repeated roots β where $k + s = n$, then the polynomial may be written

$$(\alpha - p_1)(\alpha - p_2)\dots(\alpha - p_k)(\alpha - \beta)^s = 0 \tag{8}$$

For the distinct roots

$$_iU_r = A_i p_i^r \tag{9}$$

The s repeated roots must have s arbitrary constants associated with them and lead to a solution of the form

$$U_r = \beta^r(B_0 + B_1 r + B_2 r^2 + \cdots + B_{s-1} r^{s-1}) \tag{10}$$

Hence the complementary solution is

$$Q = \sum_{i=1}^{k} A_i p_i^r + \beta^r \sum_{j=0}^{s-1} B_j r^j \tag{11}$$

In terms of the shifting operator, the particular solution may be expressed as

$$P = \frac{1}{\Phi(\mathbf{E})} f(r) \tag{12}$$

The methods used for evaluating this function depend upon the form of $f(r)$. Here, only two cases will be considered:

(*a*) If $f(r)$ has the form Ca^r, by equation (4), we may write

$$P = \frac{Ca^r}{\Phi(a)} \tag{13}$$

(*b*) If $f(r)$ is a polynomial in r, e.g. $f(r) = b_0 + b_1 r + b_2 r^2$, the particular solution may be found by introducing the difference operator $\mathbf{\Delta}$ and expanding by the binomial theorem. The difference operator is related to the shifting operator as follows:

$$\mathbf{\Delta} = \mathbf{E} - 1 \tag{14}$$

i.e.

$$\mathbf{\Delta} y_r = y_{r+1} - y_r = (\mathbf{E} - 1) y_r \tag{15}$$

Hence for $\mathbf{E}$ in equation (12) we substitute $1 + \mathbf{\Delta}$ to give

$$P = \frac{1}{\Phi(1 + \mathbf{\Delta})} f(r) \tag{16}$$

For example, to find the particular solution of the equation

$$(\mathbf{E}^2 - \mathbf{E} - 6) U_r = r^2 - 1$$

$$P = \frac{1}{(\mathbf{E} - 3)(\mathbf{E} + 2)} (r^2 - 1) \tag{17}$$

$$P = \frac{1}{(\Delta - 2)(\Delta + 3)} (r^2 - 1)$$

$$= \frac{1}{5}\left\{\frac{1}{\Delta - 2} - \frac{1}{\Delta + 3}\right\} (r^2 - 1)$$

$$= \frac{1}{5}\left\{-\frac{1}{2}\left(1 + \frac{\Delta}{2} + \frac{\Delta^2}{4} + \frac{\Delta^3}{8} + \cdots\right)\right.$$

$$\left. - \frac{1}{3}\left(1 - \frac{\Delta}{3} + \frac{\Delta^2}{9} - \frac{\Delta^3}{27} + \cdots\right)\right\} (r^2 - 1)$$

Noting that

$$\Delta r^2 = (r + 1)^2 - r^2 = 2r - 1$$

$$\Delta^2 r^2 = \{2(r + 1) - 1\} - (2r - 1) = 2$$

$$\Delta^3 r^2 = 0 = \Delta^4 r^2 \quad \text{etc.}$$

gives

$$P = -\frac{r^2}{6} - \frac{r}{18} + \frac{2}{27}$$

In this case the complementary solution is

$$Q = A_1 3^r + A_2(-2)^r$$

giving the complete solution

$$U_r = A_1 3^r + A_2(-2)^r - \frac{r^2}{6} - \frac{r}{18} + \frac{2}{27}$$

Appendix 2

Absolute acceleration in a rotating coordinate system

Consider a point moving in a rectangular coordinate system u, v which rotates with a constant angular velocity Ω relative to a stationary rectangular coordinate system x, y. Relative to the rotating reference frame let the point move with velocity components $\dot{u}$ and $\dot{v}$ and acceleration components $\ddot{u}$ and $\ddot{v}$. The absolute velocity and acceleration, that is relative to the stationary axes, may then be found as follows.

Resolving onto the x and y axes gives

$$x = u \cos \Omega t - v \sin \Omega t \tag{1}$$

$$y = u \sin \Omega t + v \cos \Omega t \tag{2}$$

Differentiation of equations (1) and (2) with respect to time gives the components of absolute velocity, that is,

$$\dot{x} = (\dot{u} - v\Omega) \cos \Omega t - (\dot{v} + u\Omega) \sin \Omega t \tag{3}$$

$$\dot{y} = (\dot{u} - v\Omega) \sin \Omega t + (\dot{v} + u\Omega) \cos \Omega t \tag{4}$$

Similarly, by differentiating equations (3) and (4) the components of absolute acceleration are found to be

$$\ddot{x} = (\ddot{u} - u\Omega^2 - 2\dot{v}\Omega) \cos \Omega t - (\ddot{v} - v\Omega^2 + 2\dot{u}\Omega) \sin \Omega t \tag{5}$$

$$\ddot{y} = (\ddot{u} - u\Omega^2 - 2\dot{v}\Omega) \sin \Omega t + (\ddot{v} - v\Omega^2 + 2\dot{u}\Omega) \cos \Omega t \tag{6}$$

In connection with the analysis presented in chapter 7 it should be noted that $\ddot{u} - u\Omega^2 - 2\dot{v}\Omega$ is the component of absolute acceleration along a stationary axis instantaneously coincident with the u-axis. Similarly $\ddot{v} - v\Omega^2 + 2\dot{u}\Omega$ is the component of absolute acceleration along a stationary axis instantaneously coincident with the v-axis.

Answers to selected exercises

Chapter 1

Page 6

1. $mL^2\ddot{\theta} + (ka^2 + mgL)\theta = 0$

2. $m\ddot{y} + \dfrac{3EI}{L^3}y = 0$

3. $J\ddot{\theta} + c\dot{\theta} + \dfrac{\pi D^4 G}{32L}\theta = 0$

4. $(m\ddot{\xi} - \omega^2\xi) + k\xi = 0$

   ξ is the displacement from the steady state position, that is when the system is rotating but not vibrating.

5. (a) $L\ddot{\theta} + g\theta = 0$

   (b) $\left\{L^2 + \dfrac{R^2}{2}\right\}\ddot{\theta} + gL\theta = 0$

6. (i) $mL^2\ddot{\theta} + (ka^2 + mgL)\theta = akX \sin \omega t$

   (ii) $mL^2\ddot{\theta} + (ka^2 + mgL)\theta = -[mL\ddot{x} + kax]$

7. $m\ddot{y} + \dfrac{3EI}{L^3}y = F(t)$

8. $J\ddot{\theta} + c\dot{\theta} + \dfrac{\pi D^4 G}{32L}\theta = T(t)$

9. $m\ddot{x} + (c_1 + c_2)\dot{x} + kx = c_2 \dfrac{\mathrm{d}}{\mathrm{d}t}f(t)$

Page 13

1. $x = x_0 \exp(-1{\cdot}57t)\{0{\cdot}07 \sin 22{\cdot}3t + \cos 22{\cdot}3t\}$
2. 6·29

3. $x = 0{\cdot}02 - \exp(-2{\cdot}5t)\{0{\cdot}002 \sin 25{\cdot}7t + 0{\cdot}02 \cos 25{\cdot}7t\}$

4. 13·6 N s/m; 8·75%; 0·38%

Page 30

1. This is an undamped system, therefore the free vibrations do not decay.

2. $$x = \left\{2\alpha\omega p \sin \omega t + p\left(\frac{k}{m} - \omega^2\right)\cos \omega t\right.$$

$$\left.-\exp(-\alpha t)\left[\alpha\left(\frac{k}{m} + \omega^2\right)\sin pt + p\left(\frac{k}{m} - \omega^2\right)\cos pt\right]\right\}$$

$$\times \frac{F}{2}\left\{\left(\frac{k}{m} - \omega^2\right)^2 + 4\alpha^2\omega^2\right\} mp$$

$$\text{where } p^2 = \frac{k}{m} - \frac{c^2}{4m^2}$$

$$\text{and } \alpha = \frac{c}{2m}$$

Chapter 2

Page 41

1. $f_1 = 0{\cdot}104$ Hz, amplitude ratio $= 2{\cdot}88$
   $f_2 = 0{\cdot}135$ Hz, amplitude ratio $= -0{\cdot}347$
   Angle $= 90°$

2. $f_1 = 11{\cdot}7$ Hz, $\theta_1/\theta_2 = 0{\cdot}28$
   $f_2 = 23$ Hz, $\theta_1/\theta_2 = -1{\cdot}78$
   $PE = \frac{1}{2}\{5400U_1^2 + 20\,860U_2^2\}$
   $KE = \frac{1}{2}\{\dot{U}_1^2 + \dot{U}_2^2\}$

Page 44

(*a*) $\mathrm{k} = \begin{bmatrix} k_1 + k_2 & -k_2 \\ -k_2 & k_2 \end{bmatrix}$; $\mathrm{m} = \begin{bmatrix} m_1 & 0 \\ 0 & m_2 \end{bmatrix}$
static coupling.

(*b*) $\mathrm{k} = \begin{bmatrix} k_1 + k_2 & -k_2 \\ -k_2 & k_2 \end{bmatrix}$; $\mathrm{m} = \begin{bmatrix} I_0/l^2 & 0 \\ 0 & m \end{bmatrix}$
static coupling.

(*c*) $\mathrm{k} = \begin{bmatrix} k_1 & 0 \\ 0 & m_2 g \end{bmatrix}$; $\mathrm{m} = \begin{bmatrix} m_1 + m_2 & m_2 l \\ m_2 & m_2 l \end{bmatrix}$
dynamic coupling.

(*d*) $k = \begin{bmatrix} kl_1 + (m_1 + m_2)g & 0 \\ 0 & m_2 g \end{bmatrix}$; $m = \begin{bmatrix} m_1 l_1 + m_2 l_1 & m_2 l_2 \\ m_2 l_1 & m_2 l_2 \end{bmatrix}$

dynamic coupling.

Page 47

2. $f_1 = 0$ Hz; $\dot{x}_1 : \dot{x}_2 : \dot{x}_3 :: 1:1:1$; $|\dot{x}_1| = 0{\cdot}576$ m/s
   $f_2 = 6{\cdot}65$ Hz; $x_1 : x_2 : x_3 :: 1 : 0{\cdot}129 : -0{\cdot}419$; $|x_1| = 0{\cdot}027$ m
   $f_3 = 10{\cdot}8$ Hz; $x_1 : x_2 : x_3 :: 1 : -1{\cdot}296 : 0{\cdot}531$; $|x_1| = 0{\cdot}00915$ m

Page 49

1. $f_1 = 3{\cdot}55$ Hz; $\theta_2/\theta_1 = \frac{1}{2}$
   $f_2 = 5{\cdot}04$ Hz; $\theta_2/\theta_1 = -\frac{1}{2}$

Page 55

1. $$x_1 = \frac{-m_1 \omega^2 F \cos \omega t}{m_1 m_2 \omega^4 - \omega^2(m_1 k_2 + m_2 k_2 + m_2 k_1) + k_1 k_2}$$

   therefore, unless k_2 is infinite, x_1 will have a finite amplitude.

Page 65

1. An increase in mass of the added mass causes a reduction in the first and second natural frequencies and an increase in the third natural frequency.

Page 68

1. 0·37 Hz; 0·89 Hz
2. 11·4 Hz

Page 74

1. $f_1 = 8{\cdot}76$ Hz; $x/\theta = 4$
   $f_2 = 18{\cdot}5$ Hz; $x/\theta = -\frac{1}{16}$
2. Zero Hertz.
3. $$m = \begin{bmatrix} \left\{M + m_1 + m_2 + 0{\cdot}5m\left(1 + \frac{k^2}{r^2}\right)\right\} & (m_1 + m_2)L_1 & m_2 L_2 \\ m_1 + m_2 & (m_1 + m_2)L_1 & m_2 L_2 \\ m_2 & m_2 L_1 & m_2 L_2 \end{bmatrix}$$

   $$k = \begin{bmatrix} K & 0 & 0 \\ 0 & (m_1 + m_2)g & 0 \\ 0 & 0 & m_2 g \end{bmatrix}$$

4.
$$\begin{bmatrix} I\mathrm{D}^2 & 0 & 0 \\ 0 & (\mathcal{J} + ML^2)\mathrm{D}^2 & -ML^2\omega \sin 2\Phi \mathrm{D} \\ 0 & 2ML^2\omega \sin 2\Phi\mathrm{D} & 2(\mathcal{J} + ML^2 \sin^2 \Phi)\mathrm{D}^2 \end{bmatrix} \begin{bmatrix} \theta \\ \phi \\ \alpha \end{bmatrix}$$

$$+ \begin{bmatrix} S & 0 & -S \\ 0 & \begin{matrix}(-ML^2\omega^2 \cos 2\Phi \\ + MgL \cos \Phi)\end{matrix} & 0 \\ -S & 0 & S \end{bmatrix} \begin{bmatrix} \theta \\ \phi \\ \alpha \end{bmatrix} = 0$$

6.
$$\begin{vmatrix} \begin{matrix} K(A^2 + B^2) + 2kB^2 \\ -(I + mB^2)\omega^2 \end{matrix} & 2kLB - mLB\omega^2 & \begin{matrix} K(A - B) - 2kB \\ + m\omega^2 B \end{matrix} \\ 2kLB - mLB\omega^2 & 2kL^2 - (\mathcal{J} + mL^2)\omega^2 & -2kL + mL\omega^2 \\ \begin{matrix} K(A - B) - 2kB \\ + m\omega^2 B \end{matrix} & -2kL + mL\omega^2 & \begin{matrix} 2K + 2k \\ -(M + m)\omega^2 \end{matrix} \end{vmatrix} = 0$$

Chapter 3

Page 92

1. $p = \alpha K\{\sin \alpha x + \cot \alpha L \sin \alpha x\} \sin \omega t$
   where $\alpha^2 = \rho\omega^2/K$

3. (*a*) $$\xi = \frac{16pL^2}{\pi^3 AE} \sin\left\{\frac{2\pi}{L}\sqrt{\frac{E}{\rho}}\,t\right\} \sum_n \frac{\cos\left\{\frac{3\pi}{8}(2n-1)\right\}}{(2n-1)[(2n-1)^2 - 16]}$$

   (*b*) $$\xi = \frac{16pL^2}{\pi^3 AE} \sin\left\{\frac{2\pi}{L}\sqrt{\frac{E}{\rho}}\,t\right\} \sum_n \frac{\cos\left\{\frac{3\pi}{16}(2n-1)\right\} - \cos\left\{\frac{5\pi}{16}(2n-1)\right\}}{(2n-1)[(2n-1)^2 - 16]}$$

Page 94

2.
$$\begin{vmatrix} AE\alpha & -K + M_q\omega^2 \\ -EA\alpha \cos \alpha L + M_p\omega^2 \sin \alpha L & AE\alpha \sin \alpha L + M_p\omega^2 \cos \alpha L \end{vmatrix} = 0$$

3. $$u_x = \sum_n A_n \sin \frac{\omega_n x}{c} \sin \omega_n t$$

where $$c^2 = E/\rho, \quad m = \rho A L, \quad \dot{u}_{L0} = \frac{\int_0^\tau P(t)\,dt}{M}$$

and $$A_n = \frac{M\dot{u}_{L0} \sin \dfrac{\omega_n L}{c}}{\dfrac{m\omega_n}{2} - \dfrac{mc}{4L} \sin \dfrac{2\omega_n L}{c} + M\omega_n \sin^2 \dfrac{\omega_n L}{c}}$$

Page 95

1. 0·0116 m, 16·8 m

3. $$\begin{bmatrix} c_1 \\ c_2 \end{bmatrix} = GJ\alpha \begin{bmatrix} \cot \alpha L & -\operatorname{cosec} \alpha L \\ -\operatorname{cosec} \alpha L & \cot \alpha L \end{bmatrix} \begin{bmatrix} \theta_1 \\ \theta_2 \end{bmatrix}$$

Page 101

2. 81·7, 226, 442, and 732 Hz
3. 73, 235, 494, and 843 Hz

Page 104

1. $2\phi^2 \sin \alpha L \sinh \alpha L + \cos \alpha L \cosh \alpha L - 1 = 0$
   where $\phi = m\omega^2/EI\alpha^3$

2. $$\begin{aligned} &\gamma \sin \alpha L \sinh \alpha L + (1 - \gamma\delta)(\sin \alpha L \cosh \alpha L - \cos \alpha L \sinh \alpha L) \\ &\quad + \delta \cos \alpha L \cosh \alpha L - \beta\{(1 + \gamma\delta)(1 - \cos \alpha L \cosh \alpha L) \\ &\quad + 2 \cos \alpha L \cosh \alpha L + (\gamma - \delta)(\cos \alpha L \sinh \alpha L + \sin \alpha L \cosh \alpha L) \\ &\quad - 2\gamma \sin \alpha L \cosh \alpha L\} = 0 \end{aligned}$$

Page 113

1. $$y_{L/2} = \sum_{r=1,3,5}^{\infty} \frac{4p \sin \dfrac{r\pi}{2}}{(EI\alpha_r^4 - \rho A\omega^2) r\pi}$$

Page 125

2. $$\frac{E}{\rho} \frac{\partial^2 \xi}{\partial z^2} + \nu^2 k^2 \frac{\partial^4 \xi}{\partial^2 t\, \partial^2 z} - \frac{\partial^2 \xi}{\partial t^2} = 0$$

$$-\frac{3n^2}{80}\%; \quad k^2 = \frac{\int r^2\, dA}{A}$$

Page 131

1. $W = (C_1 \sin \beta y + C_2 \cos \beta y + C_3 \sinh \gamma y + C_4 \cosh \gamma y) \;\sin \dfrac{p\pi x}{a}$

where $a^2\beta^2 = a^2\lambda^2 - p^2\pi^2$

$$a^2\gamma^2 = a^2\lambda^2 + p^2\pi^2$$

$$\lambda^4 = \frac{12(1 - \nu^2)\rho\omega_p{}^2}{Ed^2}$$

Frequency equations:

(a) $2(1 - \cos \beta L \cosh \gamma L) + \left(\dfrac{\gamma}{\beta} - \dfrac{\beta}{\gamma}\right) \sin \beta L \sinh \gamma L = 0$

(b) $2(1 - \cos \beta L \cosh \gamma L) + \left(\dfrac{\rho_1\rho_4}{\rho_2\rho_3} - \dfrac{\rho_2\rho_3}{\rho_1\rho_4}\right) \sin \beta L \sinh \gamma L = 0$

$$\rho_1 = -\beta^2 - \nu\frac{p^2\pi^2}{a^2}$$

$$\rho_2 = \gamma^2 - \nu\frac{p^2\pi^2}{a^2}$$

$$\rho_3 = -\beta\left\{\beta^2 + (2 - \nu)\frac{p^2\pi^2}{a^2}\right\}$$

$$\rho_4 - \gamma\left\{\gamma^2 - (2 - \nu)\frac{p^2\pi^2}{a^2}\right\}$$

(c) $\beta \cos \beta L \sinh \gamma L - \gamma \sin \beta L \cosh \gamma L = 0$

Chapter 4

Page 139

1.
$$\begin{bmatrix} I_1D^2+306K_1 & -390K_2 & 8K_1 & -24K_2 & -2K_1 & 6K_2 \\ -390K_2 & M_1D^2+642K_3 & -24K_2 & 72K_3 & 6K_2 & -18K_3 \\ 8K_1 & -24K_2 & I_2D^2+320K_1 & -432K_2 & 8K_1 & -24K_2 \\ -24K_2 & 72K_2 & -432K_2 & M_2D^2+768K_3 & -24K_2 & 72K_3 \\ -2K_1 & 6K_2 & 8K_1 & -24K_2 & I_3D^2+306K_1 & -390K_2 \\ 6K_2 & -18K_3 & -24K_2 & 72K_3 & -390K_2 & M_3D^2+642K_3 \end{bmatrix} \begin{bmatrix} \theta_1 \\ \delta_1 \\ \theta_2 \\ \delta_2 \\ \theta_3 \\ \delta_3 \end{bmatrix} = 0$$

where $K_1 = \dfrac{EI}{88L}$, $K_2 = \dfrac{K_1}{L}$ and $K_3 = \dfrac{K_1}{L^2}$

2.

$$S = 10^7 \begin{bmatrix} 0{\cdot}693 & 0{\cdot}147 & -0{\cdot}308 & 0{\cdot}046 \\ 0{\cdot}147 & 0{\cdot}274 & 0{\cdot}046 & -0{\cdot}177 \\ -0{\cdot}308 & 0{\cdot}046 & 0{\cdot}337 & -0{\cdot}232 \\ 0{\cdot}046 & -0{\cdot}177 & -0{\cdot}232 & 0{\cdot}284 \end{bmatrix}$$

$$M = \begin{bmatrix} 500 & 0 & 0 & 0 \\ 0 & 500 & 0 & 0 \\ 0 & 0 & 250 & 0 \\ 0 & 0 & 0 & 250 \end{bmatrix}$$

Page 140

Exact	$f =$	2·849	11·397 Hz
Model (i)		2·870	11·337 Hz
Model (ii)		2·858	11·325 Hz

Page 146

1.	Exact	$\lambda =$	12·36	485·5
	One element		12·48	1212
	Two elements		12·60	493·3

Chapter 5

Page 161

2. Yes, each satisfies the kinematic boundary conditions for a cantilever.

Page 165

1. $\omega_1^2 = 12{\cdot}54\ EI/\rho AL^4$
   $\omega_2^2 = 1210\ EI/\rho AL^4$

2. $\omega_1^2 = 12{\cdot}38\ EI/\rho AL^4$
   $\omega_2^2 = 494\ EI/\rho AL^4$
   $\omega_3^2 = 13\ 950\ EI/\rho AL^4$

Page 168

1. Yes, the first vibratory mode.

Page 170

1. θ vector is 0·107, 0·015, −0·081, −0·142

2. $\omega_3{}^2 = 421\ \text{rad}^2/\text{s}^2$; θ vector is 0·04, −0·12, −0·16, 0·12
   $\omega_4{}^2 = 994\ \text{rad}^2/\text{s}^2$

Page 171

2. (*a*) 7·00 Hz; 1, 0·227, 0·406, 0·577, 0·617
   (*b*) 11·59 Hz; 0, 0·245, 0·606, 0·645, 0

3. 4, 0, −2, −2 mrad

Page 174

1. 26·34 Hz, 172·4 Hz, 463·3 Hz

Page 177

1. (*a*) Lowest 7·0 Hz; 1·0, 0·227, 0·406, 0·577, 0·617
   Highest 36·24 Hz; −0·003, 0·055, 0·401, −0·641, 1·0
   (*b*) Lowest 11·59 Hz; 0, 0·245, 0·606, 0·645, 0
   Highest 30·50 Hz; 0, 0·921, −0·620, 0·116, 0

3. 26·34 Hz, 172·4 Hz, 463·3 Hz

Page 182

1. (*a*) 0·391 Hz, 1·09 Hz, 1·57 Hz, 1·88 Hz
   (*b*) 8·88 Hz, 71·5 Hz, 236 Hz
   (*c*) 0 Hz, 12·99 Hz, 17·35 Hz, 26·13 Hz

2. 4·47 Hz, 7·46 Hz, 12·02 Hz, 33·18 Hz

Chapter 6

Page 189

1. 5, 41.

2. $$\omega_n{}^2 \simeq \frac{\pi^2 G n^2}{\rho L^2}\left\{1 - \frac{\pi^2 n^2}{12N^2} + \cdots\right\}$$

Chapter 7

Page 212

1. 1·415 Hz; $|x| = \dfrac{0{\cdot}1\omega^2}{79 - \omega^2}$ mm; a forward whirl.

2. In the horizontal plane motion of mass would be $\dfrac{A \sin \Omega t}{0{\cdot}5(79 - \Omega^2)}$

and it would be independent of rotational speed of shaft. There would be no motion of the mass in the vertical plane and therefore no forced whirl.

Page 214

2. If it were possible for the shaft axis to cross the axis of rotation it would behave in the same manner as a bent shaft; at high speeds the mass centre would coincide with the axis of rotation. If it were not possible for it to cross the axis of rotation the system would diverge at the critical speed, that is, the mass would increase in radius until the shaft broke.

3. (*a*) Equivalent to a balanced mass on an initially straight shaft.
   (*b*) Equivalent to a balanced mass on an initially bent shaft.
   (*c*) Equivalent to an unbalanced mass on an initially straight shaft.

Page 220

1. $$[|MR|] = \begin{bmatrix} 134 \\ 23{\cdot}7 \\ 73{\cdot}4 \end{bmatrix} \times 10^{-4} \text{ kg m}$$

$$[\theta] = \begin{bmatrix} 150 \\ 297 \\ 24 \end{bmatrix}$$ degrees, relative to $+u$ axis rotating towards $+v$ axis

2. To balance the forces associated with a mode, at least one of the balancing masses must be fixed at a point other than a node of that mode.

3. (i) Critical speed would be reduced.
   (ii) (*a*) If the bearings were self aligning and a was less than the distance to the free-free node the critical speed would increase.
   (*b*) If the bearings were long, that is give position and direction fixity, the critical speed would increase.
   (iii) Critical speed would be reduced.

Page 222

1. $$r = \frac{mg}{\{k^2 + c_r{}^2\Omega^2\}^{1/2}}; \quad \theta = \tan^{-1}\frac{c_r\Omega}{k}$$

Page 226

1. The energy is derived from the motor or turbine which drives the shaft.

Page 230

1. No, yes, $z = 0$, oscillatory, yes.

Page 234

1. Secondary resonances much less severe than primary resonances.

Page 245

3. $\omega_u = 182$ rad/s, $\omega_v = 190$ rad/s
 $u = -0{\cdot}17$ mm, $v = -0{\cdot}22$ mm

Chapter 8

Page 252

5. $I^2p^4 - (ISa^2 + Ikb^2 + J^2\omega^2)p^2 + Skb^2a^2 = 0$

6. (i) $p^2 = \dfrac{\mu f + J^2\omega^2}{IK}$ where μ and f are viscous damping coefficients.

 (ii) Only if μ and f were equal to zero.

Page 275

1. Critical speeds at 33·85, 34·45, 248·6, 373·1 rad/s. Unstable between upper two critical speeds.

Author index

Subject index